Smart Mobile In-Vehicle Systems

Gerhard Schmidt • Huseyin Abut
Kazuya Takeda • John H.L. Hansen
Editors

Smart Mobile In-Vehicle Systems

Next Generation Advancements

Springer

Editors
Gerhard Schmidt
Christian-Albrechts-Universität
Kiel, Germany

Huseyin Abut
San Diego State University
San Diego, CA, USA

Kazuya Takeda
Nagoya University
Nagoya, Japan

John H.L. Hansen
The University of Texas at Dallas
Dallas, TX, USA

ISBN 978-1-4614-9119-4 ISBN 978-1-4614-9120-0 (eBook)
DOI 10.1007/978-1-4614-9120-0
Springer New York Heidelberg Dordrecht London

Library of Congress Control Number: 2013953325

Printed on acid-free paper

Springer is part of Springer Science+Business Media (www.springer.com)

Foreword

Part A: Interesting Facts About the Automobile and Its History

The adventure of the automobile and its place in our lives exhibit a long and fascinating history. Until the introduction of intelligence and information processing, milestones have included the invention of engines using different power sources: steam, electricity, and gasoline. Its roots, however, can be traced back to the end of the fifteenth century. At that time Leonardo Da Vinci was drawing a three-wheel vehicle. It was not really a motor-car because it was powered by strings. They had to be stretched by operators. The vehicle did not have seats and it is not entirely clear what application Da Vinci really had in mind [1].

The first car that did not need muscular strength was built in 1769 by a Frenchman named Joseph Cugnot. He took advantage of the invention of the steam engine and its further enhancements. His vehicle was bulky and consequently was very difficult to control. In fact, a "chauffeur"—the French word for fireman—was needed to heat the boiler [2]. Steam driven cars were still in use in the twenties of the last century.

January 29, 1886 is considered as the real birth day of the automobile as we know it today. On this date Carl Benz received the patent for a "car with a combustion engine". The vehicle had three wheels. The engine was placed under and behind the seats [2].

A dictionary printed in 1897 reports on the efficiency of gasoline engines in those days: One liter of gasoline was sufficient to transport two passengers over a distance of 16 km [3]. However, 1 h was needed to cover this distance. So, neglecting the low speed and the lack of comfort of the passengers, our "automobile-grandfathers" already developed 6 1/4-L-cars.

Electricity was used for powering automobiles even before the utilization of gasoline. Like the current plug-in electric vehicles, their cruising range was limited by the low capacity of available batteries [4]. Nevertheless, they gained popularity in cases, where only short distances had to be covered and where frequent stops were needed as in the distribution of cargo in large metropolitan areas. The German postal service used some battery-powered trucks up to the sixties of the last century [5].

Approximately 1900, engineers accepted the challenge to combine the advantages of gasoline- and battery-powered engines. Ferdinand Porsche, at that time a young engineer at the Lohner-Factory in Vienna, has exhibited a gasoline-electric-powered car, (i.e. the first hybrid vehicle [6]) at the 1900 Exposition Universelle in Paris. During driving, two single-cylinder gasoline engines charged a battery. The car was then driven by electric motors tied to the front wheels. Because of its unreliability, heavy weight, and high cost the car was not a commercial success.

The automobile industry has experienced a remarkable growth during the past century interrupted only by two world wars. Cars with combustion engines became easier to operate and more comfortable for the passengers as well. The introduction of sensors and digital processors in cars offered mechanisms to further enhance their performance: engines have become more reliable, produced less exhaust, and consumed less gasoline—just to mention a few. In many countries the use of passenger cars outnumbered the use of railways. This continued even in times where gasoline becomes more and more expensive. One reason may be that people "feel at home" in their cars, and driver warning and assistance systems and car infotainment units help to create this feeling. The first group helps the driver to be more prudent and safe for all the persons in the vehicle. Whereas, the second provides valuable information concerning road, traffic, and weather conditions, keeping the driver in touch with the world as well as helping to turn the car into a virtual concert hall in the middle of heavy traffic during commuting hours.

As with the widespread deployment of driver warning and assistance systems becoming the norm, passive and active control systems are emerging and finding their way in a number of applications. One example is the driver-free auto parking, available now in many mid-level to high-end cars. Along with these, the first long distance competition of driverless cars sponsored by DARPA in the US took place in 2004 [7]. Recently, Google intelligent vehicles have traveled both in metropolitan areas and long-haul trips without a single accident. In a number of states in the US and in Europe, cars without drivers are legalized and already permitted to be on the road. Many US states are moving forward with specialized laws to governing who is responsible for safety and operation of "driverless" vehicles.

The real enthusiasts of cars, however, fear that electronically loaded cars will disappear from the streets when they reach the age of old-timers because electronic components to repair them will not be available. They are also concerned that too many tools and schemes for their assistance may take away their "fahrvergnügen" (fun of driving) [8].

In addition to the electronics that go into cars, "how" we drive says a lot about who we are, and how our society incorporates personal transportation within the economic infrastructure, as well as issues of both personal and public safety. The recent best seller by Vanderbilt: "TRAFFIC: Why we drive the way we do, and what is says about us" is an interesting view of our society and the automobile. In particular, emerging countries such as India and China, in comparison to countries in Europe and North America, have fundamentally different views of personal transportation and changing views of what is acceptable in terms of safety and convenience. In spite of this, in-vehicle technology continues to progress forward with new advances appearing in high-end vehicles, as well as down to the most affordable entry level vehicles in many of these countries.

Part B: Things We Do When We Drive a Car Today

As people spend more time in their vehicles, and commuting time to and from work continues to increase as urban populations grow in this age of high-tech, drivers are attempting to perform many more tasks than simply driving their vehicle from point A to point B, which was the case in the twentieth century. The introduction of wireless technology, digital audio/music players, mobile Internet access, advanced entertainment/multimedia systems, and smart navigation technologies into the car has placed increased cognitive demands on drivers. Yet, the typical driving test all over the world continues to focus exclusively on the logistics of operating the vehicle itself and does not include the management of these outside technologies as part of the driver assessment for issuing a license.

Many countries including the US have therefore instituted laws that restrict the use of cell phones and text messaging while operating a vehicle. For instance, large, bright, and illuminated road signs saying "Click it & Ticket it," are posted along the highways in California. USA State Legislative groups and Governors have come together to bring better consistency within the US for laws addressing cell phone use and texting while driving [11, 12]. Restrictions on the use of cell phones while driving have reached worldwide acceptance at various levels (see [13] for a summary). Again, the recent book by Vanderbilt "Traffic: Why We Drive the Way We Do" offers a number of perspectives on society, culture, and government engagement on driving and drivers [10].

Driver distractions in the car are many and have been documented by countless research studies. On the average, drivers attempt to adjust their radio 7.4 times per hour of driving, turn their attention to infants 8.1 times/hour, and are generally searching for something (e.g., sunglasses, coins, etc.) 10.8 times/hour. It is further observed that the average driver looks away from the road 0.06 s every 3.4 s, i.e., 64 s/h. Mobile devices with "intense displays" such as the iPod, other smart phones, and tablets require more mental concentration to perform secondary tasks like searching for songs, pausing, or skipping a song.

While there are some differences of opinion, researchers have noted that any task that requires a driver to divert his/her attention (typically visual) away from the road for more than 1.5 s is viewed as a distraction. However, some scholars believe that this threshold is around 3.0 s. Irrespective of the exact time figure, such a guideline is important as a general rule. But it should be clear that not all drivers are equally skilled, and even advanced/experienced drivers go through periods of fatigue, or they can be unfamiliar with the vehicle they are operating at the time. As a consequence, even for brief periods of time, these could alter their driving abilities and could result incostly and fatal accidents.

Part C: Workshops on Signal Processing in Present and Future Cars

In 2011, the 5th Biennial Workshop for In-Vehicle Systems took place in Kiel, Germany. This meeting served to bring together researchers from diverse research areas to consider advancements in digital signal processing within vehicles to improve safety, comfort, and potentially contribute to reduce driver distraction. A total of 27 peer-reviewed conference papers were presented with researchers from academia, automotive and technology companies, as well as government research laboratories. The workshop included two tutorials, held by highly recognized and experienced speakers from both industry and academia:

Silvia Schuchardt and Anne Theiß at the registration desk of the workshop

First tutorial on Kalman filtering with applications to automotive speech enhancement was presented by Prof. Dr.-Ing. Eberhard Hänsler from Technische Universität Darmstadt, Germany, and Dr.-Ing. Gerald Enzner, Ruhr-Universität Bochum, Germany.

The second tutorial on car hands-free testing and optimization was presented by Dr.-Ing. Hans-Wilhelm Gierlich, head of the Telecom Division of HEAD acoustics, Germany.

Highlights of the workshop have been the four keynote addresses, two of which coming from industry, the other two from academia:

The first keynote was delivered by Dr.-Ing. Luis Arévalo, Vice President, Division of Car Multimedia, Automotive Navigation and Infotainment Systems at Robert Bosch GmbH, Germany. The title of his speech was "Navigation Systems Interacting with Other Vehicle Electronic Control Units."

Professor John H.L. Hansen, from the University of Texas at Dallas, USA, gave the second high-class keynote about "UTDrive: Advances in Human-Machine Systems to Reduce Driver Distraction for In-Vehicle Environments."

Third keynote address with the title "Intelligence in Vehicles," which was presented by Dr. Arne Bartels from the Volkswagen Group Research in Wolfsburg, Germany.

Professor Tim Fingscheidt from Technische Universität Braunschweig, Germany, gave the last keynote on ""Speech Enhancement in Car Applications—Any Specifics?"

In addition to these keynotes and 27 excellent oral presentations, which had been the basis for this book, two panel sessions were organized on "Multi-Sensor and Data Fusion" and "Driver Distraction." Several highly respected panelists from industry and academia have participated and guided the panels. Along with these panels, talks, and keynotes the workshop participants were able to have a look at the latest publications from Springer, the publisher of this book and its four predecessors, at their exhibit booth.

Dr. Baumann at the Springer booth

Following the kick-off of the workshop with tutorials on the first day, the participants were invited to a welcome reception in the "Landeshaus" of Kiel—the Parliament House of the State of Schleswig-Holstein. The welcome addresses were given by Dr. Cordelia Andreßen, Undersecretary, Ministry of Science, Economics, and Transport and by Dr. Hans-Wilhelm Gierlich, Head of the Telecom Division, HEAD acoustics, Germany, the main sponsor of this workshop. After these talks, attendees were given a tour the all parliament building of the State of Schleswig-Holstein.

The participants in the plenary hall of the parliament of Schleswig-Holstein

After the sessions on the second day there was an organized visit to the Leibniz Institute of Marine Sciences (IFM-GEOMAR). The tour was followed by a dinner on the boat "Stadt Kiel" cruising on the Baltic Sea. The captain has organized a tour of the boat including the engine room. As it can be seen from the photo below on the right, Professor John Hansen, in particular, has really enjoyed this and had a long chat with the crew. Fortunately for the sake of other guests and the crew, John did not attempt to steer the ship. The dinner was served on the two decks and the guests had discussions on automotive topics and, of course, about matters beyond that.

Motorship "Stadt Kiel" cruising on the Baltic Sea and her engine room

Best student paper award

At the end of the workshop the best student paper—based on its originality, professional merit, contribution, and presentation quality—was awarded to Philipp Heidenreich and his coauthor Professor Abdelhak Zoubir for their contribution "Computational Aspects of Maximum Likelihood Direction-of-Arrival Estimation of Two Targets with Applications to Automotive Radar." An extended version of this contribution can be found in Chap. 1 of this book. On the right you see Philipp Heidenreich (right) together with Dr. Bernd Iser (left) from SVOX who sponsored the award.

October 2012 Eberhard Hänsler and Gerhard Schmidt

Literature

1. http://wn.com/Leonardo_Da_Vinci_Automovile_1495
2. http://www.leifiphysik.de/web_ph09/geschichte/08automobil/automobil.htm
3. Meyers Konversations-Lexikon. Bibliographisches Institut, 1897
4. http://de.wikipedia.org/wiki/Elektroauto
5. http://www.museumsstiftung.de/index.php?id=777
6. http://www.spiegel.de/auto/fahrkultur/porsche-semper-vivus-das-erste-hybridauto-der-welt-a-749168.html
7. http://en.wikipedia.org/wiki/DARPA_Grand_Challenge
8. "Fahrvergnügen" was used in a commercial by VW in the 1990th in California
9. T. Vanderbilt, *"TRAFFIC: Why we drive the way we do, and what is says about us"*, (Vintage Books, Random House, Inc., New York, 2008)
10. http://www.ghsa.org/html/stateinfo/laws/cellphone_laws.html, (USA's State Governors Highway Safety Association)
11. http://www.ncsl.org/issues-research/transport/cellular-phone-use-and-texting-while-driving-laws.aspx, (USA's National Conference on State Legislatures)
12. http://www.cellular-news.com/car_bans/(Cellular News listing of Countries Worldwide that ban cell phones while driving), (Contributing authors)

Preface

The *Fifth Biennial Workshop on Digital Signal Processing (DSP) for In-Vehicle Systems* took place in Kiel, Germany, on September 4–7, 2011. The workshop was organized by the *Digital Signal Processing and System Theory* research group at Kiel University, Germany. As mentioned above, this biennial is the fifth in a series. It was organized first in 2003 in Nagoya (Japan), followed by events in Sesimbra (Portugal) in 2005, in Istanbul (Turkey) in 2007 and in Dallas (Texas, USA) in 2009. World-class experts from a wide spectrum of research fields have participated and shared cutting-edge studies on driver behavior and in-vehicle technologies just as they did in earlier workshops.

The workshop at Kiel University formed a communication platform among researchers, automotive manufacturers, government foundations, and legislators for road safety and on future in-vehicle technologies as well as focusing on driver behavior. Contributions came from signal processing, control engineering, multi-modal audio–video processing, biomechanics, human factors, and transportation engineering, which opened doors for fruitful discussions and information exchange in an exciting interdisciplinary area. The main focus areas were as follows:

- DSP technologies in automobiles,
- speech dialog, hands-free, and in-car communication systems (algorithms and evaluation),
- driver-status monitoring and distraction/stress detection,
- in-vehicle dialog systems and human–machine interfaces,
- challenges in video and audio processing for in-vehicle products,
- multisensor fusion for driver identification and robust driver monitoring,
- vehicle-to-vehicle and vehicle-to-infrastructure wireless technologies
- human factors and cognitive science in enhancing safety, and
- transportation engineering.

From this workshop, 15 papers and one additional contribution stemming from a tutorial, which was held at the start of the workshop, were selected and expanded with even newer material. These 16 chapters make up this book. Chapters are grouped into five parts, each addressing key areas within in-vehicle digital signal processing arena:

Part I: Sensor and Data Fusion,
Part II: Speech and Audio Processing,
Part III: Driver Distraction,
Part IV: Driving Behavior and User Profiling,
Part V: Driving Scene Analysis.

First, Part I consists of four chapters that cover the fusion of sensor signals or data in general. The first chapter considers the estimation of the direction of arrival in automotive RADAR. Special emphasis is put here on computational aspects. The second chapter investigates stereo camera systems for estimating three-dimensional motion fields in real time for applications such as automotive driver assistance systems, robotics, or surveillance. It is followed by an overview on vehicle-assistance systems that acquire, process, and evaluate environmental data. Several state-of-the-art systems are described here. Chapter 4 addresses the design, the perception, and decision algorithms of the so-called unmanned ground vehicles. Special focus is put on the *Otonobil*, the first autonomously driven vehicle of Turkey.

The next five chapters make up Part II of the textbook which focuses on speech and audio processing for in-vehicle systems. Chapter 5 presents an overview about testing and optimization of hands-free equipment in cars and Chap. 6 focuses on combined fast-converging echo cancellation and residual echo and noise suppression schemes for wideband automotive hands-free systems. Chapter 7 deals with the systems that improve the (speech) communication within the passenger compartment. Next, Chap. 8 discusses the acoustic concept of a *room in a room*, which allows for recording and playback of sound fields with a multitude of microphones and loudspeakers. The last chapter in this second part of the book is about a novel post-processing scheme that can be applied after a conventional filterbank. It *refines* the original short-term spectra and allows for improved pitch estimation or improved convergence speed or complexity reduction of echo cancellation filters.

Part III is on driver distraction with two chapters. Chapter 10 focuses on understanding how drivers react to various secondary tasks such as phone calls, and creating text messages. The CAN bus is used then for analyzing the distraction effect of such actions. The second chapter provides the definition of reference labels for perceptual evaluations from external evaluators, and the consistency and effectiveness of using a visual-cognitive space for subjective evaluations are investigated.

The next two chapters form Part IV concentrating on driver behavior and user profiling. Chapter 12 is about evaluation methods of save driving skills. The second chapter of this part is on the impact of emotions on driving behavior with special emphasis on pre- and post-accident situations.

The last portion of the book is Part V which addresses driving scene analysis. In Chap. 14 two driving scene analysis systems are proposed: The first system measures the similarities between driving behavior signals in driving scenes involving stops, starts, and right and left turns. The second system measures the similarities between environmental driving signals, focusing on surrounding vehicles and driving road configuration. In Chap. 15 studies are presented on algorithms that use front cameras or, in particular, motion vectors of standard video encoding algorithms to detect various driving events. The detection results can be used to gain understanding of the driving dynamics, and eventually to support driver decisions and improve driving safety. In the last chapter of the book, in Chap. 16, automotive radar systems for estimation of target shapes are described. Special focus is put on a two-stage approach for combining high-resolution techniques with conventional Fourier-based methods.

We hope that this book provides an up-to-date perspective on automotive signal processing, with novel ideas for researchers, engineers, and scientists in the field. We wish to thank all those who participated in the 2011 workshop. We wish to express our continued appreciation of Springer Publishing for a smooth and efficient publication process for this book. Specifically, we would like to thank Alex Greene and Ms. Ania Levinson of Springer Publishing for their extensive efforts to enhance the structure and content of this book, as well as providing our community a high-quality and scholarly platform to stimulate public awareness, scientific research, and technology development in this field.

Kiel, Germany — Gerhard Schmidt
San Diego, CA, USA — Huseyin Abut
Nagoya, Japan — Kazuya Takeda
Dallas, TX, USA — John H.L. Hansen

Contents

Contributors

Hüseyin Abut San Diego State University, San Diego, CA, USA

Boğaziçi University, Istanbul, Turkey

Arne Bartels Volkswagen AG, Wolfsburg, Germany

Pınar Boyraz Istanbul Technical University, Istanbul, Turkey

Stefan Brosig Volkswagen AG, Wolfsburg, Germany

Carlos Busso The University of Texas at Dallas, Richardson, TX, USA

Elias S.G. Carotti Politecnico di Torino, Torino, Italy

Markus Christoph Harman/Becker Automotive Systems GmbH, Straubing, Germany

Çagri Dikilitaş Istanbul Technical University, Istanbul, Turkey

Florian Engels A.D.C. Automotive Distance Control Systems GmbH, Lindau, Germany

Ziya Ercan Istanbul Technical University, Istanbul, Turkey

Tim Fingscheidt Technische Universität Braunschweig, Braunschweig, Germany

Uwe Franke Daimler AG, Sindelfingen, Germany

Hans-Wilhelm Gierlich HEAD acoustics GmbH, Herzogenrath, Germany

Metin Gökaşan Istanbul Technical University, Istanbul, Turkey

John H.L. Hansen University of Texas at Dallas, Dallas, TX, USA

Hasan Heceoğlu Istanbul Technical University, Istanbul, Turkey

Philipp Heidenreich ADC Automotive Distance Control Systems GmbH, Lindau, Germany

Marc-André Jung Technische Universität Braunschweig, Braunschweig, Germany

Norhaslinda Kamaruddin Universiti Teknologi Mara, Selangor Darul Ehsan, Malaysia

Norihide Kitaoka Nagoya University, Nagoya, Japan

Reinhard Koch Christian-Albrechts-Universität zu Kiel, Kiel, Germany

Mohamed Krini Nuance Communications, Ulm, Germany

Nanxiang Li The University of Texas at Dallas, Richardson, TX, USA

Yiyang Li Nagoya University, Nagoya, Japan

Christian Lüke Christian-Albrechts-Universität zu Kiel, Kiel, Germany

Enrico Masala Politecnico di Torino, Torino, Italy

Chiyomi Miyajima Nagoya University, Nagoya, Japan

Ryo Nakagawa Nagoya University, Nagoya, Japan

Masaru Noda Nara Institute of Science and Technology, Nara, Japan

Haruo Noma Advanced Telecommunications Research Institute International, Kyoto, Japan

Norzaliza M. Nor International Islamic University, Kuala Lumpur, Malaysia

Gülay Öke Istanbul Technical University, Istanbul, Turkey

Alper Öner Istanbul Technical University, Istanbul, Turkey

Matti Pouke Oulu University, Oulu, Finland

Clemens Rabe Daimler AG, Sindelfingen, Germany

Thomas Ruchatz Volkswagen AG, Wolfsburg, Germany

Amardeep Sathyanarayana University of Texas at Dallas, Dallas, TX, USA

Gerhard Schmidt Christian-Albrechts-Universität zu Kiel, Kiel, Germany

Volkan Sezer Istanbul Technical University, Istanbul, Turkey

Masahiro Tada Advanced Telecommunications Research Institute International, Kyoto, Japan

Kazuya Takeda Nagoya University, Nagoya, Japan

Anne Theiß Christian-Albrechts-Universität zu Kiel, Kiel, Germany

Abdul Wahab International Islamic University, Kuala Lumpur, Malaysia

Jochen Withopf Christian-Albrechts-Universität zu Kiel, Kiel, Germany

Yosuke Yoshida Nara Institute of Science and Technology, Nara, Japan

Abdelhak M. Zoubir Technische Universität Darmstadt, Darmstadt, Germany

Part I
Sensor and Data Fusion

Chapter 1
Computational Aspects of Maximum Likelihood DOA Estimation of Two Targets with Applications to Automotive Radar

Philipp Heidenreich and Abdelhak M. Zoubir

Abstract Direction-of-arrival (DOA) estimation of two targets with a single snapshot plays an important role in many practically relevant scenarios in automotive radar for driver assistance systems. Conventional Fourier-based methods cannot resolve closely spaced targets, and high-resolution methods are required. Thus, we consider the maximum likelihood DOA estimator, which is applicable with a single snapshot. To reduce the computational burden, we propose a grid search procedure with a simplified objective function. The required projection operators are pre-calculated off-line and stored. To save storage space, we further propose a rotational shift of the field of view such that the relevant angular sector, which has to be evaluated, is centered with respect to the broadside. The final estimates are obtained using a quadratic interpolation. An example is presented to demonstrate the proposed method. Also, results obtained with experimental data from a typical application in automotive radar are shown.

Keywords Automotive radar • Direction of arrival (DOA) • Driver assistance systems • Maximum likelihood (ML) estimation

---

P. Heidenreich (✉)
ADC Automotive Distance Control Systems GmbH, Peter-Dornier-Str. 10, 88131 Lindau, Germany
e-mail: philipp.heidenreich@continental-corporation.com

A.M. Zoubir
Signal Processing Group, Technische Universität Darmstadt, Merckstr. 25, 64283 Darmstadt, Germany
e-mail: zoubir@spg.tu-darmstadt.de

G. Schmidt et al. (eds.), *Smart Mobile In-Vehicle Systems: Next Generation Advancements*, DOI 10.1007/978-1-4614-9120-0_1,

1.1 Introduction

Ever-increasing amount of advanced signal processing algorithms is used in various automotive applications [1, 2], e.g., advanced driver assistance systems [3]. These utilize from various sensors to determine the environment of a vehicle. From an identified traffic situation, the driver assistance system regulates the behavior of the vehicle, instructs the driver, or warns the driver in dangerous situations. Often radar sensors are employed, which work reliably even in bad weather conditions, and can provide accurate measurements of the range and relative velocity of multiple targets. To also measure the lateral position of a target, an array of antennas in horizontal direction with digital beamforming can be applied. For typical applications such as collision avoidance or adaptive cruise control (ACC), it is essential to accurately estimate the lateral position and to be able to resolve multiple closely spaced targets. For the array system with limited aperture, this can be achieved with high-resolution processing, which is considered computationally intensive and numerically complex, so that real-time implementation becomes a challenging task.

A pulsed radar system with an array of receive antennas can be effectively used for target localization in terms of range, relative velocity, and direction of arrival (DOA) [4, 5]. After radar preprocessing, which consists of a pulse compression and a Fourier transform over the pulses, the received sensor data is divided into processing cells according to the range and relative velocity, each represented by a single snapshot. For more details, an exemplary radar system is described in [6].

In most practical situations of automotive radar, multiple targets can be distinguished by their range and/or relative velocity so that each processing cell contains at most one target. In the single-target case, the optimal DOA estimates can be found using the Beamformer (BF) spectrum, which is computationally simple [7]. However, there are situations in which multiple targets have similar range and relative velocity so that they are superposed in a processing cell. We consider the situation with two targets per processing cell as practically relevant. In the ACC application, this is motivated by experimental data and may occur when there is an horizontal multipath with a close guardrail, as depicted in Fig. 1.1. If the two propagation paths fall into the same processing cell and cannot be resolved, this generally results in a false localization of the observed car, which seems to be pulled towards the guardrail. To correctly localize the observed car and a ghost target, high-resolution DOA estimation is required. Note that the multipath situation can be correctly identified using the guardrail location, which can be estimated from stationary target detections.

A number of high-resolution DOA estimators are available in the literature, see, e.g., [7–9]. Among the subspace methods, there is the popular MUSIC algorithm [10], which requires an eigendecomposition of the spatial covariance matrix, and a one-dimensional search on a fine grid to obtain the DOA estimates. For particular array geometries, there are also analytic solutions, e.g., unitary ESPRIT [11]. Implementing an eigendecomposition on a practical system with real-time constraints can be numerically complex. Eigendecomposition is iterative in nature

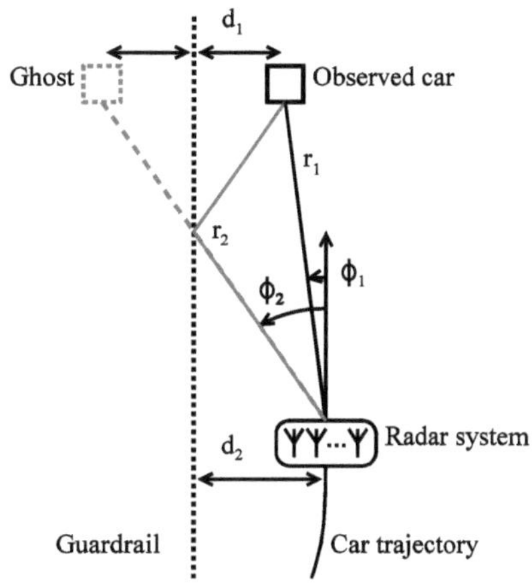

Fig. 1.1 Automotive radar situation with horizontal multipath with a close guardrail

and therefore hard to parallelize [12]. Moreover, when only a single snapshot is available, decorrelation techniques are required so that the signal subspace is fully represented. This can be achieved using the forward/backward (FB) averaging and/or spatial smoothing [13], which is suboptimal in general and can result in a reduction of the array aperture. The described drawbacks can limit the practical usage of subspace methods.

In contrast, the maximum likelihood (ML) DOA estimator of multiple targets can be directly applied with a single snapshot. It is asymptotically efficient [14] and possesses an improved threshold performance when compared to subspace methods [15]. Further, it allows resolving correlated targets [8]. Despite its good properties, the ML estimator has not enjoyed much practical application due to its high computational cost. It requires the optimization of a multidimensional objective function with a complicated multi-model shape. Computationally efficient but iterative implementations are the method of alternating projections [16] or the relaxation algorithm (RELAX) in [17].

Here, for the two-target case, we consider a global search of the two-dimensional ML objective function as practically feasible. We propose to use a simplified calculation of the objective function and a delimited search range. The required projection operators are data independent and can be pre-calculated off-line, which enables a trade-off between the computational complexity and the required storage space.

1.2 Signal Model

Let $\boldsymbol{x}$ denote the M-element array vector, or snapshot, from a pre-detected processing cell according to the range and relative velocity, whose power is significantly above the noise level. The problem formulation is posed as follows: decide between the single-target model and the two-target model

$$\begin{aligned} D=1: &\quad \boldsymbol{x} = s_0 \boldsymbol{a}(\psi_0) + \boldsymbol{n} \\ D=2: &\quad \boldsymbol{x} = s_1 \boldsymbol{a}(\psi_1) + s_2 \boldsymbol{a}(\psi_2) + \boldsymbol{n} \end{aligned} \tag{1.1}$$

and estimate the respective parameters. s_0 and ψ_0 are the target response parameter and DOA parameter in the single-target model, respectively. Likewise, s_1, s_2, ψ_1, and ψ_2 are the corresponding parameters in the two-target model.

$$\boldsymbol{a}(\psi) = \frac{1}{\sqrt{M}} \left[1, e^{j\psi}, \ldots, e^{j(M-1)\psi}\right]^T \tag{1.2}$$

but the steering vector of the considered uniform linear array (ULA) with electrical angle $\psi = \frac{2\pi}{\lambda} d \sin \phi$, where λ is the wavelength, d is the array element spacing, and ϕ is the spatial azimuth angle. The measurement noise vector $\boldsymbol{n}$ is assumed to be spatially white, circular complex Gaussian with zero mean and variance σ^2.

1.3 Optimal Processing

The optimal processing is described in the following. It consists of the ML DOA estimation for the single-target model and the two-target model and a generalized likelihood ratio test (GLRT).

1.3.1 Maximum Likelihood for One Target

The ML estimator for ψ_0 in model (1.1) for $D = 1$ corresponds to the location of the global maximum of the BF spectrum:

$$P(\psi) = \left|\boldsymbol{a}(\psi)^H \boldsymbol{x}\right|^2.$$

The inner vector product corresponds to a spatial Fourier transform at frequency ψ. Hence, $P(\psi)$ can be evaluated efficiently using a Fast Fourier transform (FFT) with zero-padding.

Let the step size of the evaluation grid be $\Delta\psi$, and let the location of the global maximum on the evaluation grid be ψ_n. A refined DOA estimate can be obtained using a quadratic interpolation in the neighborhood of ψ_n as

$$\hat{\psi}_0 = \psi_n + 0.5\Delta\psi \frac{P(\psi_{n-1}) - P(\psi_{n+1})}{P(\psi_{n-1}) - 2P(\psi_n) + P(\psi_{n+1})}$$

1.3.2 Maximum Likelihood for Two Targets

The ML estimators for ψ_1 and ψ_2 in model (1.1) for $D = 2$ correspond to the location of the global maximum of the two-dimensional ML objective function:

$$c(\psi_1, \psi_2) = \boldsymbol{x}^H \boldsymbol{P}_A(\psi_1, \psi_2)\boldsymbol{x} \tag{1.3}$$

where

$$\boldsymbol{P}_A(\psi_1, \psi_2) = \boldsymbol{A}\left(\boldsymbol{A}^H\boldsymbol{A}\right)^{-1}\boldsymbol{A}^H, \quad \boldsymbol{A} = [\boldsymbol{a}(\psi_1), \boldsymbol{a}(\psi_2)]$$

is the projection matrix onto the column span of steering matrix $\boldsymbol{A}$. An intuitive interpretation is that we seek for parameters ψ_1 and ψ_2 which maximize the projection of $\boldsymbol{x}$ onto the plane spanned by the columns of $\boldsymbol{A}$.

The optimization of $c(\psi_1,\psi_2)$ needs to be numerical and is generally computationally intensive. Below, we describe the direct calculation of the objective function and a global search procedure.

1.3.2.1 Direct Objective Function Evaluation

To determine projection matrix $\boldsymbol{P_A}(\psi_1,\psi_2)$, the matrix inverse of $\boldsymbol{A}^H\boldsymbol{A}$ is required. Using the inversion formula for a matrix of dimension two, and notation $\boldsymbol{a}_1 = \boldsymbol{a}(\psi_1)$ and $\boldsymbol{a}_2 = \boldsymbol{a}(\psi_2)$ for convenience, we have

$$\boldsymbol{P}_A(\psi_1, \psi_2) = \frac{1}{1 - |\beta|^2}\left(\boldsymbol{a}_1\boldsymbol{a}_2^H - \beta\boldsymbol{a}_1\boldsymbol{a}_2^H - \beta^*\boldsymbol{a}_2\boldsymbol{a}_1^H + \boldsymbol{a}_2\boldsymbol{a}_2^H\right)$$

where $\beta = \boldsymbol{a}_1^H\boldsymbol{a}_2$, and we have used $\boldsymbol{a}(\psi)^H\boldsymbol{a}(\psi) = 1$. This allow us to calculate directly (1.3) using

$$c(\psi_1, \psi_2) = \frac{1}{1 - |\beta|^2}\left(|y_1|^2 - 2\mathrm{Re}\{\beta y_1^* y_2\} + |y_2|^2\right) \tag{1.4}$$

where $y_1 = \boldsymbol{a}_1^H\boldsymbol{x}$ and $y_2 = \boldsymbol{a}_2^H\boldsymbol{x}$. Provided all steering vectors are available on a discrete grid of the field of view, a significant part of the computational cost, required to evaluate a single point of (1.4), constitutes the calculation of y_1, y_2, and β, which corresponds to $12M$ real-valued multiply–add operations.

1.3.2.2 Global Search

Due to the complicated multimodal shape of the objective function $c(\psi_1,\psi_2)$, a numerical search procedure, e.g., using a damped Newton method, critically depends on the initialization [18]. A fairly reliable initialization without eigendecomposition appears difficult to find, especially when the targets are not resolved in the BF spectrum. Here, we consider a global evaluation of the two-dimensional objective function on a selected grid for ψ_1 and ψ_2. Unlike numerical search procedures, this allows a non-iterative implementation.

Let $\psi_{1,m}$ and $\psi_{2,n}$ be the location of the global maximum on the evaluation grid with step size $\Delta\psi$; refined DOA estimates can be obtained using a quadratic interpolation in the neighborhood of the global maximum, as

$$\hat{\psi}_1 = \psi_{1,m} + 0.5\Delta\psi \frac{c(\psi_{1,m-1}, \psi_{2,n}) - c(\psi_{1,m+1}, \psi_{2,n})}{c(\psi_{1,m-1}, \psi_{2,n}) - 2c(\psi_{1,m}, \psi_{2,n}) + c(\psi_{1,m+1}, \psi_{2,n})}$$

$$\hat{\psi}_2 = \psi_{2,n} + 0.5\Delta\psi \frac{c(\psi_{1,m}, \psi_{2,n-1}) - c(\psi_{1,m}, \psi_{2,n+1})}{c(\psi_{1,m}, \psi_{2,n-1}) - 2c(\psi_{1,m}, \psi_{2,n}) + c(\psi_{1,m}, \psi_{2,n+1})}$$

Regarding computational cost, the global evaluation of the objective function is required only for $\psi_1 < \psi_2$. Note that the resulting triangular search range is shown in Fig. 1.2 (top right). The corresponding computational cost is

$$C = C_1 N_2, \quad N_2 = \frac{N_\psi(N_\psi - 1)}{2}$$

where C_1 represents the computational cost, required to evaluate a single point of the objective function; N_2 is the number of points in the two-dimensional search range; and N_ψ is the number of grid points in the field of view, $\psi \in [-\pi, \pi)$.

1.3.3 Generalized Likelihood Ratio Test

According to the model (1.1) for $D = 1$ and $D = 2$, let the unknown parameters be collected in vectors $\boldsymbol{\Theta}_1$ and $\boldsymbol{\Theta}_2$, respectively, and let $p_1(\boldsymbol{x}|\boldsymbol{\Theta}_1)$ and $p_2(\boldsymbol{x}|\boldsymbol{\Theta}_2)$ be the corresponding likelihood functions, i.e., the conditional probability density function of the snapshot given the unknown parameter. A GLRT for deciding between the single-target model and the two-target model is

$$T = \frac{\max_{\boldsymbol{\Theta}_2} p_2(\boldsymbol{x}|\boldsymbol{\Theta}_2)}{\max_{\boldsymbol{\Theta}_1} p_1(\boldsymbol{x}|\boldsymbol{\Theta}_1)} > \gamma$$

which involves the determination of the corresponding ML estimates. For the signal model in Sect. 1.2, and taking the logarithm, the GLRT can be simplified to [19]

$$\log T = M \log \hat{\sigma}_1^2 - M \log \hat{\sigma}_2^2 > \log \gamma$$

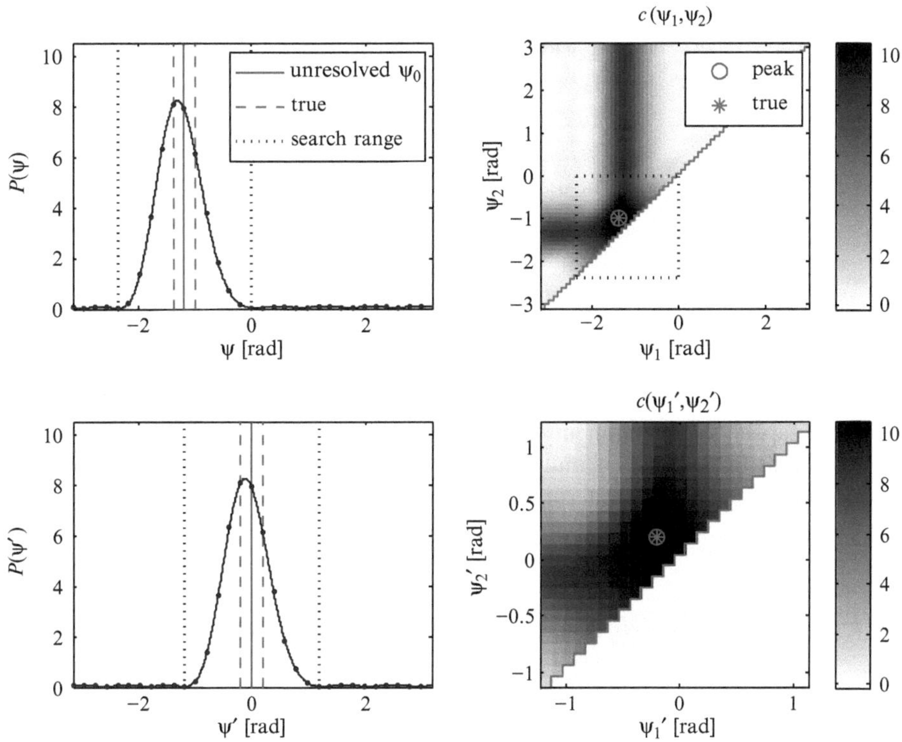

Fig. 1.2 Example with $M = 8$, two targets separated by $\psi_2 - \psi_1 = 0.5\text{BW}$, noise-free and single snapshot: BF spectrum to identify relevant angular sector (*top left*), ML objective function for full search range with $\Delta\psi = \pi/32$ (*top right*), shifted BF spectrum (*bottom left*), and ML objective function for shifted delimited search range (*bottom right*)

where

$$\hat{\sigma}_1^2 = \frac{1}{M}\left\|\boldsymbol{x} - \boldsymbol{a}(\hat{\psi}_0)\boldsymbol{a}(\hat{\psi}_0)^H\boldsymbol{x}\right\|^2, \quad \hat{\sigma}_2^2 = \frac{1}{M}\left\|\boldsymbol{x} - \boldsymbol{P}_A(\hat{\psi}_1, \hat{\psi}_2)\boldsymbol{x}\right\|^2$$

A suitable threshold value $\log\gamma$ can be determined numerically, according to the Neyman–Pearson principle.

1.4 Proposed Approach

If multiple snapshots of the same processing cell, say at different cycles n, are considered, model (1.1) extends to $\boldsymbol{x}[n]$, $n = 1, \ldots, N$, where N is the number of available snapshots. In this case, the ML objective function in (1.3) extends to

$$c(\psi_1,\psi_2) = \frac{1}{N}\sum_{n=1}^{N} \boldsymbol{x}[n]^H \boldsymbol{P}_A(\psi_1,\psi_2)\boldsymbol{x}[n] = \mathrm{Tr}\left\{\boldsymbol{P}_A(\psi_1,\psi_2)\hat{\boldsymbol{R}}\right\} \tag{1.5}$$

where

$$\hat{\boldsymbol{R}} = \frac{1}{N}\sum_{n=1}^{N} \boldsymbol{x}[n]\boldsymbol{x}[n]^H$$

is the sample covariance matrix. We remark that the single-snapshot case is of primary importance for the considered application. However, to enhance the DOA estimation accuracy, it may be desirable to combine multiple snapshots, which have been associated on a tracking procedure. Therefore, we consider the general case with N snapshots and comment on the special case with $N = 1$.

Note that in the case of multiple snapshots, one has to trade off between the evaluation of the quadratic term and the trace notation in (1.5).

1.4.1 Simplified Objective Function Calculation

The projection operator $\boldsymbol{P_A}(\psi_1, \psi_2)$ is data independent; therefore, it can be pre-calculated off-line and stored. In this case, the calculation of the trace notation in (1.5) requires $C_1 = 4M^2$ real-valued multiply–add operations (note that only the diagonal entries of the matrix product have to be evaluated). Moreover, this can be simplified since $\boldsymbol{P_A}(\psi_1, \psi_2)$ has a great deal of structure to exploit. In particular, it is centro-Hermitian, i.e., we have

$$\boldsymbol{J}_M \boldsymbol{P}_A(\psi_1,\psi_2)^* \boldsymbol{J}_M = \boldsymbol{P}_A(\psi_1,\psi_2)$$

where $\boldsymbol{J}_M$ is the exchange matrix of size M, with ones on the anti-diagonal and zeros elsewhere. This property can be easily shown [19] and directly follows from the fact that the steering vector, defined in (1.2), is Hermitian symmetric up to a complex scaling.

As a consequence, the ML objective function remains unchanged when snapshot $\boldsymbol{x}$ is replaced by $\boldsymbol{J}_M\boldsymbol{x}^*$, since

$$(\boldsymbol{J}_M\boldsymbol{x}^*)^H \boldsymbol{P}_A(\psi_1,\psi_2)\boldsymbol{J}_M\boldsymbol{x}^* = \left[\boldsymbol{x}^H \boldsymbol{J}_M \boldsymbol{P}_A(\psi_1,\psi_2)^* \boldsymbol{J}_M \boldsymbol{x}\right]^* = \boldsymbol{x}^H \boldsymbol{P}_A(\psi_1,\psi_2)\boldsymbol{x}$$

where we have used the fact that $c(\psi_1, \psi_2)$ is real-valued by definition. Likewise, it remains unchanged when $\hat{\boldsymbol{R}}$ is replaced by the forward/backward (FB) averaged sample covariance matrix

$$\hat{\boldsymbol{R}}_{\mathrm{FB}} = \frac{1}{2}\left(\hat{\boldsymbol{R}} + \boldsymbol{J}_M \hat{\boldsymbol{R}}^* \boldsymbol{J}_M\right)$$

which is centro-Hermitian by definition.

1.4.1.1 Unitary Transformation

Let $\boldsymbol{Q}_M$ be a column conjugate symmetric matrix, satisfying $\boldsymbol{J}_M \boldsymbol{Q}_M^* = \boldsymbol{Q}_M$. A sparse choice for a unitary column conjugate symmetric matrix is

$$\boldsymbol{Q}_{2m+1} = \frac{1}{\sqrt{2}} \begin{bmatrix} \boldsymbol{I}_m & \boldsymbol{0} & j\boldsymbol{I}_m \\ \boldsymbol{0}^T & \sqrt{2} & \boldsymbol{0}^T \\ \boldsymbol{J}_m & \boldsymbol{0} & -j\boldsymbol{J}_m \end{bmatrix}$$

where $\boldsymbol{I}_m$ is the identity matrix of size m. An equivalent unitary column conjugate symmetric matrix of dimension $2m$ can be obtained by deleting the center row and center column of $\boldsymbol{Q}_{2m+1}$. The main result of [20] is that any square centro-Hermitian matrix is equivalently expressed by a real-valued matrix of the same dimension so that

$$V(\psi_1, \psi_2) = \boldsymbol{Q}_M^H \boldsymbol{P}_A(\psi_1, \psi_2) \boldsymbol{Q}_M \tag{1.6}$$

and

$$\hat{\boldsymbol{C}} = \boldsymbol{Q}_M^H \hat{\boldsymbol{R}}_{\mathrm{FB}} \boldsymbol{Q}_M \tag{1.7}$$

are the real-valued projection operator and the sample covariance. The similarity transformation with unitary matrix $\boldsymbol{Q}_M$ is referred to as unitary transformation. We note that this approach has been used in [11] and [21], respectively, to derive the unitary ESPRIT and unitary root-MUSIC algorithm, where computational cost is reduced by replacing a complex-valued eigendecomposition by a real-valued one. Since we got $\boldsymbol{Q}_M \boldsymbol{Q}_M^H = \boldsymbol{I}_M$, the objective function in (1.5) can be rewritten as

$$\begin{aligned} c(\psi_1, \psi_2) &= \mathrm{Tr}\left\{\boldsymbol{P}_A(\psi_1, \psi_2)\hat{\boldsymbol{R}}\right\} \\ &= \mathrm{Tr}\left\{\boldsymbol{P}_A(\psi_1, \psi_2)\boldsymbol{Q}_M \boldsymbol{Q}_M^H \hat{\boldsymbol{R}}_{\mathrm{FB}} \boldsymbol{Q}_M \boldsymbol{Q}_M^H\right\} \\ &= \mathrm{Tr}\left\{\boldsymbol{Q}_M^H \boldsymbol{P}_A(\psi_1, \psi_2)\boldsymbol{Q}_M \boldsymbol{Q}_M^H \hat{\boldsymbol{R}}_{\mathrm{FB}} \boldsymbol{Q}_M\right\} \\ &= \mathrm{Tr}\left\{\mathbf{V}(\psi_1, \psi_2)\hat{\boldsymbol{C}}\right\}. \end{aligned} \tag{1.8}$$

To further reduce computational cost, we exploit that $\mathbf{V}(\psi_1,\psi_2)$ and $\hat{\boldsymbol{C}}$ are symmetric and remove redundant matrix entries [19]. In this case, and provided

Table 1.1 Computations required for evaluating a single point of the objective function, C_1, and storage space, in the single-snapshot case, using $N_2 = N_\psi(N_\psi - 1)/2$

	C_1	Storage space
Direct (1.4)	$\approx 12M$	$N_\psi 2M$
Simplified (1.8)	$(M + 1)M/2$	$N_2(M + 1)M/2$
Simplified (1.9)	$\approx 4M$	$N_2 2M$

all projection operators are available on a discrete grid of the two-dimensional search range, the calculation of (1.8) requires $C_1 = (M + 1)M/2$ real-valued multiply–add operations.

1.4.1.2 Single-Snapshot Alternative

In the single-snapshot case, an alternative is to employ an eigendecomposition of the real-valued projection operator in (1.6),

$$V(\psi_1, \psi_2) = v_1 v_1^T + v_2 v_2^T$$

where eigenvectors $v_1, v_2 \in \mathbb{R}^{M\times 1}$ are both functions of ψ_1 and ψ_2. Again, the projection operator eigenvectors can be pre-calculated off-line and stored. Using $y = Q_M^H x \in \mathbb{C}^{M\times 1}$, the objective function in (1.3) can be rewritten as

$$c(\psi_1, \psi_2) = y^H V(\psi_1, \psi_2) y = y^H \left(v_1 v_1^T + v_2 v_2^T\right) y = |z_1|^2 + |z_2|^2 \tag{1.9}$$

where $z_1 = v_1^T y$ and $z_2 = v_2^T y$. Provided all projection operator eigenvectors are available on a discrete grid of the two-dimensional search range, a significant part of (1.9) constitutes the calculation of z_1 and z_2, which corresponds to $C_1 \approx 4M$ real-valued multiply–add operations.

1.4.1.3 Comparison

The overall cost of a global search has been described in Sect. 1.3.2.2. A trade-off between the computations, required for evaluating a single point of the objective function C_1 (in real-valued multiply–add operations), and the required storage space (in real-valued numbers) is given in Table 1.1 for the single-snapshot case. Note that the calculation of the real-valued projection operators, or respective eigenvectors, is done off-line and does not contribute to the overall cost. Also, the preprocessing, such as the formation of the covariance matrix, has no significant effect, as it is performed only once.

For an eight-element ULA and for the single-snapshot case, the simplified objective function in (1.9) is the cheapest option, both in terms of required computations and the storage space. However, when multiple snapshots are available, we prefer the simplified objective function in (1.8), because the covariance matrix is employed and no extensions are necessary.

Regarding the storage, the simplified calculation requires the real-valued projection operators, or respective eigenvectors, on a two-dimensional search range with N_2 points, whereas the direct calculation only requires the steering vectors on a one-dimensional grid of the field of view.

1.4.2 Delimited Search Range

So far, we have reduced the computational cost by simplifying the calculation of the ML objective function. Next, we consider a delimited search range so that the number of points to evaluate on a two-dimensional search range and the storage space is reduced.

We only consider the more difficult case of closely spaced targets, which cannot be reliably resolved in the BF spectrum, i.e., $\psi_2 - \psi_1 < \text{BW}$, where $\text{BW} = 2\pi/M$ is the Rayleigh beamwidth. We remark that when the targets are widely separated so that they are reliably resolved in the BF spectrum, there exist computationally simple methods to reduce the estimation bias due to the leakage effect [22].

Let $\hat{\psi}_0 \in [\psi_1, \psi_2]$ be the peak location of the unresolved targets in the BF spectrum. Consider the shifted array output model, which is obtained by a rotational shift of the field of view:

$$\boldsymbol{x}' = \sqrt{M}\boldsymbol{a}(-\hat{\psi}_0) \odot \boldsymbol{x} = s_1\boldsymbol{a}\left(\psi_1'\right) + s_2\boldsymbol{a}\left(\psi_2'\right) + \boldsymbol{n}' \tag{1.10}$$

where $\psi_1' = \psi_1 - \hat{\psi}_0$ and $\psi_2' = \psi_2 - \hat{\psi}_0$ are the shifted DOA parameters and $\odot$ is the element-wise Hadamard product. The random characteristics of the rotationally shifted noise vector $\boldsymbol{n}'$ remain unchanged. The rotational shift allows to evaluate $c(\psi_1', \psi_2')$ on a delimited search range, e.g., $\psi' \in [-1.5\text{BW}, 1.5\text{BW}]$, which very likely contains the centered DOA parameters ψ_1' and ψ_2'. As a result, the number of points in the two-dimensional search range, N_2, and therewith the storage space of the projection operators have been reduced significantly. For the given example, the reduction corresponds roughly to $(3\text{BW}/2\pi)^2 = (3/M)^2$.

1.4.3 Example

We present an example to demonstrate the principle of the delimited search range and the rotational shift. A ULA with $M = 8$ elements, spaced by $d = \lambda/2$, is used. A noise-free single snapshot is simulated according to model (1.1) for $D = 2$, with target response parameters $s_1 = \sqrt{2}e^{-j\pi/4}s_2 = 1$, and an angular separation of $\psi_2 - \psi_1 = 0.5\mathrm{BW}$. Figure 1.2 shows the results.

The upper and lower left plots show the BF spectra of the original snapshot and the shifted snapshot, respectively. Since the targets are not resolved in the BF spectrum, the unresolved peak ψ_0 can be used to identify the relevant sector for the delimited search range, which is indicated by the dotted lines. The upper and lower right plots show the ML objective function with step size $\Delta\psi = \pi/32$ for the full search range and the shifted delimited search range, respectively, which correspond to $\psi \in [-\pi, \pi)$ and $\psi' \in [-1.5\mathrm{BW}, 1.5\mathrm{BW}]$.

1.5 Experimental Data Analysis

We present results obtained with experimental data from a typical application in automotive radar. The scenario with horizontal multipath and a close guardrail, as shown in Fig. 1.1, is considered again. The two propagation paths, corresponding to the observed car and the ghost target, fall into the same processing cell if $r_2 - r_1 < \Delta r$, where Δr is the size of a range cell. The range and DOA parameters are related by

$$\begin{aligned} r_1 \sin(\phi_1) &= d_2 - d_1 \\ r_2 \sin(\phi_2) &= d_2 + d_1 \end{aligned}$$

where d_1 and d_2 are the normal distances from the guardrail.

The employed radar system operates at carrier frequency 24 GHz and has a range resolution of 1.8 m. For DOA estimation, an array of microstrip patch antennas in the form of a ULA with $M = 7$ elements, spaced by $d = \lambda/2$, is used. In the selected recording, the car with the radar system is following another car on the left lane of the motorway. In roughly 300 cycles, the distance of the observed car increases from 25 to 50 m.

For extracting relevant processing cells an initial DOA is determined by the peak of the BF spectrum. Relevant processing cells are extracted as follows:

- Detection, to select only cells with significant energy
- Clustering of cells with neighboring range, similar relative velocity, and initial DOA
- Gating, to consider only cells of interest for a certain application, whose relative velocity and initial DOA fall into a desired gate

The proposed ML estimator for two targets from Sect. 1.4 and the GLRT from Sect. 1.3.3 are applied to all relevant processing cells. For two selected cycles, Fig. 1.3 shows the camera recording of the scene and the result of the radar target localization as a function of x- and y-position.

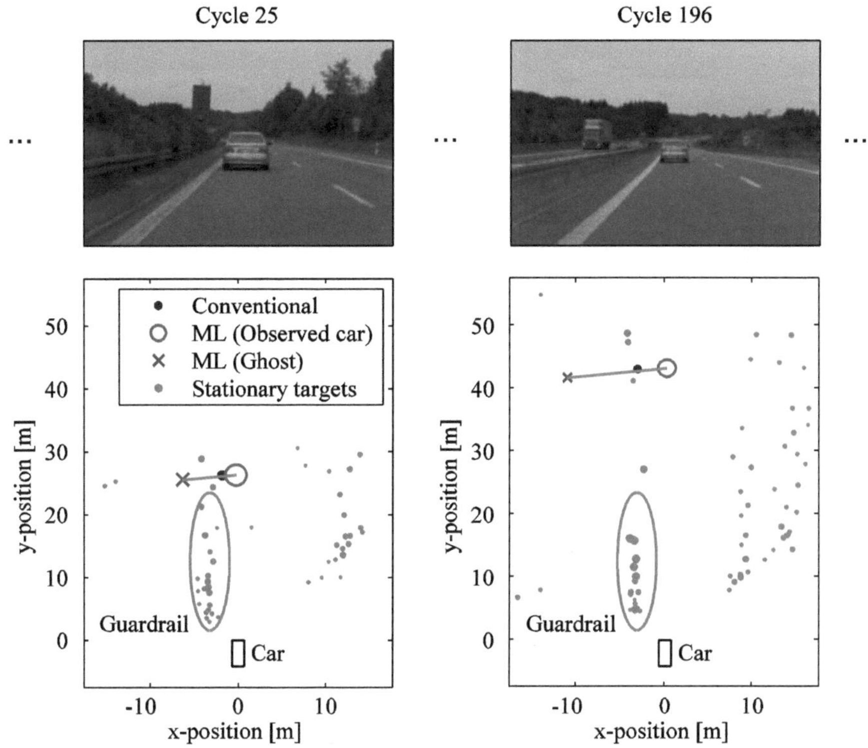

Fig. 1.3 Experimental data analysis. Scenario and all detected and clustered targets in bird's eye view for two selected cycles

Gray dots correspond to stationary targets, while black dots correspond to moving targets in the relevant gate, both for single-target DOA estimates. The result of the proposed ML estimator for two targets is indicated with a circle and cross, corresponding to the observed car and ghost target, respectively. On average, the measured power of the ghost target is roughly 6 dB smaller than the power of the observed car. The marker size of all displayed targets is proportional to the measured SNR. Note that the indicated stationary target detections can be used to localize the guardrail, which is required to identify the multipath situation.

Figure 1.4 shows the DOA estimation results of all cycles versus range. Note that the two selected cycles from Fig. 1.3 are indicated. In the upper plot, we show the conventional results with single-target DOA estimation using the BF. In the lower plot, however, we display improved results from two-target ML DOA estimation.

It can be observed from Fig. 1.4 that for the selected situation the conventional single-target DOA estimates tend to erroneously localize the observed car closer to the guardrail. When the multipath propagation is identified correctly, it is possible to apply the proposed ML estimator for two targets and adequately localize the observed car and a ghost target.

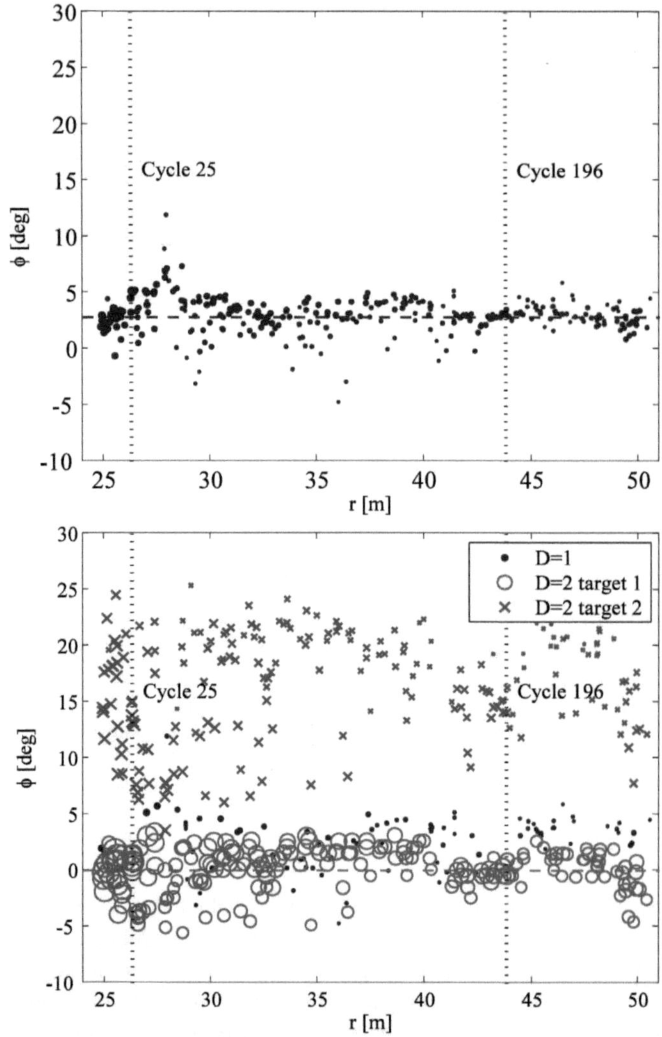

Fig. 1.4 Experimental data analysis. DOA estimates versus range: conventional results with single-target DOA estimation using the BF (*top*), improved results with two-target ML DOA estimation (*bottom*)

1.6 Conclusions

We have considered the practically relevant problem of high-resolution DOA estimation and detection of up to two targets. We have proposed a fast implementation of a grid search ML estimator, in which the ML objective function has been simplified and the required projection operators are pre-calculated off-line and stored. For saving storage space and computations, we have proposed a rotational shift of the field of view such that the relevant angular sector, which has to be evaluated, is delimited and centered with respect to the broadside. The proposed method allows a computationally simple and straightforward implementation. The principle of the proposed method has been demonstrated using an example with simulated data. Finally, we have presented results obtained with experimental data from a typical application in automotive radar, in which high-resolution DOA estimation results in enhanced target localization.

References

1. J. Hansen, P. Boyraz, K. Takeda, H. Abut, *Digital Signal Processing for In-Vehicle Systems and Safety* (Springer, New York, 2011)
2. F. Gustaffson, Automotive safety systems. IEEE Signal Process. Mag. **26**(4), 32–47 (2009)
3. H. Winner, S. Hakuli, G. Wolf, *Advanced Driver Assistance Systems Handbook* (Vieweg + Teubner, Wiesbaden, 2009) (in German: Handbuch Fahrerassistenzsysteme: Grundlagen, Komponenten und Systeme für aktive Sicherheit und Komfort)
4. M. Skolnik, *Radar Handbook* (McGraw-Hill, New York, 2008)
5. M. Richards, *Fundamentals of Radar Signal Processing* (McGraw-Hill, New York, 2005)
6. M. Wintermantel, Radar system with improved angle formation, Germany Patent Application WO 2010/000252, 2010
7. H. Krim, M. Viberg, Two decades of array signal processing research. IEEE Signal Process. Mag. **13**(4), 67–94 (1996)
8. H. van Trees, *Detection, Estimation, and Modulation Theory—Part IV Optimum Array Processing* (Wiley, New York, 2002)
9. E. Tuncer, B. Friedlander, *Classical and Modern Direction-of-Arrival Estimation* (Academic Press, New York, 2009)
10. R. Schmidt, Multiple emitter location and signal parameter estimation. IEEE Trans. Antennas Propag. **34**(3), 276–280 (1986)
11. M. Haardt, J. Nossek, Unitary ESPRIT: how to obtain increased estimation accuracy with a reduced computational burden. IEEE Trans. Signal Process. **43**(5), 1232–1242 (1995)
12. G. Golub, C. van Loan, *Matrix Computations* (The Johns Hopkins University Press, Baltimore, MD, 1996)
13. S. Pillai, C. Kwon, Forward/backward spatial smoothing techniques for coherent signal identification. IEEE Trans. Acoust. Speech Signal Process. **37**(1), 8–15 (1989)
14. P. Stoica, A. Nehorai, MUSIC, maximum likelihood, and the cramér-Rao bound. IEEE Trans. Acoust. Speech Signal Process. **37**(5), 720–741 (1989)
15. Y. Abramovich, B. Johnson, X. Mestre, Performance breakdown in MUSIC, G-MUSIC and maximum likelihood estimation, in *Proc. of the 32nd IEEE Int. Conf. on Acoustics, Speech and Sig. Proc. (ICASSP)*, Honolulu, USA, 2007
16. I. Ziskind, M. Wax, Maximum likelihood localization of multiple sources by alternating projection. IEEE Trans. Acoust. Speech Signal Process. **36**(10), 1553–1560 (1988)
17. J. Li, D. Zheng, P. Stoica, Angle and waveform estimation via RELAX. IEEE Trans. Aerosp. Electron. Syst. **33**(3), 1077–1087 (1997)
18. B. Ottersten, M. Viberg, P. Stoica, A. Nehorai, Exact and large sample ML techniques for parameter estimation and detection in array processing, in *Radar Array Processing* (Springer, Berin, 1993)
19. P. Heidenreich, *Antenna Array Processing: Autocalibration and Fast High-Resolution Methods for Automotive Radar*, Ph.D. thesis, Technische Universität Darmstadt, 2012

20. A. Lee, Centrohermitian and skew-centrohermitian matrices. Linear Algebra Appl. **29**, 205–210 (1980)
21. M. Pesavento, A. Gershman, M. Haardt, Unitary root-MUSIC with a real-valued eigendecomposition: a theoretical and experimental performance study. IEEE Trans. Signal Process. **49**(5), 1306–1314 (2000)
22. P. Heidenreich, A. Zoubir, Computationally simple DOA estimation of two resolved targets with a single snapshot, in *Proc. of the 37th IEEE Int. Conf. on Acoustics, Speech and Sig. Proc. (ICASSP)*, Kyoto, Japan, 2012

Chapter 2
Dense 3D Motion Field Estimation from a Moving Observer in Real Time

Clemens Rabe, Uwe Franke, and Reinhard Koch

Abstract In this chapter an approach for estimating the three-dimensional motion fields of real-world scenes is proposed. This approach combines state-of-the-art dense optical flow estimation, including spatial regularization, and dense stereo information using Kalman filters to achieve temporal smoothness and robustness. The result is a dense and accurate reconstruction of the three-dimensional motion field of the observed scene. An efficient parallel implementation using a GPU and an automotive compliant FPGA yields a real-time vision system which is directly applicable in real-world scenarios including driver assistance systems, robotics, and surveillance.

Keywords Computer vision • Driver assistance • Motion estimation

2.1 Introduction

Future driver assistance systems are expected to support the driver in complex driving situations. This requires a thorough understanding of the car's environment, which includes not only the perception of the infrastructure but also the precise detection and tracking of moving traffic participants.

The 3D structure of the scene can easily be obtained through a stereo camera system. To detect obstacles, this information is commonly accumulated in an evidence grid-like structure [13]. We refer to it as the bird's view map. Since stereo

---

C. Rabe (✉) • U. Franke
Research and Technology/Machine Perception, Daimler AG, Hanns-Klemm-Str. 45, 71059 Sindelfingen, Germany
e-mail: clemens.rabe@daimler.com; uwe.franke@daimler.com

R. Koch
Multimedia Information Processing, Christian-Albrechts-Universität zu Kiel, Hermann-Rodewald-Str. 3, 24098 Kiel, Germany
e-mail: rk@mip.informatik.uni-kiel.de

G. Schmidt et al. (eds.), *Smart Mobile In-Vehicle Systems: Next Generation Advancements*, DOI 10.1007/978-1-4614-9120-0_2,

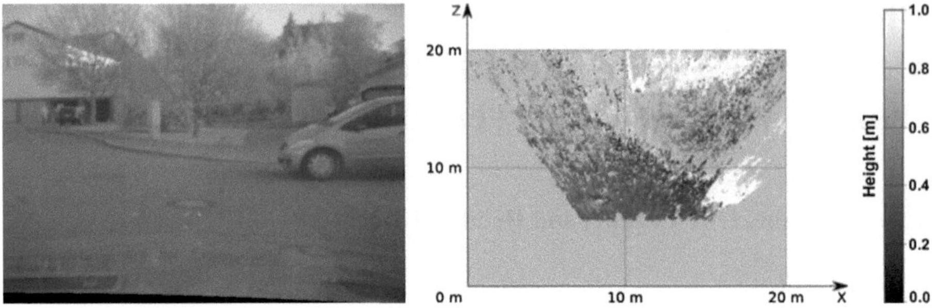

Fig. 2.1 *Left*: Typical traffic scene. The car is stationary, whereas the pedestrian runs towards the intersection. *Right*: Bird's eye view of the reconstructed 3D structure of the scene. The color encodes the height above ground (from *black* (0 m) to *white* (1.0 m)) of the observed points

does not reveal any motion information, usually this map is segmented and detected objects are tracked over time in order to obtain their motion state. The major disadvantage of this standard approach is that the performance of the detection depends highly on the correctness of the segmentation. In particular, moving objects in front of stationary ones, as illustrated in Fig. 2.1, are often bundled and therefore not detected. This causes erroneous misinterpretations and requires more powerful solutions.

In literature there are many examples using the optical flow, i.e., the apparent image motion, to detect moving objects. Argyros et al. describe a method to detect moving objects using stereo vision [1]. Comparing the normal flow of the right camera image with the normal flow between the left and the right images of the stereo cameras they detect image regions with independent object motion as inconsistencies in the flow data. Other approaches like Kellman and Kaiser [12], Mills [14], and Heinrich [8] use the geometric constraints stemming from the stereo configuration to detect independent motion. However, these approaches lack a precise measurement of the detected movements.

One of the first attempts to fuse stereo and optical flow information was studied by Waxman and Duncan [20]. They have analyzed the relationship between the optical flow fields of each camera and defined the so-called relative flow. Using this information the relative longitudinal velocity between the observer and the object is directly determined. Other approaches known as scene flow algorithms estimate the 3D motion field directly from two consecutive image pairs. The term scene flow was introduced by Vedula et al. in [19] and was defined as the three-dimensional motion of points in the world. In practice, scene flow algorithms estimate the optical flow and disparity change, optionally also the disparity map, in a combined approach. From this information, the 3D motion field is then reconstructed. Although computationally expensive, dense scene flow algorithms running in real time are available, as shown by Rannacher in [18].

Fig. 2.2 *Left*: Typical traffic scene. *Right*: Motion field, estimated by the Dense6D algorithm proposed in this chapter. The color encodes the velocity of the observed points

Although scene flow analysis provides fast detection results, it is limited with respect to robustness and accuracy due to the immanent measurement noise caused by its differential character. To get more reliable results, an integration of the observations over time is necessary. This was done in the 6D-Vision system that was first presented in [5]. The basic idea is to track points with depth known from stereo vision over two and more consecutive frames and then to fuse the spatial and temporal information using Kalman filters [11]. The result is an improved accuracy of the 3D position and an estimation of the 3D motion of the point under study at the same time. Since we get a rich 6D-state vector for each point this method is referred as 6D-Vision. Taking into account the motion information, the detection of moving objects can be carried out significantly easier and more robust than using bird view maps. In addition, using the 3D motion information a prediction of the object movement is possible. This allows a driver assistance system to warn and react to potential collisions in time.

The improved system presented in [16] is able to track up to 10,000 points in real time on modern hardware devices. However, in safety relevant applications, robustness, density, and volume of information are of utmost importance. Inspired by the recent progress of dense optical flow algorithms, Rabe et al. applied the 6D-Vision principle to dense optical flow and dense stereo data [17]. Implemented on the Graphics Processing Unit (GPU), this system is able to analyze 640×480 pixels at a frame rate of 20 Hz. A typical result for a traffic scene is shown in Fig. 2.2. Since the 3D motion information is estimated for nearly every pixel of the image, this system is called Dense6D.

In this contribution, the Dense6D system is described in more detail, focusing on the implementation aspects for a real-time system. The chapter is organized as follows: Section 2.2 describes the core elements of the system, namely, the dense stereo and dense optical flow algorithms, as well as the Kalman filter-based fusion to obtain the 3D motion estimates. In Sect. 2.3, the quantitative evaluation results on synthetic ground-truth image sequences and qualitative results on real-world sequences are presented, followed by the conclusion in Sect. 2.4.

2.2 Estimation of the 3D Motion Field

The Dense6D algorithm estimates a 3D motion field based on a dense disparity map and a dense optical flow field that can be computed by any stereo and optical flow algorithms, respectively. However, automotive applications demand robust and real-time capable algorithms. Therefore, we use two state-of-the-art algorithms meeting these requirements: the Semi-Global-Matching (SGM) stereo algorithm and the TV-L^1 optical flow algorithm. In the following, both algorithms are presented, followed by a detailed description of the Dense6D algorithm.

Throughout this chapter, we assume that the camera system is calibrated, and the captured images are preprocessed by a rectification module that performs a lens-correction and establishes a standard stereo configuration.

2.2.1 *Dense Stereo*

In [9] Hirschmüller presented an algorithm to obtain dense disparity maps by using mutual information as a cost function and minimizing the energy functional

$$\begin{aligned} E = &\sum_{\mathrm{x}\in\Omega} C(\mathrm{x}, d_{\mathrm{x}}) \\ &+ \sum_{\mathrm{x}\in\Omega}\sum_{\mathrm{y}\in N_{\mathrm{x}}} p_1 T\left[\left|d_{\mathrm{y}} - d_{\mathrm{x}}\right| = 1\right] \\ &+ \sum_{\mathrm{x}\in\Omega}\sum_{\mathrm{y}\in N_{\mathrm{x}}} p_2 T\left[\left|d_{\mathrm{y}} - d_{\mathrm{x}}\right| > 1\right] \end{aligned} \tag{2.1}$$

with $C(\mathrm{x},d_{\mathrm{x}})$ being the matching cost function of the disparity d_{x} at the image position x of the image domain $\Omega = \{\mathrm{x}\} \subset \mathbb{R}^2$, N_{x} the neighborhood of x, p_1 and p_2 the smoothness penalties, and $T[\cdot]$ a function returning 1 if the inner expression is true and 0 otherwise. The first term simply sums all matching costs and can be interpreted as the data term. The second and third terms act as smoothness terms: Small deviations of neighboring disparities are penalized by a penalty p_1, and large deviations are penalized by the (constant) penalty p_2. Since the penalty p_1 is smaller than the penalty p_2, slanted or curved surfaces are preferred over disparity discontinuities.

Since a global solution to the energy minimization problem is computationally expensive, Hirschmüller proposed to resolve it approximately using dynamic programming, thus giving it the name SGM. The recursive scheme for the costs of the applied dynamic programming is defined as

$$L_{\mathbf{r}}(\mathrm{x},d) = C(\mathrm{x},d) + \min \begin{cases} L_{\mathbf{r}}(\mathrm{x}-\mathbf{r},d) \\ L_{\mathbf{r}}(\mathrm{x}-\mathbf{r},d-1)+p_1 \\ L_{\mathbf{r}}(\mathrm{x}-\mathbf{r},d+1)+p_1 \\ L_{\mathbf{r}}(\mathrm{x}-\mathbf{r},i)+p_2 \quad i<d-1 \\ L_{\mathbf{r}}(\mathrm{x}-\mathbf{r},i)+p_2 \quad i>d+1 \end{cases} - \min_k L_{\mathbf{r}}(\mathrm{x}-\mathbf{r},k) \tag{2.2}$$

along a path defined by the step vector $\mathbf{r}$. The costs over multiple paths of different directions (horizontal, vertical, and diagonal) are accumulated, and the resulting disparity is then found as the one with the minimum accumulated cost. Since paths of different directions are used, the typical streaking effects known from stereo algorithms evaluating only one path can be removed almost completely.

The resulting disparity map is only pixel-discrete, because the costs are calculated for discrete disparity values only. To obtain sub-pixel accuracy, the costs near the obtained minimal disparity are taken and the refined disparity is then found at the minimum of the parabola passing through these points.

Although the original algorithm used mutual information to determine the costs of a disparity match, the above energy minimization scheme can be used with almost any matching score. In practice, the zero-mean sum of absolute differences (ZSAD) and the Census operator [23] proved to be most robust even in situations of large illumination changes and are also less computationally expensive compared to the mutual information measure.

Solution of the minimization problem still remains computationally heavy, and a straightforward implementation takes about 2 s to compute a VGA disparity map on a state-of-the-art computer. Since the calculation of the cost cube requires access to the predecessors, each path must be computed sequentially, rendering it unsuitable for massively parallel machines like GPUs. However, Gehrig el al. proposed an implementation on an FPGA, which is able to calculate the disparity map in about 30 ms [6]. The massive speedup was achieved by calculating the disparity map on a sub-sampled image of half the resolution (overview image) and then combining it with the disparity map calculated for a portion of the image computer at the full resolution (fine image). This strategy assumes that distant objects of interest mainly occur in the center of the image, which is typical for traffic scenes. The implemented engine allows the computation of 64 disparity steps at each level, which leads to a total disparity range of 128 pixels. The implementation supports the ZSAD and Census matching costs, and the sub-pixel refinement is performed using an equi-angular fit. The sub-pixel accuracy of the engine is 1/16 pixel due to the use of fixed-point arithmetic. To remove mismatches, especially for partially occluded pixels, a right–left verification step is performed additionally. Using a CPU implementation with similar optimizations, the algorithm runs at 14 Hz [7]. The results for a typical traffic scene are displayed in Fig. 2.3.

Fig. 2.3 *Left*: Traffic scene. *Right*: Three-dimensional visualization of the corresponding scene

2.2.2 Dense Optical Flow

Dense optical flow algorithms calculate an optical flow vector for every pixel in the reference image by introducing a regularization term and formulating the problem as an energy minimization problem. Inspired by the seminal work of Horn and Schunck [10], a diverse range of such techniques has been developed. For a detailed review, the reader is referred to the surveys [2, 4, 22].

The method presented by Horn and Schunck solves the aperture problem by introducing a smoothness constraint, with the assumption that nearby optical flow vectors are similar in direction and magnitude. The optical flow field $\mathbf{u} = \left(u_x, u_y\right)^{\mathrm{T}} : \Omega \rightarrow \mathbb{R}^2$ is found by minimizing the energy function

$$E = \int_{\Omega} \left\{ \lambda |\rho(\mathrm{x}, \mathbf{u}(\mathrm{x}))|^n + \sum_{i=x,y} |\nabla u_i(\mathrm{x})|^n \right\} d\mathrm{x} \tag{2.3}$$

with $n = 2$. The parameter λ defines the weight of the data term with respect to the regularization term. The data term $\rho(\mathrm{x},\mathbf{u})$ is the constant brightness assumption, which is defined as

$$\rho(\mathrm{x}, \mathbf{u}) = I_1(\mathrm{x} + \mathbf{u}) - I_0(\mathrm{x}) \tag{2.4}$$

with $I_{\{0,1\}} : \Omega \rightarrow \mathbb{R}^2$ as the intensity function of the previous and current images, respectively.

The resulting optical flow field yields very encouraging results in regions of constant optical flow. However, due to the quadratic penalization in the smoothness term, the algorithm tends to over-smooth the optical flow field at flow boundaries and also over-weights the outliers. Therefore, Zach et al. presented in [24] a computational method to solve the energy function for the case $n = 1$, with significantly improved results. However, in automotive scenarios the constant brightness assumption is often violated, e.g., shadow casts or changes of the exposure time, that results in incorrect optical flow estimates (Fig. 2.4b). Wedel et al. has proposed a structure–texture decomposition of the input images to overcome this problem [21]. Excellent results were obtained for image pairs containing only small displacements. However, this decomposition removes vital image information essential in estimating large displacements correctly (Fig. 2.4c).

Instead of a computationally expensive preprocessing step, we use a modification proposed by Müller et al. in [15]. The key idea here is to replace the original data term based on the constant brightness assumption with a robust data term $\rho_r(\mathrm{x},\mathbf{u})$ based on the Census operator:

$$\rho_r(\mathrm{x},\mathbf{u}) = h(c_1(\mathrm{x}+\mathbf{u}), c_0(\mathrm{x})) \tag{2.5}$$

Here, $c_{\{0,1\}}(\mathrm{x})$ gives the Census signature around the center pixel x in the previous and current images, respectively. $h(c_1,c_0)$ denotes the Hamming distance of the two signatures. To solve the energy minimization problem, a variation of the framework of Zach et al. is utilized.

Implemented on the GPU, this version runs in real time (25 Hz) with a Census window of 3×3 pixels. Due to the reduced entropy caused by the Census operator, the subpixel accuracy is slightly worse compared to the original version on ideal images, but the increased robustness against illumination changes outweighs this fact easily (Fig. 2.4d).

In practice, the algorithm is implemented using an image resolution pyramid of five levels to estimate large displacement vectors, and 25 iterations are performed on each pyramid level. In each iteration step, the data term is firstly minimized using a first-order Taylor approximation, followed by a median filter and the smoothing step.

2.2.3 *Temporal Integration of the Motion Field*

Having established a correspondence over time for an observed image point by an optical flow or feature tracking algorithm, the 3D motion can be calculated directly from the reconstructed 3D points and the known time interval. However, such techniques suffer heavily from the immanent measurement noise, and the results are not robust. Therefore, we use a Kalman filter to estimate the 3D position and 3D motion of a point [5, 16, 17]. Due to the recursive nature of Kalman filters, the estimation keeps improving continuously with each measurement and by updating the state vector and its associated covariance matrix. This eliminates the need to save a history of measurements and is computationally very efficient. Additionally, Kalman filters take advantage of measurement uncertainties which can be considered when the flow field is evaluated for subsequent applications.

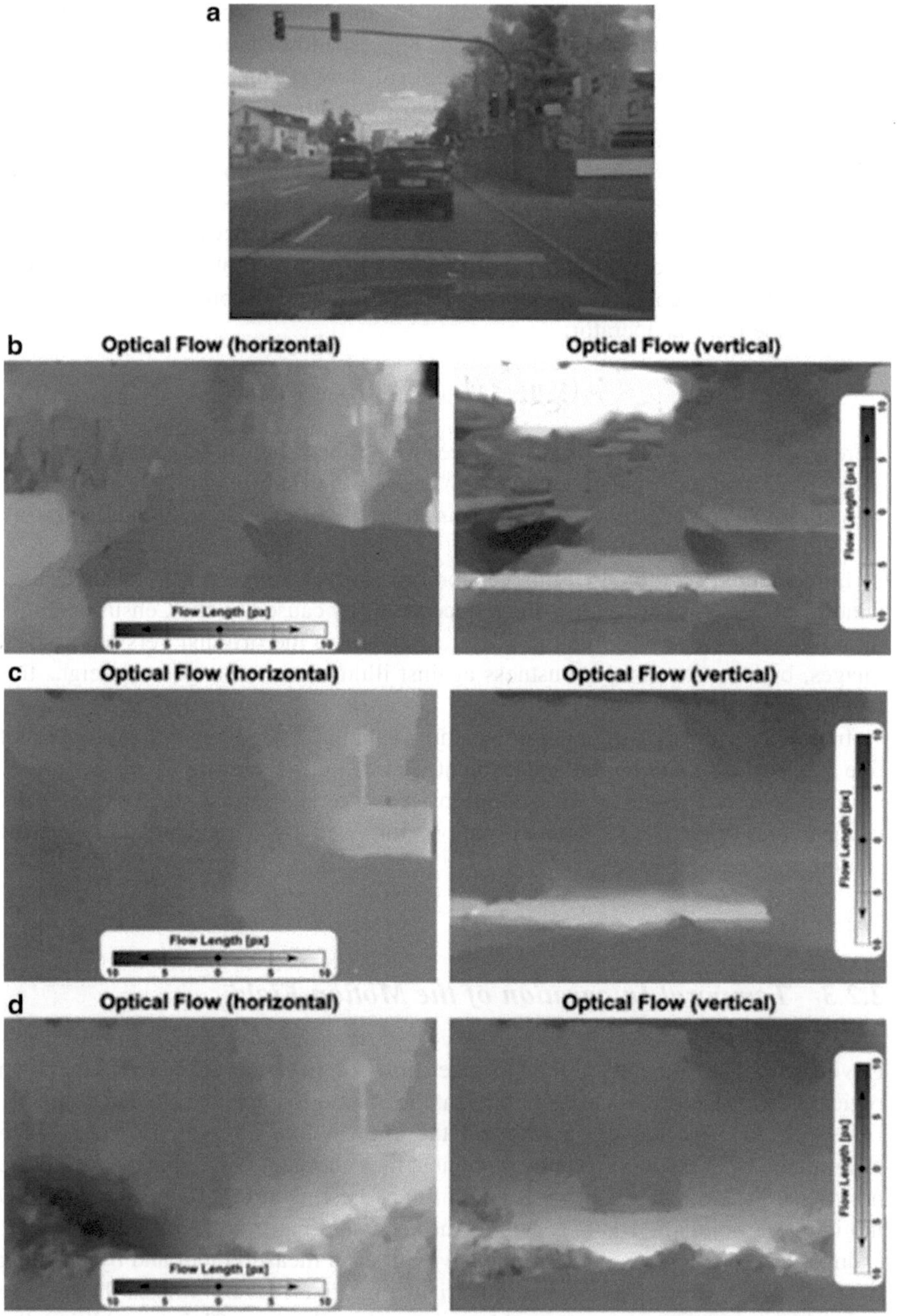

Fig. 2.4 Estimated optical flow fields under strong illumination changes. (**a**) Original image. (**b**) Without compensation. (**c**) With structure–texture decomposition. (**d**) Modified data term based on the Census operator

Using a stereo camera system, the 3D structure of the observed scene is readily reconstructed by the stereo algorithm. Here, the left image point $\mathrm{x} = (x,y)^{\mathrm{T}}$ is the projection of a world point $\mathrm{X} = (X,Y,Z)^{\mathrm{T}}$. Expressed in homogeneous coordinates, this relation is given by

$$\begin{pmatrix} x \\ y \\ d \\ 1 \end{pmatrix} \simeq \mathbf{\Pi} \cdot \widetilde{X} \tag{2.6}$$

with the positive disparity $d \equiv d(\mathrm{x})$, related to the left image. The extended projection matrix $\mathbf{\Pi}$ is written as

$$\mathbf{\Pi} = \begin{pmatrix} f_x & 0 & c_x & 0 \\ 0 & f_y & c_y & 0 \\ 0 & 0 & 0 & b \cdot f_x \\ 0 & 0 & 1 & 0 \end{pmatrix} \cdot \begin{pmatrix} \boldsymbol{R}_c & \mathbf{t}_c \\ \mathbf{0} & 1 \end{pmatrix} \tag{2.7}$$

with f_x and f_y as the focal lengths in pixel, $(c_x,c_y)^{\mathrm{T}}$ as the principal point in pixels, and b as the base width of the stereo camera system. The rotation matrix $\mathbf{R}_c$ and the translation vector $\mathbf{t}_c$ describe the extrinsic orientation of the camera system to the world and to the car coordinate system, respectively. To determine the 3D position for an observed image point x with known disparity d, (2.6) has to be inverted.

The state vector of the Kalman filter is defined as $\xi = \left(X, Y, Z, \dot{X}, \dot{Y}, \dot{Z}\right)$, the combination of the 3D position and the 3D velocity vector. The system model describes the propagation of the state vector ξ at the previous time step $t - 1$ to the current time t, assuming a linear motion, which is described by the linear equation

$$\xi_t = \begin{pmatrix} \mathbf{R}_e & \Delta t \cdot \mathbf{R}_e \\ \mathbf{0} & \mathbf{R}_e \end{pmatrix} \xi_{t-1} + \begin{pmatrix} \mathbf{t}_e \\ \mathbf{0} \end{pmatrix} \tag{2.8}$$

with $\mathbf{R}_e$ and $\mathbf{t}_e$ denoting the rotation and the translation component of the inverse motion of the observer, also called the ego-motion. Δt is the time between any two time steps.

The measurement model of the Kalman filter describes the relation between the measurement vector $\mathrm{z} = (x,y,d)^{\mathrm{T}}$, consisting of the current image position and the disparity of the analyzed world point, and the state vector ξ. Here, only the position components of the state vector are directly measured, and the relation between the measured projection and the reconstructed 3D point is given by (2.6). Since the measurement model must be formulated in Euclidean space rather than in projective space, the measurement model is nonlinear:

$$\tilde{\mathrm{z}} = w \begin{pmatrix} x \\ y \\ d \\ 1 \end{pmatrix} = \mathbf{\Pi} \cdot \begin{pmatrix} X \\ Y \\ Z \\ 1 \end{pmatrix} \tag{2.9}$$

$$\mathrm{z} = \frac{1}{w} \begin{pmatrix} 1 & 0 & 0 & 0 \\ 0 & 1 & 0 & 0 \\ 0 & 0 & 1 & 0 \end{pmatrix} \cdot \tilde{\mathrm{z}} \tag{2.10}$$

The current measurement vector z_t is given as

$$\mathrm{z}_t = (\mathrm{x}_t, d_t)^{\mathrm{T}}, \quad \mathrm{x}_t = \mathrm{x}_{t-1} + \mathrm{u}_t \tag{2.11}$$

with u_t denoting the optical flow related to the previous position x_{t-1} of the image point and the corresponding disparity d_t at the new image position x_t. It is worth noting that x_{t-1} in (2.11) depicts the old measured image position at the previous frame, not the projection of the filtered state ξ_{t-1}. That means the image position of the features is only determined by the optical flow algorithm, while the filtering only influences the velocity and the disparity estimation. This way, undesired low-pass filtering effects of the Kalman filter are avoided.

To estimate the 3D motion field for all pixels, we associate every pixel $\hat{\mathrm{x}}$ on the discrete pixel grid with one Kalman filter $\mathcal{K}(\mathrm{x}_t)$. In addition to the internal state of the Kalman filter, the image position x is stored with subpixel accuracy in the same data structure. At each time step, the prediction step of all Kalman filters is performed first. Here, we use the inertial sensors of the vehicle—the speed and yaw rate information—to calculate the ego-motion required for the state transition. For the optical flow field u_t, the Kalman filter field $\mathcal{K}_{t-1}(\hat{\mathrm{x}}_{t-1})$ is warped to the filter field $\mathcal{K}_{t-1}(\hat{\mathrm{x}}_t)$, with $\hat{\mathrm{x}}_t$ as the new pixel discrete image position, which is derived from x_t, i.e., calculated according to (2.11). After this resampling step, the Kalman filters are updated according to the given measurement model, resulting in the updated Kalman filter field $\mathcal{K}_t(\mathrm{x}_t)$.

During the resampling step it is possible that not every pixel x_t of the current image is referred by a flow vector u_t. In this case, a new filter has to be created with initial values and connected to the empty pixel. An initialization based on the states and the covariances of the surrounding pixels is beneficial.

If one pixel x_t of the current image is referred by more than one flow vector set u_t, one either has to decide which one of the filters to use with the corresponding pixel on the next frame or has to combine them into a new one. In this case, the covariances of the concurring filters can be used as weights to generate the new filter state. It is also reasonable to use the depth information so that the nearest filter survives, while the others are reset. In our implementation, the filter which is assigned as the last one remains alive. A more complex solution decreases the real-time capability significantly and thus outweighs the benefit in practice.

To achieve maximum performance, this Dense6D system is implemented on a GPU. The Kalman filter code is based on an implementation according to Bierman [3] and was automatically generated by a code generator that exploits the sparse structure of our matrices. The overall system achieves real time (20 Hz) on VGA images and consists of the following phases: image acquisition, rectification, stereo computation on an FPGA, optical flow calculation and Kalman filter calculation on the GPU, and visualization using a 3D viewer.

2.3 Evaluation

2.3.1 Evaluation on Ground Truth Sequences

In the first experimental phase, we study our vision system on a synthetic stereo image sequence of 250 frames rendered with the ray tracing software Povray. The images have a resolution of 640×480 pixels with an intensity resolution of 12 bits. In our test sequence (Fig. 2.5), the camera moves through an artificial traffic scene containing crossing and turning vehicles.

Figure 2.6a shows the ground truth motion field for a single image of this sequence. The vectors point from the current position of the world point to the position in 0.250 s. The velocity is in gray scale: white corresponding to 0.0 m/s, whereas black encodes a velocity of 8.0 m/s. Figure 2.6b shows the result of a direct combination of the stereo information and the optical flow. It is obvious that this naive approach performs very poorly due to the temporal noise of the depth estimation. Also, it should be noted that the prediction horizon had to be reduced to 0.050 s since the calculated motion vectors are extremely noisy.

The result from the proposed Dense6D approach is shown in Fig. 2.6c. Here, the prediction horizon is again 0.250 s. As it can be seen, the static points on the street are correctly estimated, and the moving vehicles are clearly visible. The estimation of the motion fields of the preceding and the crossing car is quite accurate. Only the motion of the turning vehicle on the right seems to be underestimated. This is primarily caused by the violation of the linear motion model and the integrated

Fig. 2.5 Stereo image pair of the ground truth stereo sequence used in this evaluation

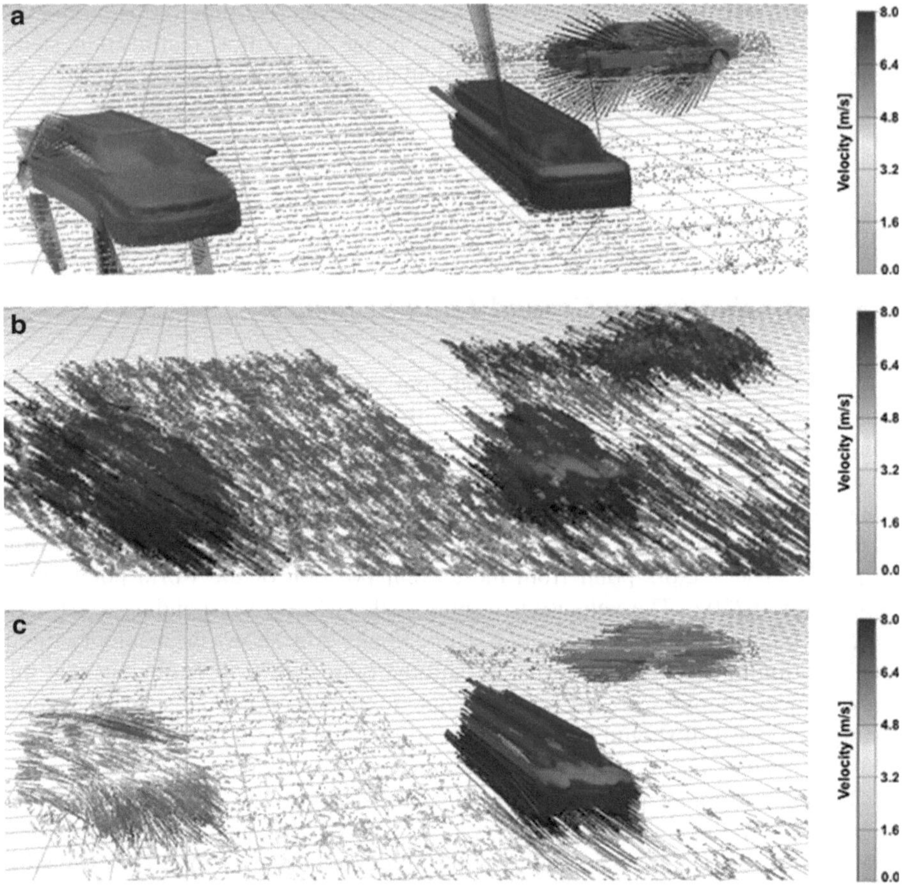

Fig. 2.6 Estimated motion field of the described methods. (**a**) Ground truth. (**b**) Direct combination of optical flow and stereo. (**c**) Dense6D. The vectors point to the predicted 3D position in 0.250 s (**a**) and (**c**) respectively 0.050 s (**b**)

outlier detection method (3σ-test). All estimates violating the motion model are rejected, which results in constantly re-initialized filters for the turning car. Hence, the visible motion vectors of the turning car correspond mainly to filters that have not yet reached their steady states.

The three-dimensional ground truth position and the motion field are used for the calculation of the error distributions $\rho[\chi]$ with $\chi = Z, \dot{X}, \dot{Y}, \dot{Z}$ as the quantities to analyze and χ^* as the corresponding ground truth. The error distributions, accumulated over the whole image Ω and the whole sequence $[0,T]$, are shown in Fig. 2.7. Again, the superiority of the Dense6D system compared to the direct approach is clearly visible.

In addition, the median of the error distribution of χ (ME) and the root mean squared error (RMS).

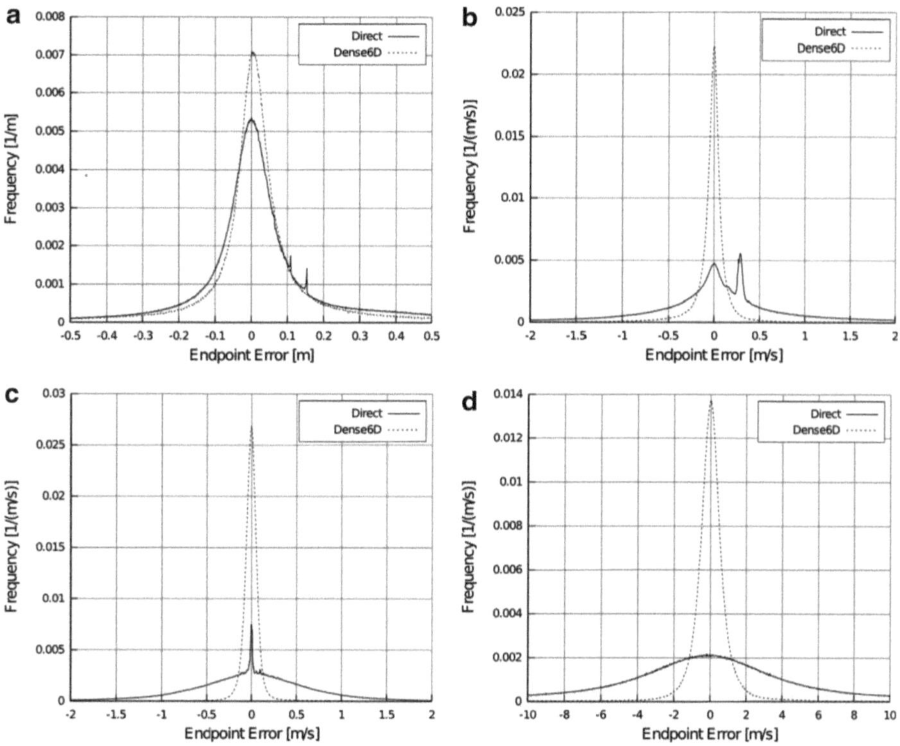

Fig. 2.7 Error distributions of the *Z*-position and the velocity components calculated from the direct combination of optical flow and stereo (*solid*) and the Dense6D method (*dashed*). (**a**) Position component *Z*. (**b**) Velocity component *Z*. (**c**) Position component y. (**d**) Velocity component *Z*

$$E_{\text{RMS}} = \sqrt{\frac{1}{N}\sum_{t=0}^{T}\int\left[\chi_t\left(\vec{x}\right) - \chi_t^*\left(\vec{x}\right)\right]^2 d\vec{x}} \tag{2.12}$$

are computed and given for the individual components in Table 2.1.

2.3.2 *Real-World Results*

The Dense6D system is deployed in our instrumented research car to evaluate its performance in real-world scenarios. Figure 2.8 shows the estimated motion field for a turning vehicle at a distance of about 30 m. Here, the observer was moving at a speed of about 3 m/s. Besides the turning car, a pedestrian and the shopping cart of the pedestrian behind are visible. In Fig. 2.9 multiple moving pedestrians are visible. The observer was again moving at a speed of about 3 m/s while performing

Table 2.1 Median error (ME) and root mean square error (RMS) of the Z-position and the velocity components calculated by a direct combination of the optical flow and the stereo information (direct) and the proposed Dense6D algorithm

	Z (m)		$\dot{X}$ (m/s)		$\dot{Y}$ (m/s)		$\dot{Z}$ (m/s)	
	ME	RMS	ME	RMS	ME	RMS	ME	RMS
Direct	0.0010	2.749	0.0462	42.0093	0.0004	15.370	0.4374	141.442
Dense6D	0.0104	1.068	−0.0065	0.3623	−0.0044	0.339	0.0107	2.538

Fig. 2.8 *Left*: Typical traffic scene. *Right*: Corresponding 3D motion field, estimated by the Dense6D algorithm proposed in this chapter

Fig. 2.9 *Left*: Typical traffic scene. *Right*: Corresponding 3D motion field, estimated by the Dense6D algorithm proposed in this chapter

a strong turning maneuver. Since the Kalman filters estimate the motion with respect to the fixed world coordinate system, the motion induced by the moving observer is completely compensated.

Obviously, the proposed Dense6D method is directly applicable in real-world scenarios and the excellent results on synthetic sequences shown in the previous section are validated.

Future work will include a multi-filter implementation and an image-based ego-motion estimation on the GPU as already known from [16]. In addition, methods to estimate the uncertainties of the stereo and optical flow information are currently under investigation.

2.4 Conclusions

In this chapter, we have presented the Dense6D system for dense, robust, accurate motion field estimation operating in real time. We have combined the state-of-the-art dense stereo and variational optical flow estimation techniques with Kalman filters under a linear motion model assumption. Evaluation of the relevant error quantities compared to simulated ground truth data shows that this approach shows far better results in real time compared to what is known in the literature so far. In real-world scenarios the technique shows its potential rather well. The Dense6D system is currently implemented in an instrumented research vehicle and is anticipated to become a key feature in emerging driver assistance systems.

Future work will include a multi-filter implementation and an image-based ego-motion estimation on the GPU as already known from [16]. In addition, methods to estimate the uncertainties of the stereo and optical flow information are currently under investigation.

References

1. A.A. Argyros, M.I. Lourakis, P.E. Trahanias, S.C. Orphanoudakis, Qualitative detection of 3D motion discontinuities, in *Proceedings of the IEEE/RSJ International Conference on Intelligent Robots and Systems (IROS'96)*, vol. 3, pp. 1630–1637, Nov 1996
2. S.S. Beauchemin, J.L. Barron, The computation of optical flow. ACM Comput. Surv. **27**, 433–466 (1995)
3. G.J. Bierman, *Factorization Methods for Discrete Sequential Estimation* (Academic Press, Inc., New York, 1977)
4. D.J. Fleet, Y. Weiss, Optical flow estimation, Chapter 15, in *Handbook of Mathematical Models in Computer Vision*, ed. by N. Paragios, Y. Chen, O. Faugeras (Springer, Berlin, 2006), pp. 239–258
5. U. Franke, C. Rabe, H. Badino, S. Gehrig, 6DVision: fusion of stereo and motion for robust environment perception, in *Proceedings of the 27th DAGM Symposium*, pp. 216–223, 2005
6. S. Gehrig, F. Eberli, T. Meyer, A real-time low-power stereo vision engine using semi-global matching, in *Proceedings of the 7th International Conference on Computer Vision Systems*, Liège, Belgium, Oct 2009
7. S.K. Gehrig, C. Rabe, Real-time semi-global matching on the CPU, in *Proceedings of the IEEE Workshop on Embedded Computer Vision*, 2010
8. S. Heinrich, Fast obstacle detection using flow/depth constraint, in *Proceedings of the IEEE Intelligent Vehicles Symposium 2002*, vol. 2, pp. 658–665, Jun 2002

9. H. Hirschmüller, Accurate and efficient stereo processing by semi-global matching and mutual information, in *Proceedings of the IEEE Computer Society Conference on Computer Vision and Pattern Recognition*, no. 2, San Diego, CA, USA, pp. 807–814, Jun 2005
10. B.K.P. Horn, B.G. Schunck, Determining optical flow. Artif. Intell. **17**, 185–203 (1981)
11. R.E. Kalman, A new approach to linear filtering and prediction problems. Trans. ASME J. Basic Eng. **82**(Series D), 35–45 (1960)
12. P.J. Kellman, M.K. Kaiser, Extracting object motion during observer motion: combining constraints from optic flow and binocular disparity. J. Opt. Soc. Am. A **12**, 623–625 (1995)
13. M.C. Martin, H. Moravec, Robot evidence grids. Tech. Rep. CMU-RI-TR-96-06, Robotics Institute, Carnegie Mellon University, Pittsburgh, PA, Mar 1996
14. S. Mills, Stereo-motion analysis of image sequences, in *Proceedings of the first joint Australia and New Zealand conference on Digital Image and Vision Computing: Techniques and Applications, DICTA'97/IVCNZ'97*, Dec 1997
15. T. Müller, C. Rabe, J. Rannacher, U. Franke, R. Mester, Illumination robust dense optical flow using census signatures, in *Proceedings of the 33th DAGM Symposium*, 2011
16. C. Rabe, U. Franke, S. Gehrig, Fast detection of moving objects in complex scenarios, in *Proceedings of the IEEE Intelligent Vehicles Symposium*, pp. 398–403, Jun 2007
17. C. Rabe, T. Müller, A. Wedel, U. Franke, Dense, robust, and accurate motion field estimation from stereo image sequences in real-time, in *Proceedings of the 11th European Conference on Computer Vision*, ed. by K. Daniilidis, P. Maragos, N. Paragios. Lecture Notes in Computer Science, vol. 6314 (Springer, Berlin, 2010), pp. 582–595
18. J. Rannacher, Realtime 3D motion estimation on graphics hardware, Bachelor thesis, Heidelberg University, 2009
19. S. Vedula, S. Baker, P. Rander, R. Collins, T. Kanade, Three-dimensional scene flow, in *Seventh International Conference on Computer Vision (ICCV'99)*, vol. 2, pp. 722–729, 1999
20. A.M. Waxman, J.H. Duncan, Binocular image flows: steps toward stereo-motion fusion. IEEE Trans. Pattern Anal. Mach. Intell. **8**(6), 715–729 (1986)
21. A. Wedel, T. Pock, C. Zach, H. Bischof, D. Cremers, Statistical and geometrical approaches to visual motion analysis, in *An Improved Algorithm for TV-L1 Optical Flow*, ed. by D. Cremers, B. Rosenhahn, A.L. Yuille, F.R. Schmidt (Springer, Berlin, 2009), pp. 23–45
22. J. Weickert, A. Bruhn, T. Brox, N. Papenberg, A survey on variational optic flow methods for small displacements, in *Mathematical Models for Registration and Applications to Medical Imaging*, ed. by O. Scherzer (Springer, New York, 2006)
23. R. Zabih, J. Woodfill, Non-parametric local transforms for computing visual correspondence, in *Proceedings of the Third European Conference on Computer Vision*, pp. 151–158, May 1994
24. C. Zach, T. Pock, H. Bischof, A duality based approach for realtime TV-L1 optical flow, in *Proceedings of the 29th DAGM Symposium on Pattern Recognition*, pp. 214–223, 2007

Chapter 3
Intelligence in the Automobile of the Future

Arne Bartels, Thomas Ruchatz, and Stefan Brosig

Abstract One trend that has been apparent in the automobile market over recent years is that more and more new vehicles from an ever-increasing number of manufacturers have been available with driver assistance systems. Most of these driver assistance systems relieve the driver of simple measuring and control tasks. Intelligence in the vehicle, however, means more than just measuring and controlling. The vehicle has got to acquire information, arrive at an interpretation and establish contextual interconnections. To do this, it needs contextual information and action options. This chapter presents an architecture for vehicle assistance systems in order to acquire, process, and evaluate environmental data, thereby bringing the objective within reach. Also, it presents certain selected projects by Volkswagen Group Research in the area of automated driving which are based on this architecture for environment perception.

Keywords Intelligent automobile • Driver assistance systems • Environment perception • Sensor data fusion • Situation interpretation • Automated driving • DARPA urban challenge

3.1 Definition of "Intelligence"

First of all, it is helpful to briefly address the concept of intelligence, because it is through intelligent actions that future driver assistance systems will differentiate themselves from current ones:

A. Bartels (✉) • T. Ruchatz • S. Brosig
Group Research, Driver Assistance and Integrated Safety, Volkswagen AG, 1777, Wolfsburg D-38436, Germany
e-mail: arne.bartels@volkswagen.de; thomas.ruchatz@volkswagen.de; stefan.brosig@volkswagen.de

G. Schmidt et al. (eds.), *Smart Mobile In-Vehicle Systems: Next Generation Advancements*, DOI 10.1007/978-1-4614-9120-0_3,

Definition from a psychological perspective:

. . . an ability which makes it possible to deal with novel situations. It is expressed in the acquisition, implementation, interpretation and establishment of relationships and contextual interconnections. . .

(Bertelsmann Lexicon)

. . . a collective term for the cognitive capacity of human beings, i.e. ability to understand, abstract, solve problems, apply knowledge and use language.

(wikipedia.de)

Definition from an information technology perspective:

. . . artificial intelligence (AI) refers to the emulation of human intelligence within information technology.

Artificial intelligence is being increasingly used in engineering sciences and medical technology. Possible application scenarios are: dealing with natural signals (understanding images and detecting patterns).

(wikipedia.de)

The definition from a psychological perspective indicates that an intelligent driver is essential in order for a vehicle to be guided on the public road. Even simple driving maneuvers such as adjusting the vehicle's speed to a speed limit as well as complex maneuvers such as turning off at a busy inner city intersection demand the acquisition, interpretation, and establishment of relationships and contextual interconnections between the subject vehicle, other road users, and the transport infrastructure (lane, traffic lights, road signs, etc.).

3.2 Perception Instead of Measurement

What significance do these definitions have for application in the automobile of the future? First of all, the terms "acquisition" and "interpretation" in the definition indicate the topic of "perception."

3.2.1 Status of Current Driver Assistance Systems

The vehicles illustrated in Fig. 3.1 feature some of the highly developed driver assistance systems that are currently available on the market. What all functions have in common is that they have environment sensors which perform measurements outside the vehicle.

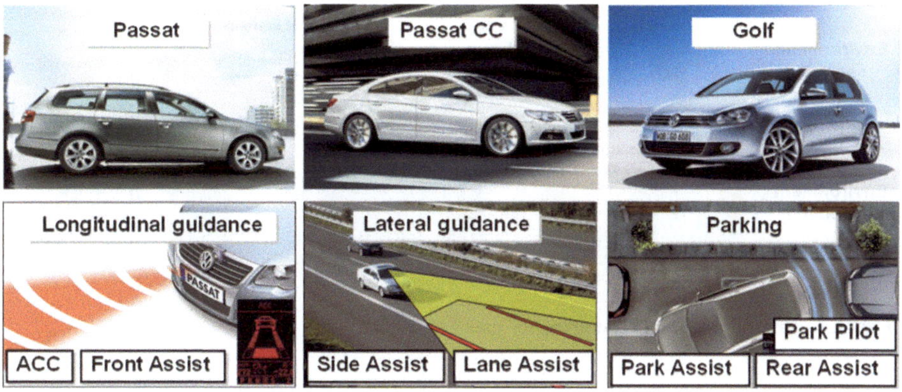

Fig. 3.1 Status of current driver assistance systems

	Source: Bosch 	Source: Bosch 	Source: Bosch 
Sensor type	Radar	Camera	Ultrasound
Opening angle Range	20-70° 50-250 m	50-70° 50-100 m	30-50° 2-5 m
Measuring parameters	Distance, angle, relative speed	Contour, road markings	Distance
Peripheral field model	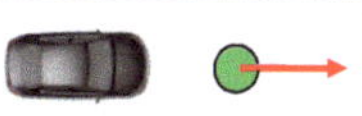	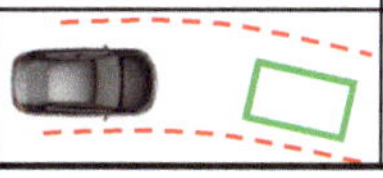	

Fig. 3.2 Environment sensors in current driver assistance systems

3.2.2 *Environment Perception*

3.2.2.1 Environment Sensors

In most cases, one environment sensor or environment sensor type is used for each function and delivers a small number of measurements. Figure 3.2 shows some examples of this.

- Radar sensors measure the positions (distance, angle) and speeds of objects relative to the subject vehicle. For instance, these objects are represented by a point model.

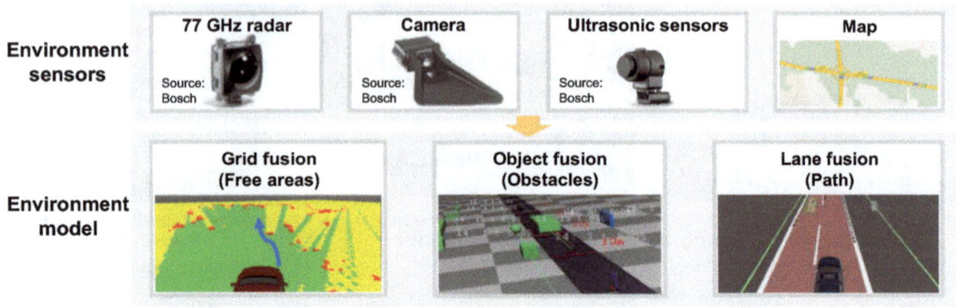

Fig. 3.3 Environment model with sensor data fusions

- Cameras measure the extent (contour) of objects and the lateral position of road markings relative to the subject vehicle. Objects are represented by a rectangle, for example, and the lane by two side lines.
- Ultrasound sensors measure the radial distance to the next object in each case.

The measuring parameters generated in this way, together with set point specifications from the functions, are sufficient for carrying out simple distance control or warning functions. In these particular situations, we can assume that the systems are measuring and controlling rather than persevering and acting intelligently.

3.2.2.2 Environment Modeling

In the near future, several environment sensors will be deployed in the vehicle and their measurements can be bundled together (fused) to give more complex knowledge on environment. In particular, the different aspects of the environment are acquired in a suitable way by measurements and are represented in environment models. An example of this is shown in Fig. 3.3: in the upper layer, environment sensors including camera, radar, and ultrasonic sensors acquire features from the environment. This information is supplemented by a digital road map.

In the subsequent layer, i.e., "environment model," this information is fused to represent certain features of the environment in this model:

- An occupancy array represents stationary objects, such as a parked car, the face of a house, or a traffic island. The great diversity of geometries of stationary objects means that an occupancy array is much better suited to this than, for example, representing objects using rectangles. The features of a cell in this occupancy array are, in the most simple case, "occupied" or "unoccupied." This input data is provided by camera, radar, and ultrasound sensors and supplemented by a digital road map.
- Information about moving objects, such as vehicles or pedestrians, is merged in an object fusion process. Representative features of an object include its distance, speed, movement direction, size, and others.

- Information about the current lane is merged in a lane fusion process. Representative features of the lane include the number and width of lanes, the type of marking lines, and the presence of marginal structures such as crash barriers. This input data is provided by camera and radar sensors, supplemented by a digital road map.

This fusion of unstructured features with objects and infrastructure data results in an environment model to represent the vehicle's environment very effectively.

3.2.2.3 Situation Interpretation

By using this environment model, it is now possible to establish contextual interconnections among the individual features of the environment. In accordance with Sect. 3.2, this is a definition of intelligence.

Now, actions can be derived on the basis of this understanding of the situation. In principle, this can be done based on rules or knowledge. Rules can be derived from the dynamics of the vehicle or accepted regulations of conduct such as the road traffic regulations. Knowledge can be represented using learned data and Bayesian networks, neural networks (NN), or support vector machines (SVM).

Three examples for intelligent, rule-based actions by driver assistance systems:

- The vehicle's speed is reduced before a tight bend: the course of the road is known from the lane fusion. Prior to the entry to the bend, this enables the vehicle's speed to be reduced so that the lateral acceleration in the bend does not exceed a maximum value.
- The prohibition on overtaking on the right is observed (in countries where driving is on the right). The object and lane fusion means that the road type (in this case, motorway) and the locations and positions of other vehicles in the lane to the left are known. It is possible to decelerate in good time when approaching these vehicles in order to avoid overtaking on the right.
- Responding in good time to vehicles cutting in: the object and lane fusion means that it is known that the vehicle on the right in front of the subject vehicle is located on the acceleration strip of a motorway entry ramp. As a result, a change of lane is about to take place. A safe distance can be set.

Figure 3.4 shows the architecture for environment perception that has been expanded with the "situation interpretation" block.

3.2.3 Driver Modeling

3.2.3.1 Motivation

A driver model can sensibly supplement the aforementioned architecture for environment perception. Three examples of this are:

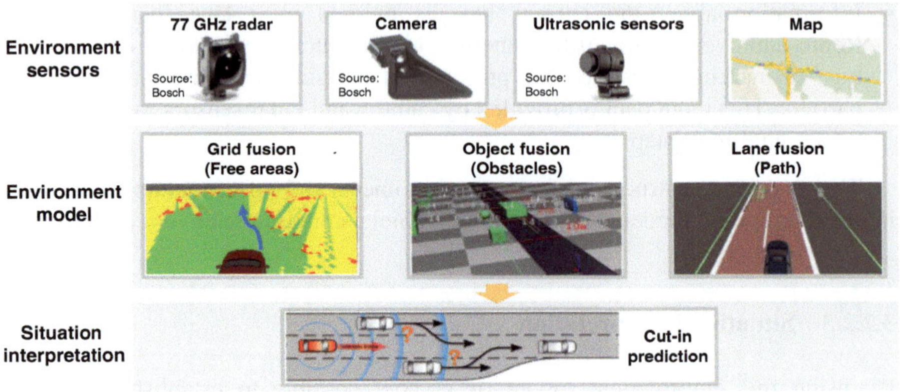

Fig. 3.4 Situation interpretation

- Advance LDW control via distraction recognition: if the driver is distracted, a warning sound comes on prior to lane departure.
- Advance ACC control by means of intention recognition of overtaking: if the driver intends to overtake the vehicle in front, ACC brakes later when approaching this vehicle.
- Lane assist function deactivation with hands off: the system switches to passive mode if the driver takes his/her hands off the steering wheel during active lane assist.

3.2.3.2 Sensors and Driver Model

Based on the representation of the environment model, Fig. 3.5 shows the draft for a driver model. The sensors used are, for example:

- Driver monitoring camera for detecting eyelid movements, pulse, and head position
- Biometric sensors for detecting pulse (electrocardiography, ballistocardiography, photoplethysmography, radar sensors, etc.) and hand position (capacitive, pressure sensors, etc.)
- Vehicle sensors (speed, steering angle, yaw rate, brake pedal, accelerator pedal, radio navigation system, hands-free system, etc.) for detecting the hand position, operating actions including ancillary activities and driving behavior
- Environment sensors (see Sect. 3.2) for detecting the driving behavior and driving context

The data from these sensors is either used directly or after preprocessing for determining driver state including fatigue, distraction, and driving ability as well as for deciding on the driver's intentions concerning maneuvers (lane change, turning off, stopping, etc.) and route selection.

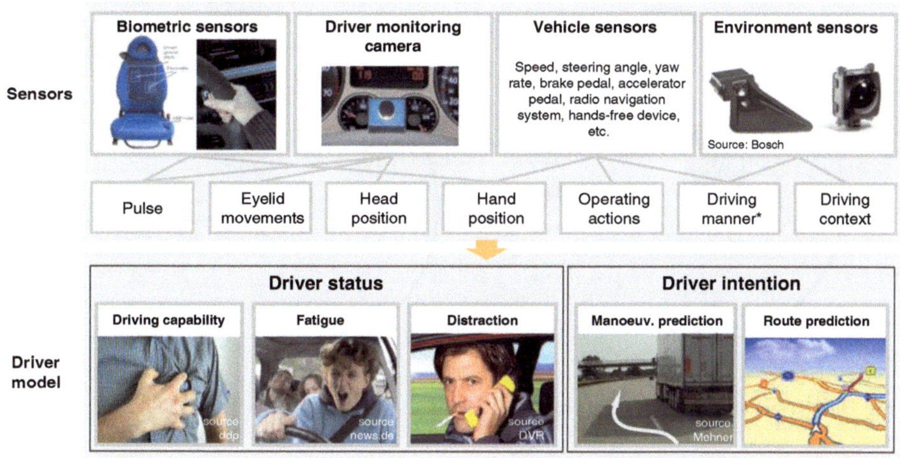

Fig. 3.5 Driver model with sensors

3.3 The Intelligent Automobile

3.3.1 Trajectory Planning and Following

However, it is not sufficient simply to represent a situation interpretation and action options in the vehicle in a suitable way. In order to allow an action to be performed automatically, it is also necessary to plan an action, e.g., in the form of trajectory planning. In the following section, the procedural principles for this are presented taking the example of the "Golf GTI 53 + 1" research project. This is composed of three phases (see Fig. 3.6): (a) environment detection, (b) trajectory planning, and (c) automated driving along/following the planned trajectory.

(a) Phase 1: Environment detection

During an initial drive at a slow speed, the named vehicle measures a course laid out by traffic cones with the help of a laser scanner. The course in this case can be laid out however required/desired and also rapidly modified.

(b) Phase 2: Trajectory planning

Next, an ideal trajectory (in terms of completing the lap in optimum time) is calculated for the course measured based on the physical driving properties of the vehicle. The lateral planning function calculates the ideal line with regard to the edges of the course. The longitudinal planning function calculates the optimum cornering speeds and braking points.

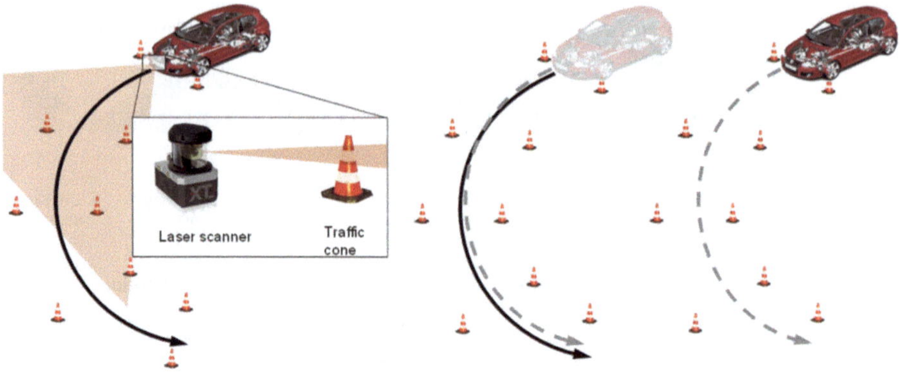

Fig. 3.6 Trajectory planning and following taking the example of the "Golf GTI 53+1"

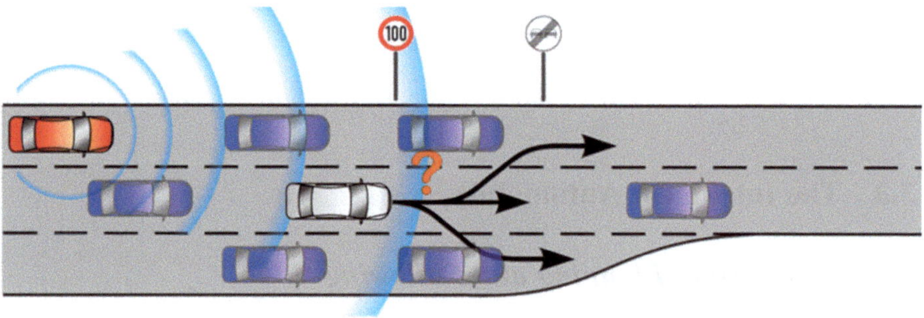

Fig. 3.7 Maneuver prediction

(c) Phase 3: Automated driving

Finally, the course is driven in a reproducible manner within the limit area of driving dynamics—the planned trajectory is provided as a set point specification for lateral and longitudinal control. Firstly, this makes it possible to reproduce the driving behavior of a professional driver; secondly, reproducibility of the driven trajectory means that it is possible to conduct an objective vehicle evaluation on handling tracks with regard to the lateral dynamics.

3.3.2 *Maneuver Prediction*

The scenario represented above seems to function very well as long as nothing unforeseen takes place, such as another vehicle cuts in suddenly to the subject vehicle's own lane. Therefore, intelligent driving also requires predicting the behavior of other users on the road. In order to deal with the situation shown in Fig. 3.7 independently, it is necessary for the subject vehicle to consider at least four things:

(a) Vehicle status: The physical properties of other road users are shown here, e.g., in terms of longitudinal and lateral acceleration when changing lanes.
(b) Lane assignment: This shows the positioning of other road users in the adjacent lane.
(c) Traffic rules and compulsory requirements: These show traffic regulations on the road which other users must adhere to.
(d) Lane change intention: This shows the prediction and recognition of an imminent lane change by other road users.

If these are taken into account, lane changes by other road users can be predicted in principle and a special action kept in readiness.

3.4 Intelligent Vehicle Functions

As an example of intelligent vehicle functions, the following section presents a number of research projects undertaken by the Volkswagen Group in the area of automated driving. This is initiated with a brief motivation for functions which support the automated driving.

3.4.1 Motivation for Automated Driving

Depending on the level of difficulty of the driving task, people today are more or less capable of controlling a vehicle effectively (see Fig. 3.8). If the driving task becomes very complex—e.g., when filtering into flowing traffic at motorway on

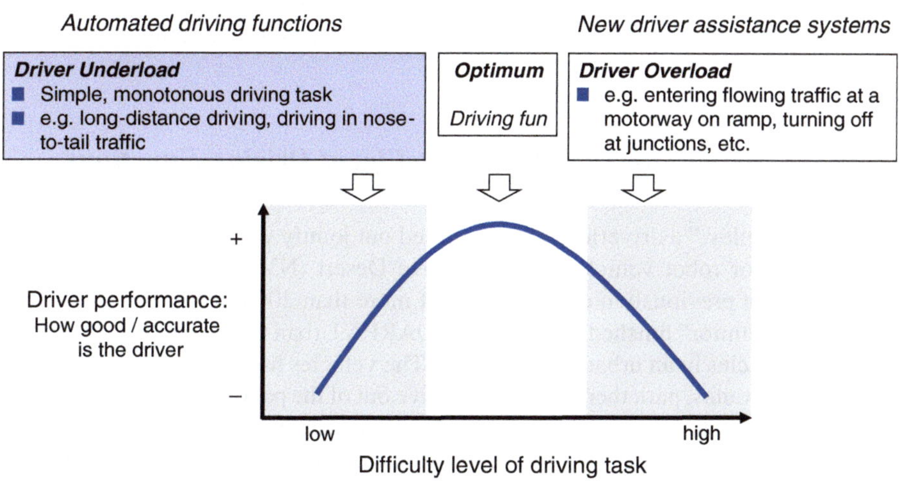

Fig. 3.8 Applications for automated driving functions and driver assistance systems

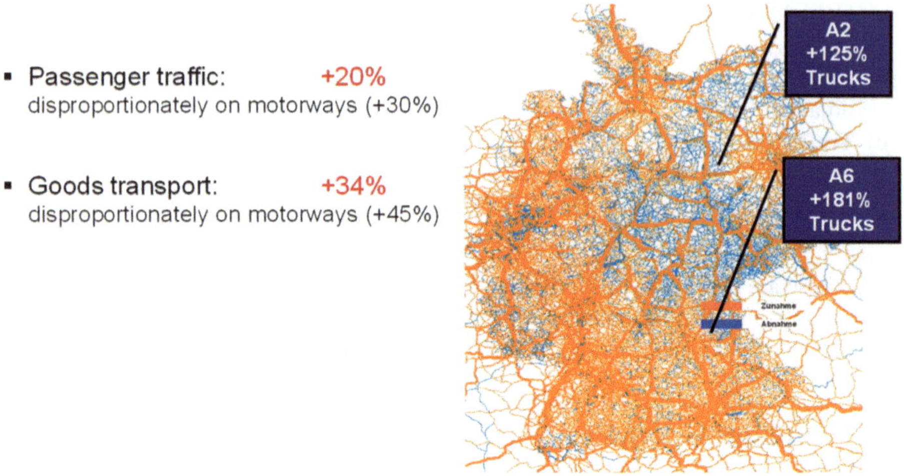

Fig. 3.9 Trend in road transport 2002–2020

ramps, or turning off at junctions—then the support of new, intelligent driver assistance systems can be helpful to the driver by providing targeted information, assistance, or protection mechanisms. At the other end of the scale, there is the issue of a driver underload when very simple driving tasks are involved—such as driving in nose-to-tail traffic or on long haul. Here, research community believes that support from automated driving functions can be sensible.

Another argument in favor of automated driving functions concerns the increasing density of traffic on German and European motorways. The ACATECH study shown in Fig. 3.9 shows that passenger traffic on German motorways is expected to increase by 30 % and goods traffic by 45 %. Indeed, traffic volume on the German motorways A2 or A6 will more than double or even almost triple. This means the frequency of monotonous driving situations will increase in the future.

3.4.2 Examples of Automated, Intelligent Driving Functions

In 2005, "Stanley," a driverless Touareg fitted out jointly with Stanford University, won a race for robot vehicles in the Mojave Desert (NV, USA) which involved driving over a previously unknown route of more than 200 km [1] (Fig. 3.10).

In 2007, "Junior" finished second in the DARPA Urban Challenge, a competition for robot vehicles in an urban environment. The vehicles had to observe traffic rules on unknown routes, park themselves, and drive out of the parking space; join into the lanes with moving traffic; and overtake other robot vehicles autonomously [2].

Also, the Intelligent Car was presented to the press at a VW test facility close to Wolfsburg in 2007. The iCar is capable of driving automatically on the motorway in both smoothly flowing and stop-and-go traffic. It stays in its lane and can

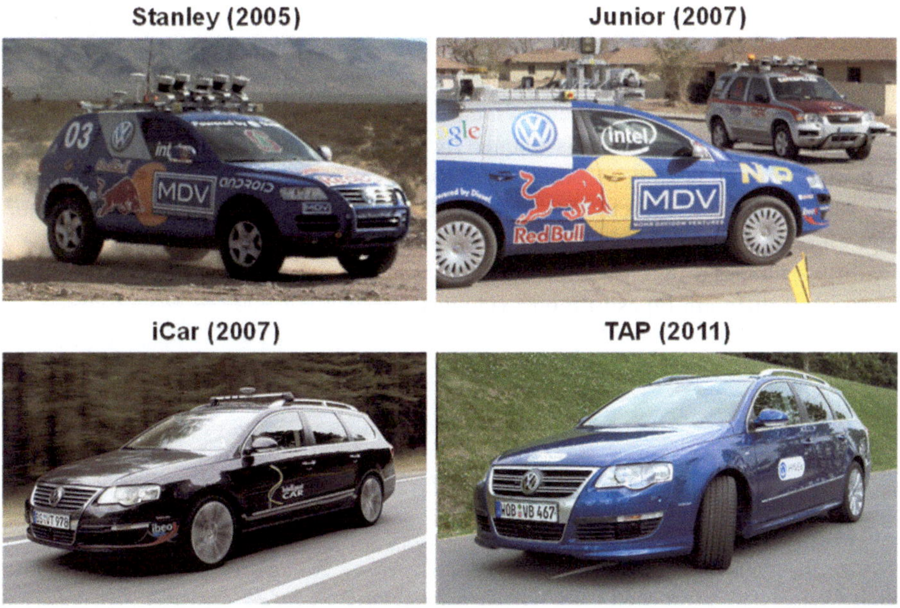

Fig. 3.10 Examples of automated, intelligent driving functions

autonomously perform lane changes, including overtaking maneuvers that have been authorized by the driver. The iCar is based on a set of environment sensors which are relatively simple in comparison to Stanley and Junior [3].

In June 2011, the Temporary Auto Pilot (TAP) was presented to the press at Hällered in Sweden. Its environment sensors have been further simplified and the environment model has been refined. The TAP can drive automatically on motorways both in traffic jams and in flowing traffic at up to 130 km/h. Furthermore, it adapts the vehicle's curve speed prior to the curve entry; it complies with the prohibition on overtaking on the right and keeps to the speed limits; and it monitors the driver's staten (fatigue, distraction) and performs an automated emergency stop if the driver does not respond adequately to an overtake request of the system [4–6].

Research plans are underway to offer vehicles to the customers in the near future that can drive automatically on selected routes, and thereby, they are expected to improve safety on these roads.

3.5 Conclusions

Intelligence in the vehicle means more than just measuring and controlling.

The automobile of the future has got to acquire a number of pieces of information, arrive at an interpretation, and establish the necessary contextual interconnections. It needs contextual information and action options. A foundation is laid for this by the usage of several environment sensors and key infrastructure knowledge, as well as by the enforcement of rules. If a goal is formulated, then planning can take place—generally with the help of a trajectory. As the last step, namely, acting, nominal parameters are provided to one or more controllers. This process is run through permanently and is adapted by updating the environment model as well as predicting the behavior of other road users.

All the elements depicted have already been implemented successfully in different embodiments of semiautomated driving. Automated driving could become a reality for the end user with the availability and economy of several environment sensors of different types to future vehicles.

References

1. S. Thrun et al., Stanley: the robot that won the DARPA grand challenge. J Field Robot **23**(9), 661–692 (2006)
2. M. Montemerlo et al., Junior: the Stanford entry in the urban challenge. J Field Robot **25**(9), 569–597 (2008)
3. A. Weiser et al., in *Intelligent car, Teilautomatisches Fahren auf der Autobahn, Proceedings of 10th symposium on automation systems, assistance systems and embedded systems for transportation (AAET)*, Braunschweig (2009), p. 11–26, ISBN 978-3-937655-19-2
4. A. Weiser, in *A probabilistic lane change prediction module for highly automated driving,* Proceedings of 7th international workshop on intelligent transportation (WIT), Hamburg, 2009
5. S. Steinmeyer, in *Modulare Sensordatenfusions-Architektur am Beispiel des "Temporary Auto Pilot", Proceedings of 11th symposium on automation systems, assistance systems and embedded systems for transportation (AAET)*, Braunschweig (2010), p. 89–104, ISBN 978-3-937655-23-9
6. S. Steinmeyer et al., in *Dynamische Bestimmung von Existenz- und Relevanzwahrscheinlichkeiten für Umfeldobjekte, 26. VDI/VW joint symposium on driver assistance and integrated safety*, Wolfsburg (2010), ISBN 978-3-18-092104-4

Chapter 4
Unmanned Ground Vehicle *Otonobil*: Design, Perception, and Decision Algorithms

Volkan Sezer, Pınar Boyraz, Ziya Ercan, Çagri Dikilitaş, Hasan Heceoğlu, Alper Öner, Gülay Öke, and Metin Gökaşan

Abstract Unmanned ground vehicles (UGV) have been the subject of research in recent years due to their future prospective of solving the traffic congestion and improving the safety on roads while having a more energy-efficient profile. In this chapter, the first UGV of Turkey, *Otonobil*, will be introduced detailing especially on its hardware and software design architecture, the perception capabilities and decision algorithms used in obstacle avoidance, and autonomous goal-oriented docking. UGV *Otonobil* features a novel heuristic algorithm to avoid dynamic obstacles, and the vehicle is an open test-rig for studying several intelligent-vehicle technologies such as steer-by-wire, intelligent traction control, and further artificial intelligence algorithms for acting in real-traffic conditions.

Keywords Autonomous car • Dynamic obstacle avoidance • Sensor fusion • Unmanned ground vehicle

4.1 Introduction

The unmanned ground vehicle (UGV) *Otonobil*[1] (Fig. 4.1) is essentially an urban concept small electric vehicle (EV) which is mechanically converted to modify the driver–vehicle interfaces for autonomous operation. Hardware conversion process

[1] *Otonobil* is one of the major projects at Mechatronics Research and Education Centre of Istanbul Technical University supported by Turkish Ministry of Development since 2008.

V. Sezer • P. Boyraz (✉) • Z. Ercan • Ç. Dikilitaş
Mechanical Engineering Department, Istanbul Technical University, İnönü Cd. No: 65, Beyoğlu, Istanbul 34437, Turkey
e-mail: pboyraz@itu.edu.tr

H. Heceoğlu • A. Öner • G. Öke (✉) • M. Gökaşan
Control Engineering Department, Istanbul Technical University, Ayazağa, Istanbul, Turkey
e-mail: gulay.oke@itu.edu.tr

G. Schmidt et al. (eds.), *Smart Mobile In-Vehicle Systems: Next Generation Advancements*, DOI 10.1007/978-1-4614-9120-0_4,

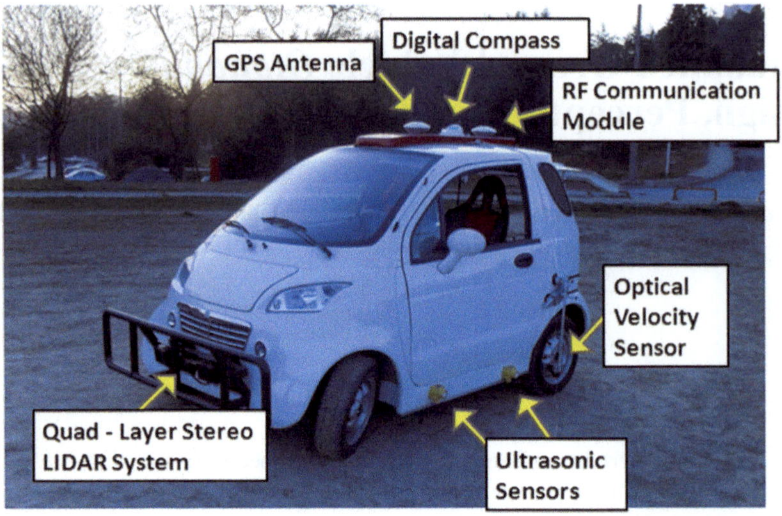

Fig. 4.1 Unmanned ground vehicle *Otonobil* at Istanbul Technical University, TR [1]

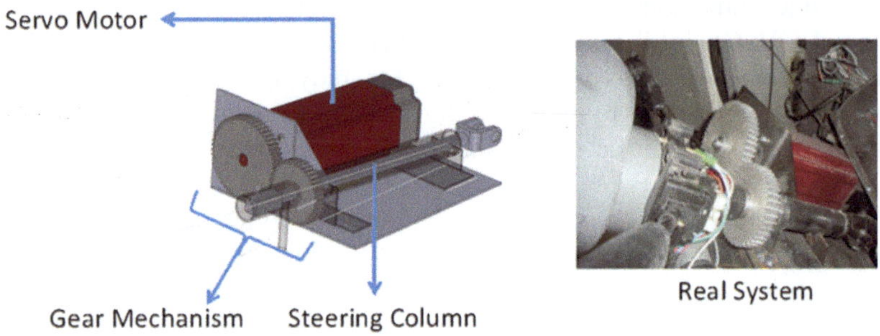

Fig. 4.2 Automatic steering system for *Otonobil* [2]

is divided into two main parts as mechanical and electrical modifications. Interface circuit, interface software, additional power system, selection of the sensors, and computer hardware are given in electrical modifications part. The multisensory perception capabilities of the UGV are given in Sect. 4.2, and the decision algorithms representing the artificial intelligence of the car is mentioned in Sect. 4.3.

The mechanical modifications on the vehicle are mainly performed on the brake and steering wheel for autonomous operation. Two separate external servo motors with a gear box are used to activate the steering wheel column and brake pedal, each according to the commands sent by the main computer onboard. The modified steering system can be seen in Fig. 4.2.

Additional mechanical modifications are in the form of extra components such as the top unit for carrying the GPS, IMU, and RF unit and the cage design in front of the car for supporting the LIDAR units.

4.2 Multisensor Data Acquisition, Processing, and Sensor Fusion

In this section, the multisensor platform in the UGV is examined together with data processing and sensor fusion strategies and algorithms. The system architecture is especially an important design consideration since the real-time performance of the applications depends heavily on the communication and general architecture such as distributed versus central structure and the communication protocols used. Another aspect of multisensor platform on the vehicle is that the platform can be used in both data acquisition/logging mode and real-time processing mode.

4.2.1 System Architecture

The sensors used in *Otonobil* are mainly for localization and state estimation purposes. The list of the sensors is given in Table 4.1.

The full system with their connection schematics and communication paths is given in Fig. 4.3. The computational components are mainly NI-PXI box used for localization, mapping, and path planning; DSpace MicroAutoBox used for local trajectory planning and tracking including low-level control of steering, braking, and wireless communication; and IBEO ECU used for object and raw data of the obstacles in front of the car. The related processing hardware and software structure is given in Fig. 4.4. Using this structure, several software pieces work in their own cycle time and the computation results are sent finally to local controller for the operation of the vehicle. These pieces are mainly image processing, LIDAR processing, vehicle state estimation, local mapping, motion planner, local trajectory planner, trajectory tracking, and wireless communication.

For digital signal processing applications used in UGV *Otonobil*, it is essential to have accurate, synchronized, and real-time logged dataset. This dataset can give the opportunity to perform the implemented algorithms in simulations to observe the performance and errors in the algorithm. First, we need the dataset in order to be able to simulate the motion. Without obtaining the data first, it would not be

Table 4.1 Sensor list of *Otonobil* for state estimation and mapping

	Sensor type	Quantity	Brand/model
Mapping sensors	Laser scanner	2	IBEO-LUX
	Laser scanner	1	SICK LMS 151
	Camera	1	SONY-XCI-SXI100
	Ultrasonic sensor	6	Banner-QT50ULB
State estimation sensors	Differential GPS	1	Trimble SPS851 & SPS551H
	Digital compass	1	KVH Azimuth 1000
	IMU	1	Crossbow VG700AB-201
	Optic speed sensor	1	Corrsys–Datron LF IIP

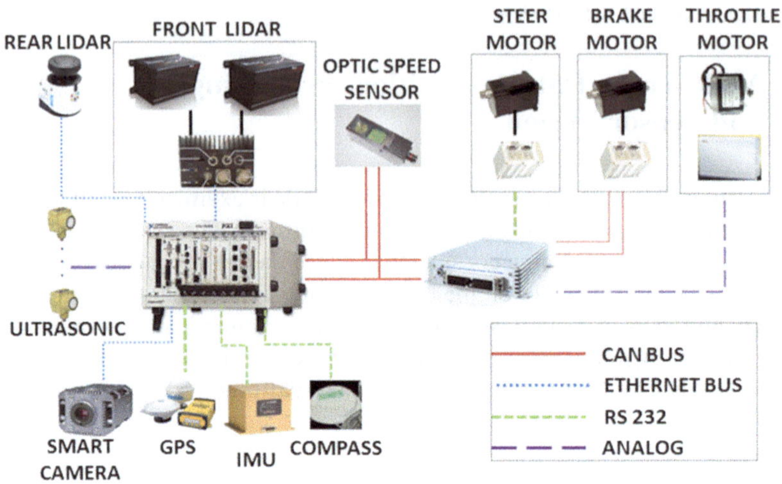

Fig. 4.3 Sensor connection and communication schematic view

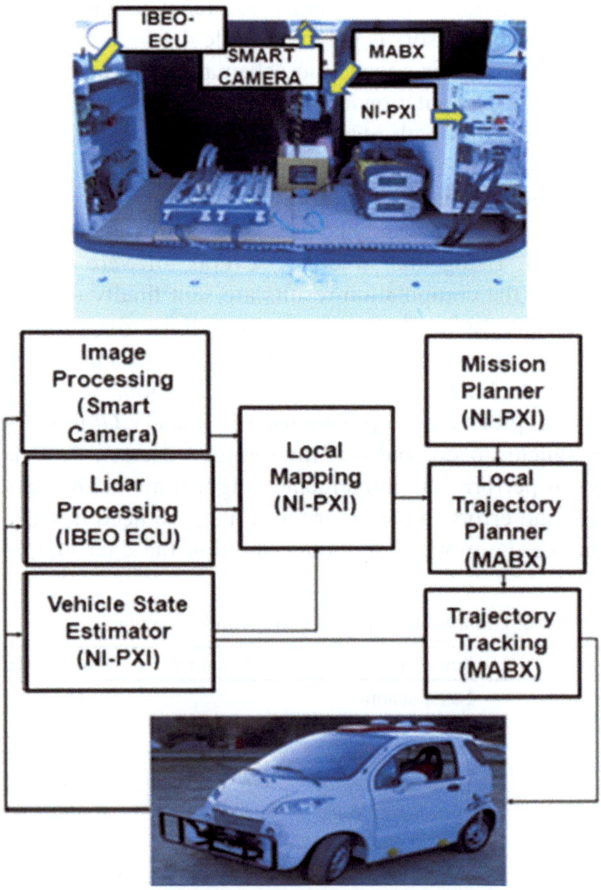

Fig. 4.4 Processing hardware and software structure of *Otonobil* [2]

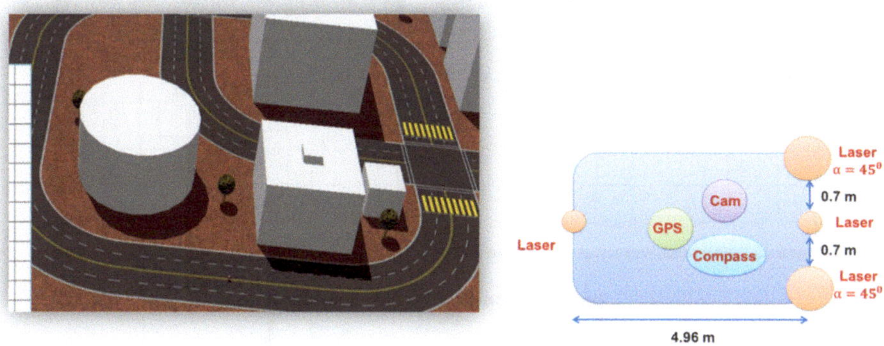

Fig. 4.5 The simulation environment in *Webots* and UGV with virtual sensors

possible to simulate the vehicle dynamics and kinematics with precision. The acquisition, processing, and logging of the sensors are performed in real-time module of *LabVIEW* since these applications require deterministic real-time performance that general purpose operating systems cannot guarantee. After acquisition and processing are performed, the data should be transferred via global variables since the data logging and the data acquisition processes are performed in different loops so that data wiring is impossible. However, this may cause race conditions because global variables could violate the strict data flow structure of *LabVIEW*. To overcome such a condition like this, functional global variable (FGV) or semaphores could be used. After the data is acquired and processed, it is transferred via an FGV to the data-logging loop which runs parallel to the acquisition and processing loop [3].

4.2.2 Sensor Fusion and Mapping

The sensor fusion and mapping algorithms are crucial in designing of a UGV because all the perception and decision making depends on the results from such calculations. For example, for path planning, an obvious map must be constructed. In context of intelligent vehicles, on board obstacle detection is an essential part. Obstacles must be detected in a correct and fast manner because the obstacles might be dynamic quickly changing their location and velocities with respect to UGV. Here, a sample algorithm for obstacle detection using laser scanner, camera, GPS, gyro, and compass is detailed. First, the vehicle is modeled as a platform with multiple sensors, and it is embedded in a simulation program (*Webots*) for simulating real driving scenarios as seen in Fig. 4.5.

Since the UGV has multiple sensors and none of these sensors are adequate for a precise measurement of the obstacle locations, a Bayesian inference method is used for detecting and tracking the obstacles. Some simulation results are given in Fig. 4.6 showing the obstacle map learning using different number of sensors.

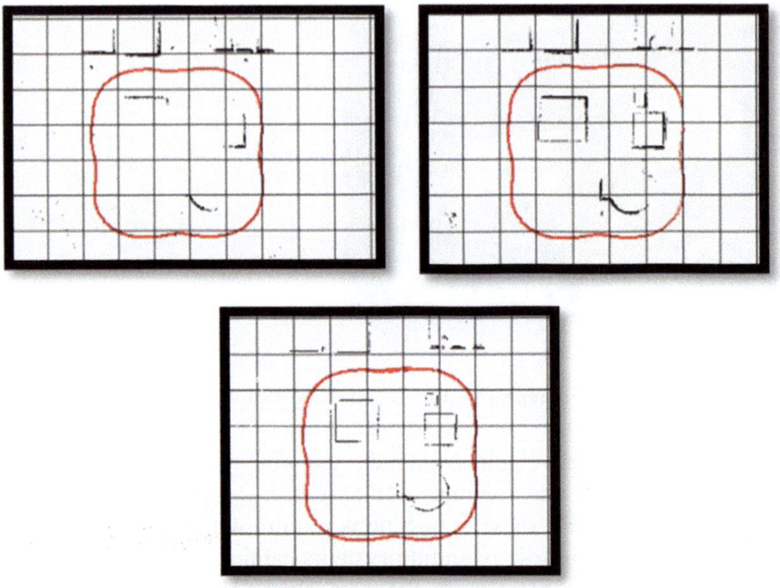

Fig. 4.6 Bayesian learning results using one, three, and four laser sensors

4.2.3 Localization and Navigation

A fundamental capability of a UGV is navigation. Using the information from various sensors, a UGV should be capable of determining vehicle's kinematic states, path planning, and calculating the necessary maneuvers to move between desired locations. To reach this objective with desired reliability, multisensor data fusion of various sensors is essential. For this objective, *Otonobil* is equipped with sensors like IMU, GPS, motor encoder, digital compass, and optic speed sensor. All the information from these sensors is fused using EKF algorithm given in Fig. 4.7. Also a stand-alone orientation estimation algorithm is proposed in order to have an accurate transformation of the information measured in vehicle's body frame into navigation frame. Full details of the work can be referred to for further information [4].

4.3 Dynamic Obstacle Avoidance

In UGV Otonobil project, a novel obstacle avoidance method called "follow the gap" is designed and it has been tested in the real scenarios. The method is easy to tune and considers the practical constraints in real vehicle such as limited field of

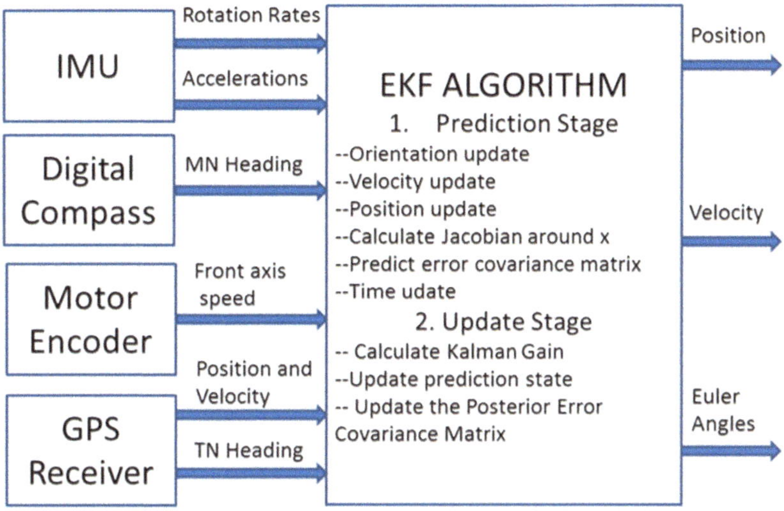

Fig. 4.7 Multisensor data fusion algorithm used in *Otonobil*

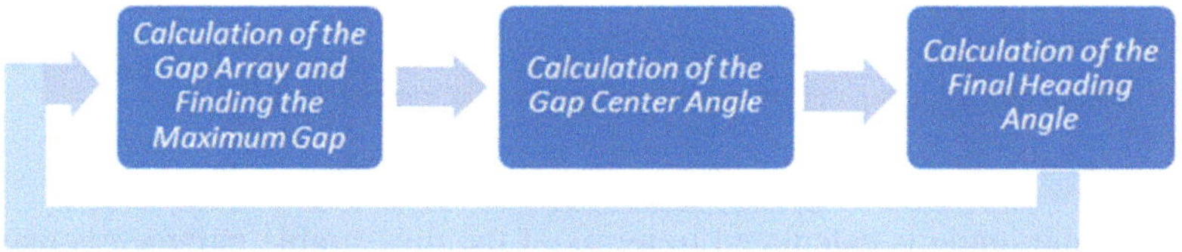

Fig. 4.8 Steps of the "follow the gap" method

view for sensors and nonholonomic motion constraints. A brief explanation and experimental results of the developed method is detailed here; however, the full details of the work can be referred to for further information [5].

4.3.1 *"Follow the Gap" Method*

The method assumes that both the UGV and the obstacles are circular objects with minimum diameter covering all the physical extensions of the real objects. "Follow the gap" method depends on the construction of a gap array around the vehicle and calculation of the best heading angle for heading the robot into the center of the maximum gap around, while at the same time considering the goal point. The algorithm can be divided into three main parts as illustrated in Fig. 4.8.

Maximum gap, the gap center angle, and the goal angle can be understood in Fig. 4.9.

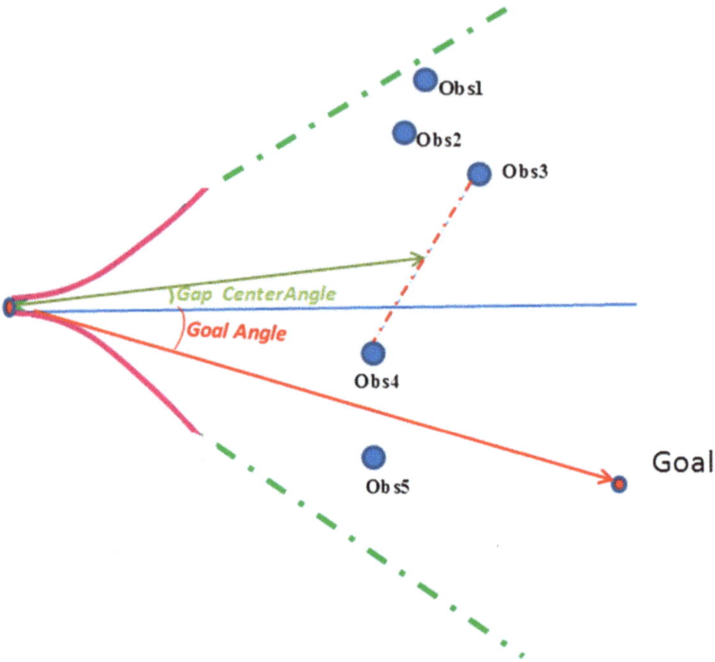

Fig. 4.9 Maximum gap, gap center angle, and goal angle together with obstacles

Gap center angle is calculated in terms of the measurable parameters using the trigonometric relations as illustrated in (4.1)

$$\phi_{gap\_c} = \arccos\left(\frac{d_1 + d_2\cos(\phi_1 + \phi_2)}{\sqrt{{d_1}^2 + {d_2}^2 + 2d_1 d_2 \cos(\phi_1 + \phi_2)}}\right) - \phi_1 \quad (4.1)$$

In this function, the variables are defined as follows: $\phi_{gap\ \_\ c}$ is the gap center angle, and d_1, d_2 are the distances to obstacles of maximum gap. ϕ_1, ϕ_2 are the angles of obstacles of the maximum gap.

The final heading angle of the vehicle is computed using both center angle and goal angle. A fusion function for the final angle calculation is given in (4.2):

$$\phi_{final} = \frac{\frac{\alpha}{d_{\min}}\phi_{gap\_c} + \beta\phi_{goal}}{\frac{\alpha}{d_{\min}} + \beta} \quad where \quad d_{\min} = \min_{i=1:n}(d_n) \quad (4.2)$$

In this function, the variables are defined as follows: ϕ_{goal}, goal angle; α, weight coefficient for gap; β, weight coefficient for goal; n, number of obstacles; d_n, distance to n-th obstacle; and $d_{mi\ n}$, minimum of d_n distance values.

The α value defines how much the vehicle is goal oriented or gap oriented.

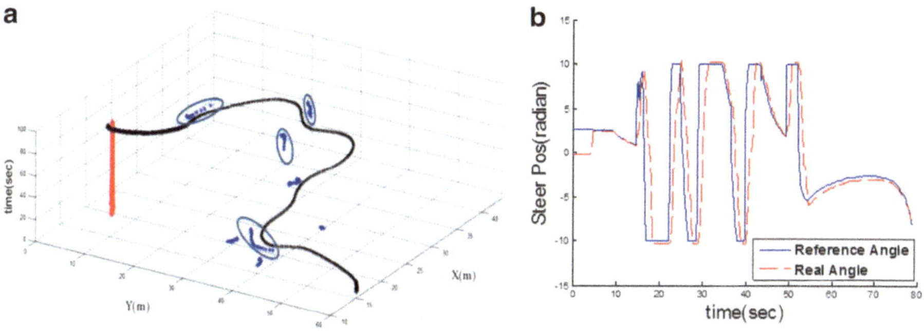

Fig. 4.10 Results of dynamic obstacle avoidance experiment. (**a**) Vehicle trajectory with obstacle measurement. (**b**) Steering wheel reference and real angle

4.3.2 Experimental Results

Obstacle avoidance algorithm based on "follow the gap method" is coded using C programming language. The real-time code runs in MicroAutoBox hardware. The only tuning parameter, alpha, is selected as 20 in experimental tests as in the simulations. The field of view of the LIDAR is 150° and its measurement range is restricted with 10 m. The first test configuration is composed of seven static obstacles with a goal point. In the second test scenario, "follow the gap" method is tested using dynamic obstacles. The results of the dynamic obstacle tests are given in Fig. 4.10 with a 3D graph showing the time-dependent locations of the obstacles and their perceived trajectories by the vehicle.

4.4 Conclusions

An *UGV* named Otonobil is introduced in this chapter detailing on its mechatronics design, perception, and decision algorithms. First the modifications on an urban EV are mentioned briefly to convert the vehicle for autonomous operation. Then the multisensory structure with several processing units is given to emphasize the importance of data acquisition and real-time processing needs of such applications. Finally a novel dynamic obstacle avoidance algorithm developed for Otonobil is explained briefly. The vehicle will be used in similar research projects in the field of active vehicle safety and intelligent vehicles in future work.

Acknowledgement The researchers in *Otonobil* project would like to thank Prof. Ata Muğan, Director of Mechatronics Education and Research Center, for his invaluable support and constant encouragement during this project.

References

1. V. Sezer, C. Dikilitas, Z. Ercan, H. Heceoglu, P. Boyraz, M. Gökaşan, "*Hardware and Software Structure of Unmanned Ground Vehicle Otonobil*", 5th biennial workshop on DSP for in-vehicle systems, Kiel, Germany, 2011
2. V. Sezer, C. Dikilitas, Z. Ercan, H. Heceoglu, A. Öner, A. Apak, M. Gökasan, A. Mugan, "Conversion of a Conventional Electric Automobile Into an Unmanned Ground Vehicle (UGV) IEEE – international conference on mechatronics, pp. 564–569, Istanbul, Turkey, 13–15 April 2011
3. Z. Ercan, V. Sezer, C. Dikilitas, H. Heceoglu, P. Boyraz, M. Gökaşan, "*Multi-sensor Data Acquision, Processing and Logging using Labview*", 5th biennial workshop on DSP for in-vehicle systems, Kiel, Germany, 2011
4. Z. Ercan, V. Sezer, H. Heceoglu, C. Dikilitas, M. Gökasan, A. Mugan, S. Bogosyan, "*Multi-sensor data fusion of DCM based orientation estimation for land vehicles*" IEEE – international conference on mechatronics, pp. 672–677, Istanbul, Turkey, 13–15 April 2011
5. V. Sezer, M. Gokasan, A novel obstacle avoidance algorithm: "follow the gap method". Robot Auton Syst **60**(9), 1123–1134 (2012)

Part II
Speech and Audio Processing

Chapter 5
Car Hands-Free Testing and Optimization: An Overview

Hans-Wilhelm Gierlich

Abstract Testing of car hands-free systems needs to take into account the specific environment in the car. This chapter describes a variety of test methods and requirements applicable to car hands-free systems. The test scenarios describe the test setup in the car and the different considerations for microphone and loudspeaker placement. The performance parameters and test methods to achieve a good conversational speech quality are introduced and described. Parameters relevant in the single-talk situation, which mainly influence the speech quality in the listening situation, are discussed. Special consideration is given to the background noise presence where additional perception-based methods are described. Furthermore, tests and test conditions targeting the echo canceller performance in different conversational situations are presented. Besides the narrowband case, a special focus is given on wideband. The users' expectations with regard to wideband are high, and the requirements for wideband car hands-free systems are high as well. The differences from the narrowband implementations are discussed. Finally, a method is given which allows a summary overview of a variety of results in one view, helping to better visualize a variety of test results and giving an easy-to-interpret overview of systems.

Keywords Car hands-free systems • Performance parameter • Noise cancellation • Echo cancellation • Testing and optimization procedures • Speech quality

H.-W. Gierlich (✉)
HEAD acoustics GmbH, Ebertstr. 30a, Herzogenrath D-52034, Germany
e-mail: h.w.gierlich@head-acoustics.de

G. Schmidt et al. (eds.), *Smart Mobile In-Vehicle Systems: Next Generation Advancements*, DOI 10.1007/978-1-4614-9120-0_5,

5.1 Introduction

Car hands-free systems were the first systems where advanced signal processing, such as adapted echo cancellation, noise cancellation,, and similar other speech enhancement techniques based on DSP implementations, was feasible due to the willingness of customers to bear the cost of such systems and due to the fact that space and power consumption for such implementations were not an issue. Consequently, the need of qualifying these nonlinear and time-variant implementations came up soon. Based on previously available analysis techniques [1–3], the first advanced test specification—the so-called VDA Specification for car hands-free [4]—was already created in 2001. Advanced tests using speech-like and speech signals were used to evaluate the quality parameters of car hands-free systems. Since that time new testing technologies—adapted to the more advanced signal processing techniques—have been created and included in the standards step by step. In 2008 this topic was also taken up by ITU-T. Besides the creation of an own question within the Study Group 12 [5], a new focus group [6] was established especially dealing with testing of speech technologies, mainly hands-free, in cars.

Car hands-free testing has to respect the user's perception of speech in the conversational situation as well as the complexity of the in-car environment. This includes the car acoustics as well as the different systems in the car used in a car hands-free setup. In modern cars, the different systems can be regarded as subsystems and are typically distributed in various units within the car.

5.2 Car Hands-Free Systems

Modern car hands-free systems can range from a simple "one box" design to a complex distributed system including microphone array techniques, distributed signal processing, and multi loudspeaker setup integrated in cars. Furthermore, an additional level of complexity is added by the fact that for the built-in car hands-free systems different components are combined in the same car depending on the car configuration ordered by the customer. The need to ensure compatibility and quality of all the different configurations increases the test time as well as the test complexity. From the customer perspective, built-in car hands-free systems are part of the car and the same quality is expected just like other components in the car. Moreover, it is expected that car hands-free systems work seamlessly with every mobile phone used by the customers. This requires universal interfaces which decouple the long product cycles in the car industry (typically 6–8 years) from the very short and dynamic product innovation cycles in the mobile terminal industry (typically less than half a year).

Figure 5.1 shows a block diagram of a typical car hands-free configuration as found in vehicles. A microphone or a microphone subsystem is the first block. The microphones may either be connected directly to the signal processing (hands-free

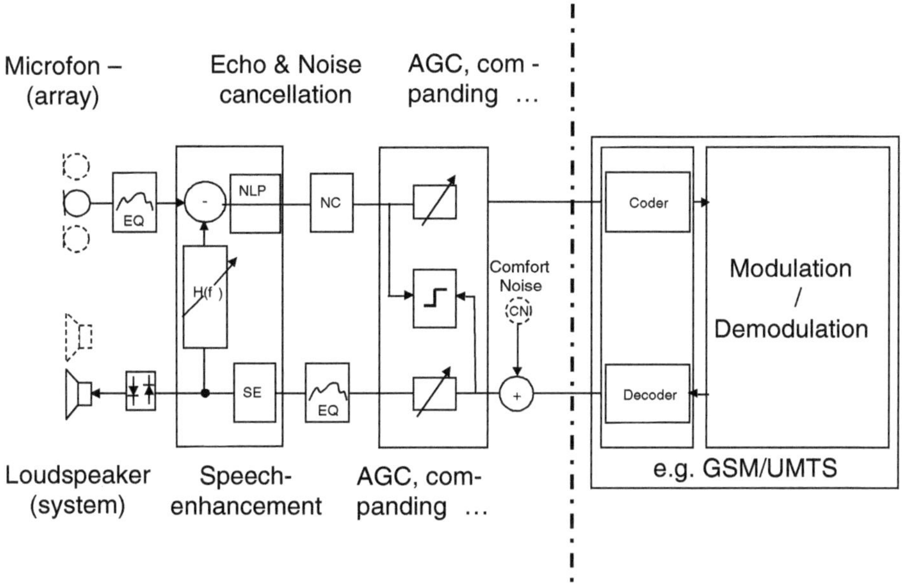

Fig. 5.1 Components in car hands-free systems

telephone system) or via the bus systems to the audio bus (e.g., MOST-BUS) in the car. On the receiver side, the car audio system is utilized. Car audio systems range from simple amplified speaker systems to multichannel multi-speaker playback systems. More advanced playback systems furthermore integrate sophisticated signal processing for audio playback of music and other content. The audio playback system may be connected either analog, digital, or by the car built-in bus system to the hands-free telephone system which includes the hands-free signal processing capability.

The connection to the radio network is typically made by the customer's mobile phone. The most common connection here is wireless (e.g., Bluetooth®). Other possibilities are built-in radio units combined with the access of the user data by special profiles (e.g., SIM-access profile with Bluetooth). The decoupling of the radio network connection from the car hands-free system by means of a mobile phone generally allows the highest flexibility in order to adapt to the fast development cycles of mobile phones. However, due to the lack of sufficient specification of performance parameters for audio characteristics on the wireless link between hands-free system and the mobile phone, numerous problems may arise from exchanging the mobile phone by the car user. The mobile phone connected to the car hands-free system may impair all quality parameters as perceived by the user in a conversational setting.

When assessing and optimizing car hands-free systems with regard to the speech quality, various system configurations have to be taken into account. The VDA Specification [4] and the ITU-T Recommendations P.1100 [7] describe in their

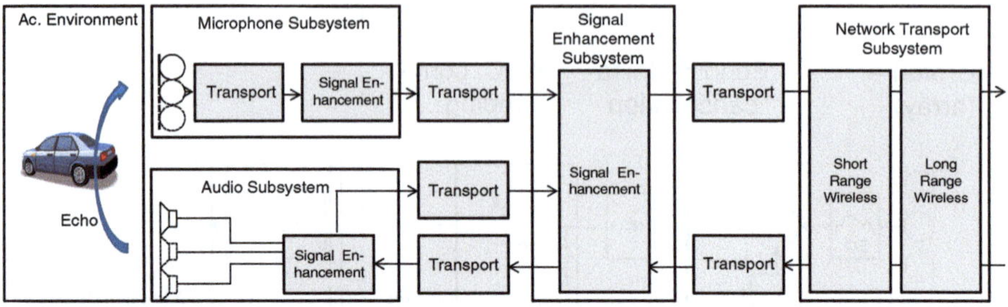

Fig. 5.2 Subsystems in car hands-free according to [6]

main sections the complete measurement of the hands-free system from microphone to the network termination point (NTP) and from the NTP to the loudspeaker. However, in case of subsystems, it is often not sufficient to determine the quality over the complete system. Often, it is useful to determine the speech quality performance of the subsystems present in the car. Any subsystem may degrade the speech quality if not properly implemented and not properly adapted to the associated subsystems. Therefore, special sections for the measurements of microphones and for the determination of the speech-related properties of the transmission between the mobile phone and the car hands-free system are found already in [7] and [8]. Since this problem seems to be more complex, a more fundamental approach on the definition of subsystems and their influence on the different speech quality aspects are needed. A new specification dealing with car hands-free subsystems is currently under development in the ITU-T Focus Group CarCom [6]. The system configuration as considered by this group is shown in Fig. 5.2.

The subsystem approach can be used in order to diagnose different components of a car hands-free system and to check the individual performance parameters of the individual subsystems contributing to the overall quality of the car hands-free system.

5.3 Speech Quality from the User's Point of View

In general, the user's perception of speech quality is independent of the communication device used. Whether the communication is made from the car, from mobile to mobile, or from an office-type environment—the user perception generally does not change. However, it should be noted that some degradations may be accepted by the user in case he/she receives other benefits, such as mobility or having the possibility of making phone calls legally in a car. Being able to communicate while driving is certainly a clear benefit from the user's point of view—if the communication is possible easily and without impairments. Nevertheless, the main task of the user in the car is driving the car. Any distraction from his driving task, i.e., secondary cognitive loads, must be avoided. Initiatives to reduce the driver distraction can be found in various entities. The ITU-T Focus Group "Driver Distraction" [9] is dealing specifically with this topic. It can be stated that a well-designed car hands-free system which allows a relaxed communication with low listening and low talking effort certainly contributes to less driver distraction. From the user's point of view, the main quality parameters can be considered as follows:

- *Delay*
  Delay or latency is a very critical parameter in human communication. As already described in ITU-T Recommendation G.114 [10], the conversational quality degrades in case the mouth-to-ear delay is higher than 150 ms. Higher delays result in increased conversational effort, increased unintended double talk, and increased echo perceptibility.
- *Speech Sound Quality*
  Speech sound quality has a variety of aspects and can be attributed to both the send side and the receive side of the car hands-free system. While it is very difficult to achieve a high speech quality in sending (due to the adverse car environment, background noise, vehicular echo, and other impairing factors), the user of the car hands-free system does not benefit from any improvements achieved in sending directly. The user of the car hands-free system would only profit indirectly by high speech quality in sending due to the increased conversational quality and less complaints of the far-end partner. The speech sound quality in receiving is mainly determined by the quality of the acoustical/audio components in the car as well as by speech coding and signal processing attributed to the receiving side.
- *Loudness*
  Loudness of the perceived speech is predominantly contributing to customer's satisfaction. As speech sound quality, loudness is influenced by the complete transmission chain between the sending and receiving processes. The part contributed by the terminal is to be divided into the sending and the receiving loudness. The telephone network is calibrated and adjusted to the nominal loudness based on the ITU-T loudness ratings [11]. Therefore, it is essential that under nominal conditions the Sending Loudness Rating SLR of the car hands-free terminal is conforming to the nominal requirements cited in [4, 7, 8]. The loudness of the received sound must be adjustable in such a way that a sufficient loudness level is guaranteed under all driving conditions, it is not exceeding limits impairing the user's ear and in such a way that always a sufficient quality of the transmitted speech is achieved. Different techniques to enhance speech loudness in receiving such as AGC (automatic gain control), noise dependent equalization, and several other techniques may be applied in order to achieve user satisfaction.
- *Intelligibility*
  Sufficient speech intelligibility is the key factor of all speech communication. All elements in a transmission chain may contribute to the degradation of speech intelligibility: on the sending side, insufficient noise cancellation, other signal processing, and speech coding; on the receiving side, speech coding and all types

of speech enhancement algorithms. Furthermore, the quality of the acoustic components (microphones, microphone arrays, and loudspeakers) may contribute significantly to the speech intelligibility issues. Although speech intelligibility is the basic requirement of the speech communication systems, no adequate testing methodology exists up to now which may predict the speech intelligibility of modern communication systems in a proper and reliable way.

- *Echo Performance*
  The echo performance of car hands-free systems in all driving situations is very critical. Any impairment due to the vehicular and the far-end echo in a car hands-free system will always be perceived by the user. Different from the other impairments mentioned above, echo is a talking-related impairment. Furthermore, echo perception depends highly on the latency or the delay of the connection (typically unknown since network delays of an individual connection are unknown). Therefore, echo performance parameters always has to be targeted to the worst case situation (high delay in the connection). Due to the complex acoustic environment in the car, the driving noise, the wind noise, and the large distance between the talker (driver or passenger) and the microphone, vehicular echo cancellation still remains a major challenge.
- *Double-Talk Behavior and Switching*
  Any switching (the remote communication partner stops speaking and the local one starts, or vice versa) in the car hands-free terminal may result in reduced speech quality and intelligibility as well as in a reduced conversation quality. Switching may result in the suppression of syllables or even words and may occur in sending or in the receiving direction. Switching may also occur in double-talk situations where sending and receiving is present simultaneously. Again, the impairments range from front end clipping of syllables—mostly unnoticeable—to the suppression of complete words—objectionable to incomprehensible. Furthermore, especially in double-talk, additional echo may occur which is not observed in single-talk instances. However, in double-talk, listeners are found to be less sensitive towards echo [3, 12]. Echo produced in double-talk situations often can be attributed to speech echo cancellers which may provide insufficient echo loss in such situations.

5.4 Test Setup

The general test setup for car hands-free systems has been known for a long time and did not change substantially over the years. It is shown in Fig. 5.3.

Whenever possible, the setup is made in the target car, i.e., the car where finally the hands-free system will be installed. All components of the car hands-free system remain intact. The driver is substituted by an artificial head according to the ITU-T Recommendation P.58 [13]. The test is conducted in a lab-type quiet environment. In order to simulate driving conditions, background noise is simulated by a simulation block (Fig. 5.3). This background noise is prerecorded by driving the car in different conditions. Typical driving conditions can be found in [7]. Often

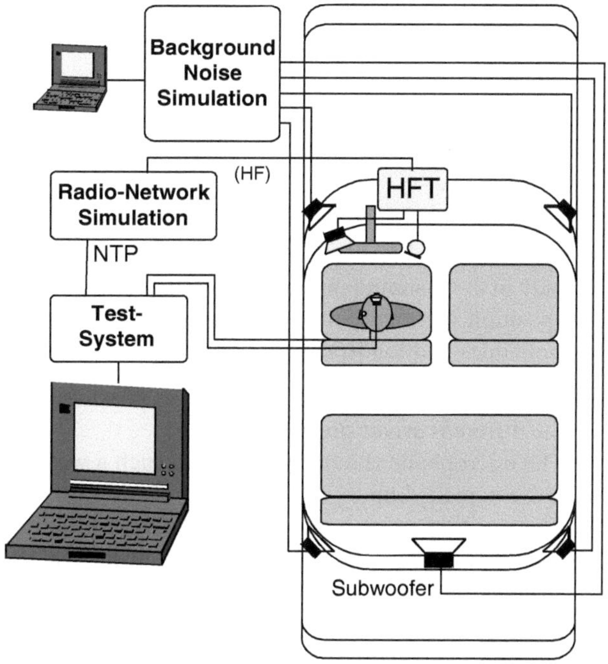

Fig. 5.3 Test setup for car hands-free systems

constant speed driving conditions are simulated at a minimum of 130 km/h but also at different driving speeds. The background noise is captured from the car hands-free microphone. It is recognized that this background noise simulation technique may be limited to single microphone scenarios. When microphone arrays are used, the simulation may be not of sufficient accuracy. In such an event, it is possible to disconnect the microphone array, record the background noise picked up by the individual microphones of the microphone array at their outputs, and fuse them later electrically with calibrated levels into the microphone inputs jointly with the test signals picked up by the microphone.

In order to simulate the radio network, a simulator—system simulator—is used which connects the mobile phone or the RF unit of the hands-free system. The test signals are inserted at the radio network simulators side—network termination point (NTP)—as well as by the artificial mouth of the artificial head. In receiving, the signal is picked up by one or both artificial ears of the artificial head placed at the driver's position. In sending, the signal is picked up at the network simulator (at the NTP).

The test signals used for conducting the tests are as speech-like as possible. New developments in ITU-T [14] define the acceptable speech signals for the measurement of such devices. In addition, speech-like signals such as artificial voice [15], *C*omposite *S*ource *S*ignals (CSS), and others are used [14]. When applying these test signals, it is always the goal to use a test signal which is as similar as possible to the signal under real-world conditions but allows repeatable and reproducible measurement as well. The basic analysis techniques used in conjunction with these test signals are defined in [16].

5.5 Microphone and Loudspeaker Positioning

The positioning of loudspeaker(s) and microphones in the car is an essential prerequisite to achieve a good overall quality. The placement should be considered already in the planning phase of new cars because it is extremely difficult to change the position of microphones or loudspeakers at a later stage. When positioning microphones and loudspeakers, the following rules should be followed:

- The microphone should be as close as possible to the driver taking into account the different driver positions.
- The microphone should be placed in such a way that a minimum of background noise is picked up by the microphone.
- The positioning of loudspeaker relative to the microphone should be made in such a way that the highest possible acoustical decoupling between the loudspeaker and the microphone in the car in question is achieved.
- Loudspeakers should be placed in such a way that a high speech level is received by the driver (and codrivers if required) but less energy is transmitted by the microphone.

For all parameters, it should be considered that the effects mentioned above are frequency dependent and should be validated across the complete frequency range. The individual performance characteristics of the signal processing (echo cancellation, noise cancellation) depending on frequency should be taken into account when optimizing the loudspeaker and microphone positions. An example for the evaluation of microphone in terms of the spectral content of the signal picked up at different microphone positions is shown in Fig. 5.4. Based on spectral level constraints for a given hands-free implementation, an optimum microphone position can be found using such types of analysis.

5.6 Test Procedures for Assessing Speech Quality

All test procedures used for car hands-free testing have to consider various conversational scenarios and the user's perception of speech quality as described beforehand. An overview about testing procedures used can be found in [2, 17] and [18]. The relevant specific standards which form the basis of all car hands-free testing procedures are as follows:

- VDA Specification [4]
- ITU-T P.1100 for narrowband car hands-free systems [7]
- ITU-T P.1110 for wideband car hands-free systems [8]

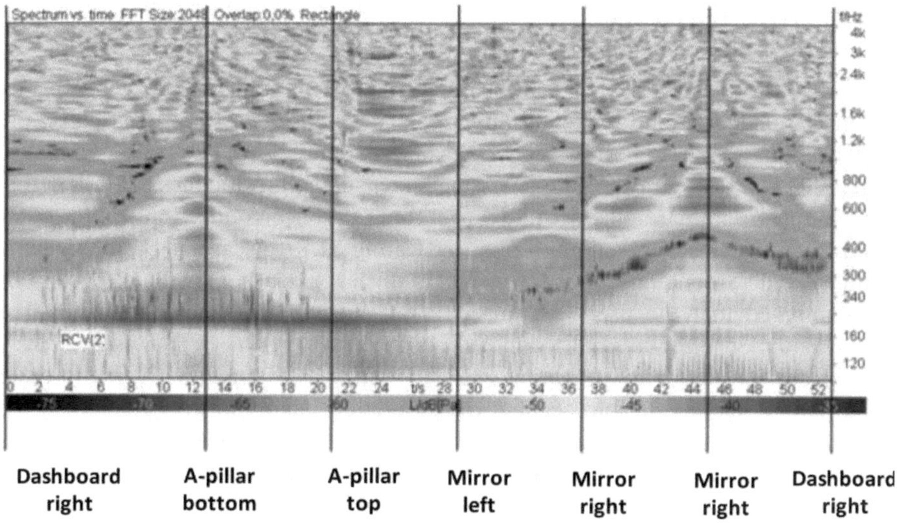

Fig. 5.4 Spectral content of the microphone signal picked up at different microphone positions

Table 5.1 Tests as defined in the VDA car hands-free specification [4] and in ITU-T Recommendations P.1100 [7] and P.1110 [8]

One-way tests	One-way tests with background noise	Echo canceller (EC) and double-talk (DT) tests
Delay	Background noise transmission after the call setup	Echo performance with and without background noise
Loudness ratings	Speech quality in the presence of background noise	Initial convergence with and without background noise
Variation of loudness rating	Quality of background noise transmission (with far-end speech)	Echo performance with time-variant echo path
Sensitivity frequency responses	"Comfort noise" injection	Switching characteristics
Speech quality during single-talk		Double-talk performance
Listening speech quality stability		
Idle channel noise		
Out-of-band signals		
Distortion		
Switching characteristics		

All tests need to take into account the different conversational situations. For each conversational scenario, a set of tests and the corresponding requirement set are defined which are needed to cover the important aspects of speech quality instrumentally. All tests should take into account the impact of signal processing expected in car hands-free systems and the environmental conditions where the system is typically used. Based on these facts, the typical tests can be grouped as tabulated in Table 5.1.

5.6.1 One-Way Tests

The basis for a setup of a car hands-free system is the adjustment of the correct loudness in sending and receiving direction. The loudness setting is based on the widely used "loudness ratings" concept which is defined in ITU-T Recommendation P.79 [11]. In contrast to simple loudness measurements, loudness ratings define a deviation from the loudness with respect to a reference connection. As for other hands-free systems, the Sending Loudness Rating SLR for car hands-free systems is higher (corresponding to lower sensitivity) than for handset terminals because it is anticipated that the user will raise his/her speech level during a hands-free conversation. The required SLR typically is 13 dB.

In receiving, the loudness rating RLR is defined as the nominal value in quiet condition. When taking into account the correction factors in the calculation, the required loudness rating in receiving is also higher (lower sensitivity) compared to a handset telephone. The reasons for this higher requirement are the loudness increase due to binaural listening as well as diffraction effects from the human body which contributes to higher loudness in the receiving direction. In order to compensate high driving noise, the loudness range variation is higher than for conventional terminals. Typically, it is required to provide a volume control with a loudness increase of at least 15 dB referred to the nominal value of RLR. In general, the control range of the volume control should allow higher than 6 dB signal-to-noise ratio for all signal and noise conditions.

A minimum prerequisite for achieving a sufficient speech sound quality is the definition of sending and receiving frequency responses allowing a sufficient margin for different implementations on the one side but providing the basis for a good sound quality perception on the other side. The objective here is to have a mostly flat response characteristic in sending, which allows for some high-pass characteristics to reduce the ever present low-frequency background noise. Similarly, in the receiving direction, the goal is a mostly flat frequency response typically averaged over the left and right ear signal of the artificial head. The artificial head in general is free-field equalized; however, more recent developments take into account a more diffuse character of the sound field in the car when multiple loudspeakers are used. In such an event, the equalization of the artificial head should be diffuse-field.

To cover artifacts from advanced signal processing techniques, such as speech coding and adaptive signal enhancing processing techniques, the perception-based speech quality measurements should be applied. Different methodologies and models are available to determine the speech sound quality based on hearing models. The most widely known method is PESQ, standardized in ITU-T Recommendation P.862 [19], which is widely used in networks for speech quality monitoring. However, PESQ does not consider the frequency response effects and their impact on the sound quality which may be significant. As a consequence, this measurement cannot be used for acoustical systems including car hands-free systems. For car hands-free systems, mostly TOSQA [20] is employed which is suitable for different types of acoustical interfaces. More recent developments in ITU-T led to a new standard for speech quality evaluation and it is described in ITU-T Recommendation P.863 [21]: POLQA. POLQA, in general, should be applicable to acoustical interfaces. However, in the characterization phase of this methodology, it was found that the prediction accuracy of this model is too limited for hands-free systems, and the current version is not recommended for use in car hands-free systems (see [22]).

Besides the test of speech quality and loudness, the measurement of delay is one of the most important tests. Delay is introduced by various components of the car hands-free system, mainly by the radio transmission, the Bluetooth® transmission, and the delay introduced by the car hands-free signal processing itself. As described above, one-way delay should be less than 150 ms for a good conversational quality communication in the car. This figure should cover all delay components measured in the car hands-free system as well as the delay introduced by the connection and the delay introduced by the far-end terminal. Since any of those components will introduce delay impairing the conversational quality, low-delay implementation should be preferred in each stage. For car hands-free systems, the delay recently was defined as the round trip delay including the sending and receiving delays but excluding the delay introduced by the radio network transmission. ITU-T Recommendations P.1100 and P.1110 require a round trip delay TRTD of less than 70 ms.

If any type of wireless connection between the mobile phone and the hands-free system is used, a round trip delay of less than 120 ms is required taking into account wireless transmission delay and including both the delay introduced by the car hands-free system and the delay introduced wireless transmission.

The distribution of the delay is up to the manufacturer. Perceptually, it does not matter whether the delay is introduced in the send or receive segment of the car hands-free system. Therefore, depending on the type of signal processing used, e.g., more delay can be spent on noise and echo cancellation in sending compared to the signal processing in the receiving side. The delay measurement is carried out separately in sending and receiving direction. The measurement is based on the cross correlation between the input signal and the measured signal at the NTP (for sending direction) or the artificial ear of the artificial head (in receiving direction). Test signals used are typically Composite Source Signals as defined in ITU-T Recommendation P.501 [14].

Often the signal processing inside the car hands-free systems is faced with switching. This may be due to speech detection algorithms or artifacts of noise cancellation. It is usually the effect of the non-linear processor of the echo canceller. From the user's point of view, it is important that also low-level speech signals are transmitted completely; in particular, initial syllables should not be suppressed. Therefore, the measurement of switching characteristics is required which is typically performed by using a sequence of Composite Source Signals slowly increasing in level. A minimum threshold is defined at which the signal should be transmitted completely. This threshold is 20 dB_{Pa} measured at the Mouth Reference Point. Although less critical, the same type of switching behavior can be measured in receiving.

5.6.2 Talking Related Impairments: Echo Canceller Tests

Echo is the main talking-related impairment found in hands-free systems. The perception of echo depends on the delay inserted in the overall connection. The higher the delay, the more sensitive human ear becomes against the echo. ITU-T Recommendation G.131 [23] describes this relationship. From those experiments, it can be seen that for delays higher than about 250 ms, a Talker Echo Loudness Rating (TELR) of more than 55 dB is required. The measurement of echo loss in car hands-free systems is based—as for any other terminal—on TCLw (Terminal Coupling Loss weighted) according to ITU-T Recommendation P.79 [11]. The measurement is performed by inserting a test signal into the receive channel at the NTP of the test setup and measuring the echo at the send output (at the NTP). The test signal used is typically artificial voice [15] followed by a highly compressed speech and noise signal which allows to convey a sufficient amount of energy into the system under test in order to measure reliably a TCLw of more than 55 dB. It is known that this type of test signal is not speech-like. Therefore, new developments in ITU-T Recommendation P.501 introduce a test signal which is a compressed real-speech signal [14]. Therefore, this test signal is used for the determination of TCL_w in newer standards.

The determination of the steady-state echo loss TCL_w achieved by the echo canceller is certainly not sufficient to characterize the echo cancellation performance and to guarantee a sufficient echo performance in different situations. A very important test to evaluate the adaptive performance of echo cancellers is the test of the initial convergence. Ideally, no echo should be perceived during the complete initial convergence phase of an echo canceller. Requirements as defined in ITU-T P.1100 and P.1110 specify an echo loss of minimum 6 dB during the first 200 ms and an echo return loss (ERL) of at least 40 dB which has to be achieved after 1.2 s as depicted in Fig. 5.5.

Another important echo canceller evaluation test is performed with background noise. Background noise—as the near end speech signal—impairs significantly the performance of echo cancellers. Since in almost all cases, noise is present, it is important to verify the initial convergence of the echo canceller with background noise especially at the beginning of a call. ITU-T P.1100 and P.1110 require that the echo signal level shall not exceed the background noise level by more than 10 dB within the first 200 ms of the initial convergence. After 1.5 s any echo signal has to be below the background noise level as shown in Fig. 5.6.

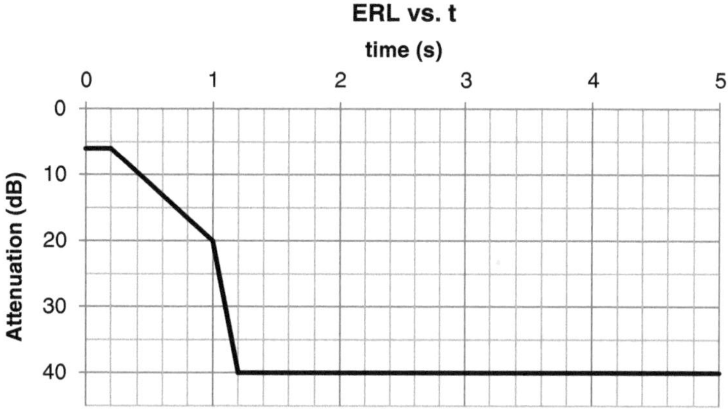

Fig. 5.5 Initial convergence, ERL versus time (from [7])

The echo performance with a time-variant echo path is another important parameter to be investigated. A time-variant echo path can be simulated in different ways. The simplest test here is to just quietly open a door during the measurement and observe the echo loss as a function of time during this period. Another more precisely defined echo path variant can be found in [7] and [8]. Here, the time-variant echo path is realized by rotating a reflecting surface positioned on the codriver's seat. Regardless of what type of time-variant echo path simulation is used, it is advisable that the measured echo attenuation should not decrease by more than 6 dB.

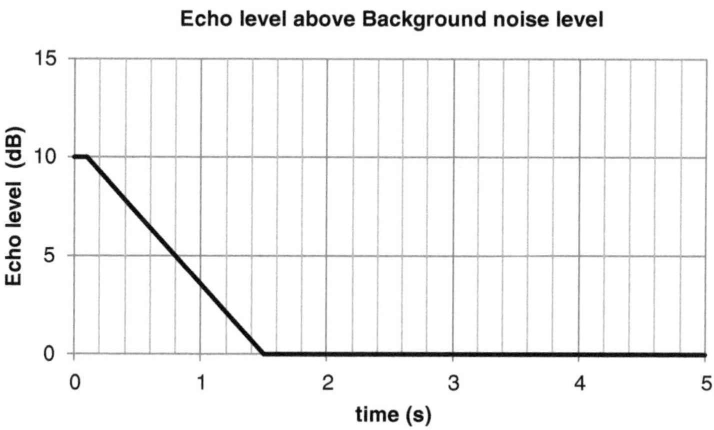

Fig. 5.6 Initial convergence with background noise (from [7])

Besides the methods described above, also a perception-based model for predicting the annoyance of echo impairments is available. The details of the method named EQUEST are given in [24] and [25]. The model takes into account the masking effects by speakers' voice and it detects echo structures in time and/or in frequency which contribute to echo annoyance. As for other perceptual models, the procedure determines an E-MOS score where E-MOS $= 5$ represents no echo impairment and E-MOS $= 1$ represents a highly annoying echo.

5.6.3 One-Way Tests with Background Noise

Besides good echo canceller performance, one of the biggest challenges in car hands-free systems is the handling of background noise. Even with optimum microphone positioning, the distance between the user's mouth and the microphone is significantly higher than for mobile phone applications. As a consequence, the SNR may be quite poor and some signal-enhancement technologies need to be applied to reduce the background noise on the one hand and to enhance the speech quality and preferably also the speech intelligibility on the other. It is well known that for all signal-enhancement techniques there exists a trade-off between reducing the amount of background noise and preserving or even better enhancing the speech sound quality. Therefore, the preferred way of testing and optimizing such techniques is perception-based models which separately determine the speech quality and the quality of the transmitted background noise. Up to now the only validated model for this purpose can be found in ETSI [26, 28]. The ETSI model (also named 3QUEST) is perception-based and determines separately the speech quality (S-MOS), the intrusiveness of the transmitted noise (N-MOS), and the overall quality (G-MOS). The prediction performance of this model for hands-free and car hands-free systems is described in [18] and [17]. The test setup for determining those parameters is shown in Fig. 5.7.

Besides the undistorted speech signal, the speech plus noise signal, as picked up by the car hands-free microphone, is required. Therefore, either a measurement microphone is positioned close to the car hands-free microphone or alternatively an equivalent electrical signal is used for processing. In order to achieve a sufficient speech sound quality, S-, N-, and G-MOS should be as high as possible. A balanced implementation leading to high values for all three parameters under different types of background noise situations is preferred. It should be avoided from increasing just the amount of noise reduction to such an extent that the speech sound quality is degraded. ITU-T Recommendation P.1100 and P.1110 require an S-, N-, G-MOS value of more than 3.0 for driving speeds under 80 km/h. For background noises produced at driving speeds between 80 km/h and 130 km/h, the requirement is relaxed to MOS ≥ 2.5.

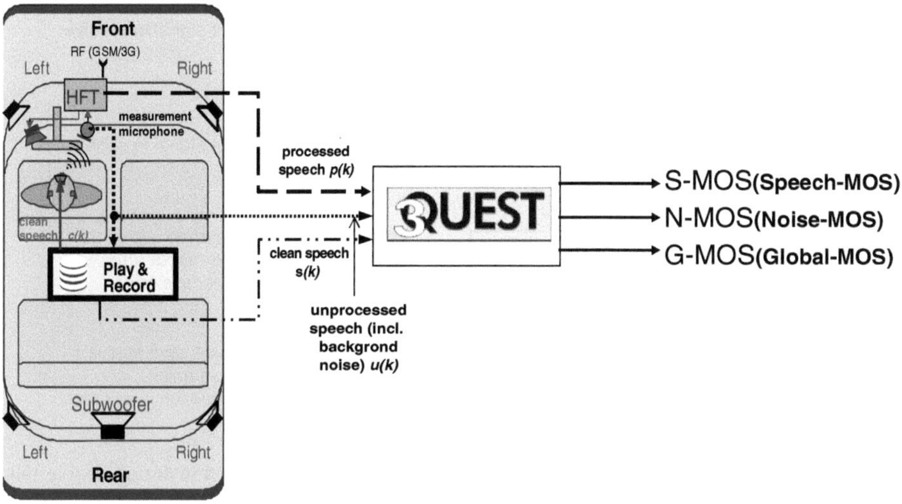

Fig. 5.7 Test setup for determining the speech quality in background noise using 3QUEST

5.6.4 *Double-Talk Performance*

Double-talk may occur in any conversational situation, especially during times where exciting or controversial information is exchanged. The amount of double talk in a conversation certainly depends on the emotional status and the temperament of the users. Furthermore, the amount of double-talk increases if the end-to-end delay is increased. High delay in conversation results in unintended double talk. Since in modern communication scenarios, the delay is often unpredictable and/or unknown, double-talk performance of car hands-free systems should be as good as possible. The double-talk characterization framework can be found in ITU-T Recommendation P.340 [12] and it is categorized into five different classes:

- 1—Full duplex
- 2a—partially duplex
- 2b—partially duplex
- 2c—partially duplex
- 3—non duplex

The double-talk behavior according to [3] is determined primarily by three parameters:

- Attenuation range during double-talk in sending
- Attenuation range during double-talk in receiving
- Echo during double-talk

To determine the attenuation range during the double-talk, a sequence of overlapping Composite Source Signals (CSS) inserted simultaneously in sending

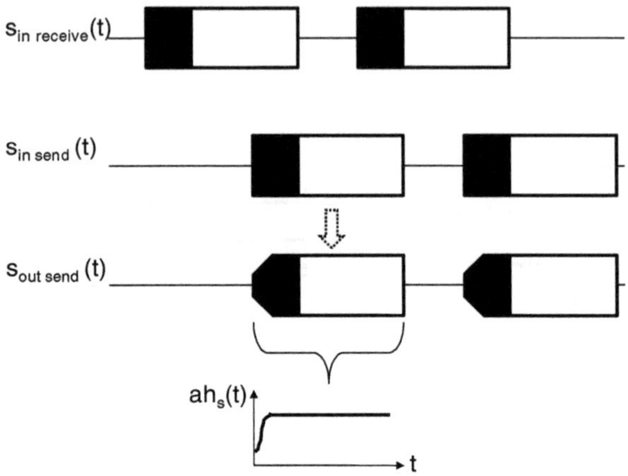

Fig. 5.8 Test signal to evaluate of the attenuation range in sending direction during double-talk

and receiving directions of the car hands-free systems under study. The test is configured to focus on the transmission of the complete signal in sending while a signal is present at the receive direction—attenuation range in sending—and the opposite in receiving direction. More information on this can be found in [16]. A typical test signal for this test is illustrated in Fig. 5.8. As shown there, the attenuation range can be determined by a level versus time analysis of the transmitted CSS component which is referred as the CSS component inserted to the hands-free system under test.

The determination of echo components during the double-talk requires the insertion of test signals in sending and receiving simultaneously. Again, the basic principle can be found in [16]. The employed test signals are voice-like signals, i.e., any voiced sound which has a comb-filter type spectrum. The signals inserted simultaneously in sending and receiving sides are constructed in such a way that they do not overlap spectrally. Thus, it is possible to separate echo components from the near-end signal in the double-talk situation e.g., by applying a comb filter. The principle of the test procedure is shown in Fig. 5.9.

The general characterization of the double-talk performance is defined in [12] and is based on subjective evaluation of different types of attenuation range and echo during double-talk instances. In Table 5.2 the requirements to be met are given in order to claim the type of double-talk behavior achieved.

5.7 Wideband: Special Requirements

In wideband, the transmission bandwidth is increased to 8 kHz to cover almost the complete frequency range of human voice. In general, a car is a great place for the deployment of wideband systems because the user (receiving side) may benefit

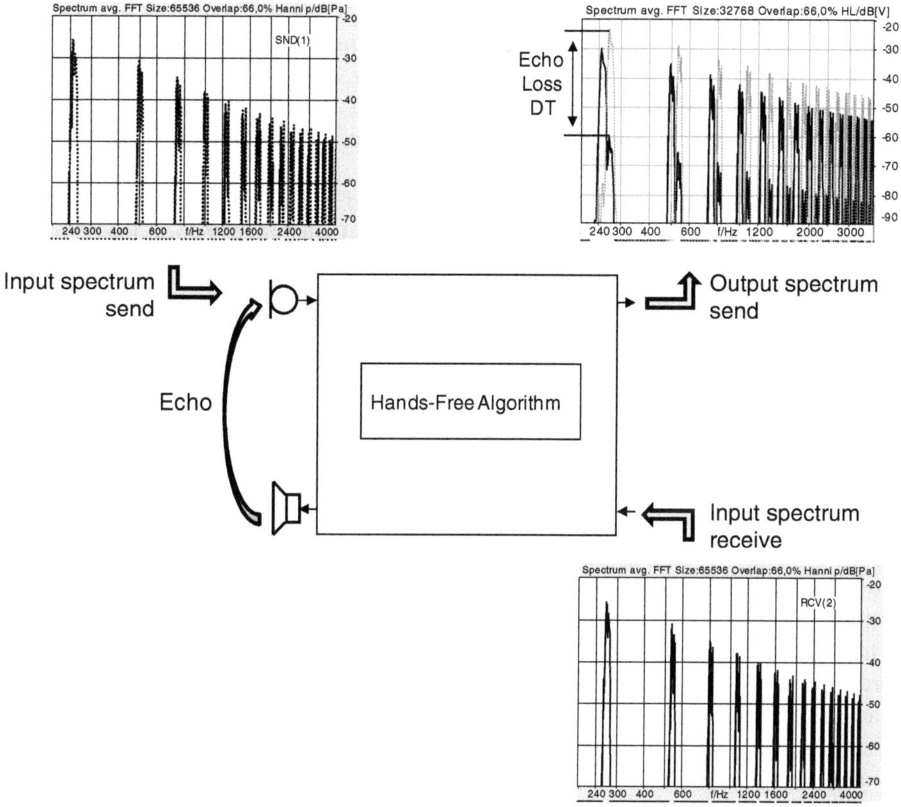

Fig. 5.9 Principle of echo evaluation during double-talk

Table 5.2 Hands-free terminal behavior as described in ITU-T recommendation P.340

	Type 1	Type 2: Type 2a	Type 2: Type 2b	Type 2: Type 2c	Type 3
$TELR_{DT}$ [dB]	≥ 37	≥ 33	≥ 27	≥ 21	<21
A_{Hsdt} [dB]	≤ 3	≤ 6	≤ 9	≤ 12	>12
A_{Hrdt} [dB]	≤ 3	≤ 5	≤ 8	≤ 10	>10

from the high-quality audio playback systems available in most modern cars. In wideband, the speech quality is increased significantly but also the intelligibility is enhanced up to 15% compared to narrowband platforms. The most important fact, however, is the reduced listening effort of users when using the wideband transmission in the presence of background noise in particular. In order to achieve a sufficient quality in wideband, all relevant speech quality parameters have to be reconsidered [29].

The minimum requirement to achieve a good speech sound quality in wideband is the extension of the frequency response requirements. The testing methodologies are exactly the same as in narrowband. However, frequency tolerance masks are extended up to 7.6 kHz and down to 100 Hz (at least) in sending as well as in receiving to guarantee a wideband user experience. Minimum requirements can be found in [8].

In order to cover processing and codec-related impairments, it is advisable to investigate also the speech sound quality by an adequate perception-based model (also known as perceptually weighted model in some circles). As in narrowband, TOSQA [20] is used up to now for this purpose. However, the new ITU-T methodology POLQA [21] could be applied in the near future. However, the same constraints apply as described in the narrowband section. So currently this is not yet the method of choice.

Due to the fact that a favorable user experience requires an extension of the frequency range not only in the upper-frequency region (8 kHz) but also in the low-frequency range (minimum 100 Hz). Hence, the impact of background noise becomes more severe. While in narrowband a high-pass filter in the transmitting side already helps to reduce the impact of the background noise significantly, however, these technologies cannot be applied in wideband. Aggressive high-pass filtering would critically impact the speech quality perceived by the user [17]. Therefore, advanced microphone and noise canceling techniques are required in order to achieve a good speech sound quality and to reduce simultaneously the background noise in wideband. For testing, the above mentioned S-, N-, G-MOS tests are available in ETSI EG 202 396-3 (3QUEST) [26]. This perception-based model allows the measurement of speech, noise, and the overall quality in the same way as it is done in narrowband systems but taking into account the special requirements in wideband.

The performance of echo cancellation in wideband also deserves special considerations. The extension of the frequency range up to 8 kHz includes a frequency range where the human ear is most sensitive. The hearing threshold shown in Fig. 5.10 has a maximum sensitivity in the frequency range around 3–5 kHz. As a consequence, any high-frequency echo contributes to echo perception in a significantly stronger way than the low-frequency echoes. Subjective investigations reported in [17] show that in this frequency range an echo loss of at least 46 dB is required to ensure a good echo performance with transmission delays up to 300 ms. The required echo loss versus frequency is shown in Fig. 5.11.

Further investigations are needed in order to define new wideband-related testing technologies for echo versus time, initial convergence, and convergence in the presence of background noise. Currently, the same testing techniques are applied as in narrowband but with extended analysis frequency range.

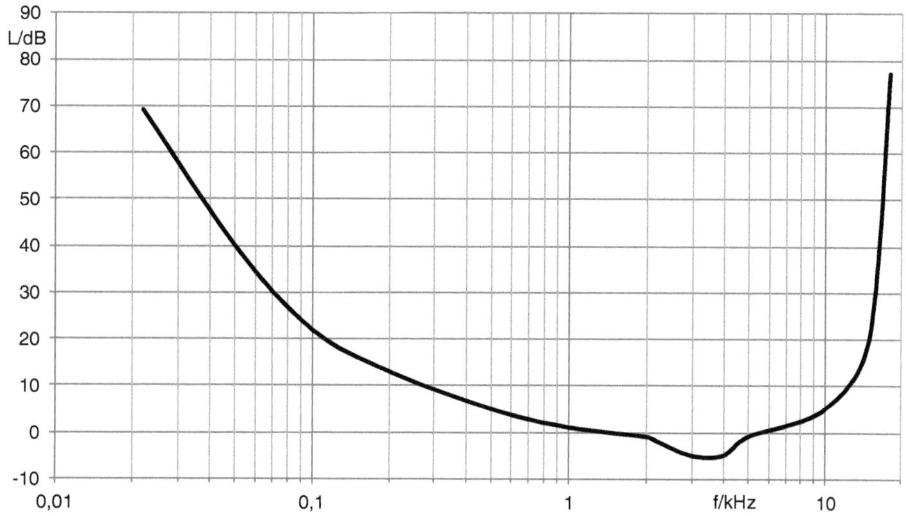

Fig. 5.10 Human hearing threshold

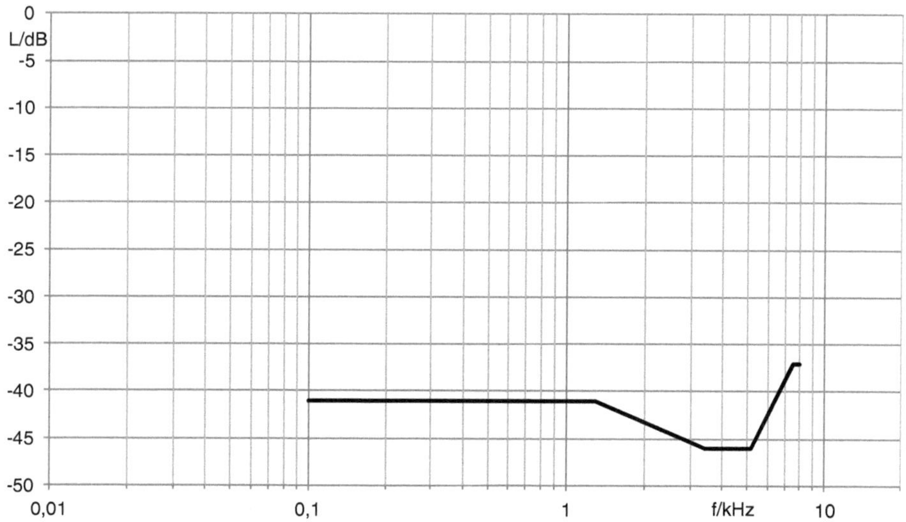

Fig. 5.11 Spectral echo loss requirements in wideband systems

The same observation is true for testing of the double-talk performance in wideband systems. Due to the lack of new subjective experiments, testing at present is based on narrowband technologies with extended frequency range.

5.8 Overall Quality Representation: "One View Visualization"

In the testing and qualification phase of car hands-free systems, a variety of parameters need to be optimized and checked. Only experts in the field are in a position to fully understand and interpret the obtained results. Although often required, due to the complexity of the problem, it is impossible to qualify a car hands-free system by a single number. A compromise which allows an overview quality representation also for non-experts and an easy-to-perform comparison of different implementations is the "one view visualization" as standardized in ITU-T Recommendation P. 505 [27]—sometimes also called "quality pie." In this Recommendation a general principle is defined on how to arrange and display the results of various speech quality measurements. This principle can be applied to car hands-free systems as well and is also standardized in [4]. Figure 5.12 shows an example of a representation taken from [4].

On the right hand-side of the diagram in Fig. 5.12, the most relevant quality parameters for sending and receiving in single-talk are displayed. Basic information on echo loss (TCLw) can be found in the lower left part of the diagram followed by

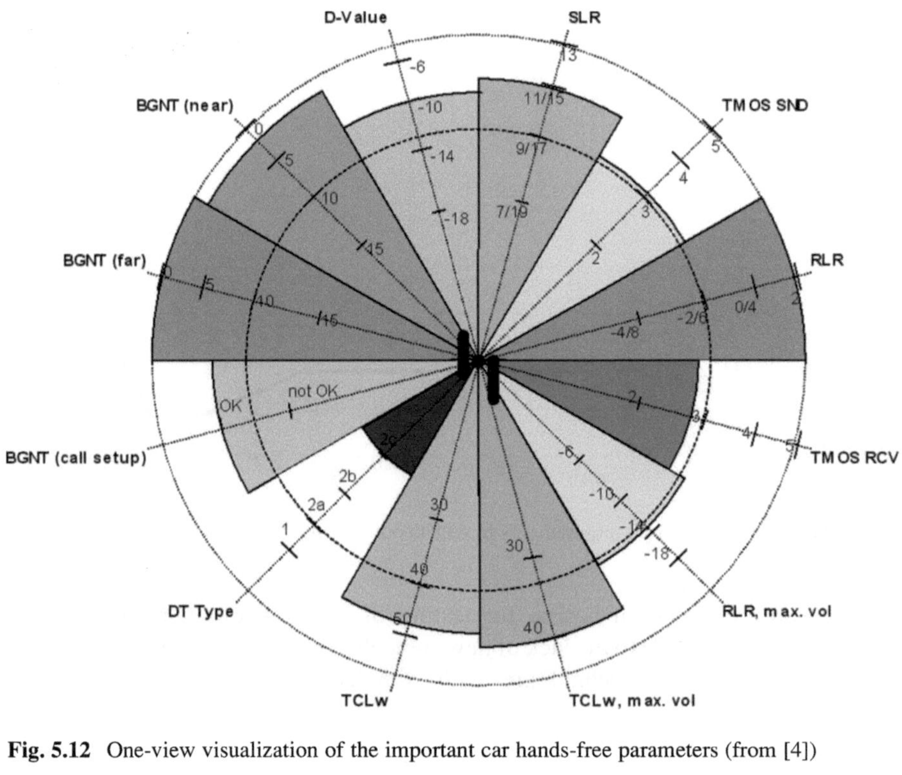

Fig. 5.12 One-view visualization of the important car hands-free parameters (from [4])

the type of double-talk performance. In the upper left part of the diagram, however, information about the car hands-free performance with background noise is shown. Often these diagrams are complemented by the 3QUEST test results under different driving conditions.

It is clear that such representation does not catch each and every quality aspect of car hands-free implementations. However, if needed additional diagrams can be created focusing on different quality aspects. It is also possible to condense different measurement parameters in one diagram slice if a method for perceptually correct combination can be found. As an example, this is done for double talk.

5.9 Conclusions

Car hands-free testing and optimization remain a challenging topic. Achieving the best possible speech quality is only possible if all components of a given car are optimized to the specifics of the car under consideration and the different conditions of use. Algorithms have to be highly flexible and adaptive and the amount of testing and optimization work remains high despite the availability of sophisticated testing and optimization procedures.

Additional challenges arise if wideband systems are to be deployed. Mainly for noise and echo cancellation, the requirements to achieve a good user experience are quite demanding and careful testing and optimization are required to achieve a noticeably better quality compared to narrowband.

References

1. H.W. Gierlich, A Measurement technique to determine the transfer characteristics of hands-free telephones. Signal Processing **27**(3), 281–300 (1992)
2. H.W. Gierlich, The auditory perceived quality of hands-free telephones: auditory judgements, instrumental measurements and their relationship. Speech Communication **20**(1996), 241–254 (1996)
3. H.W. Gierlich, F. Kettler, E. Diedrich, Proposal for the definition of different types of hands-free telephones based on double talk performance, ITU-T SG 12 meeting, Geneva, 21 Sept–01 Oct 1999, COM 12–103
4. VDA-Specification for Car Hands-Free Terminals, Version 1.6, VDA, 2010
5. ITU-T SG 12, http://www.itu.int/ITU-T/studygroups/com12/index.asp
6. ITU-T Focus Group CarCom, http://www.itu.int/en/ITU-T/focusgroups/carcom/Pages/De-fault.aspx
7. ITU-T Recommendation P.1100, *Narrow-band hands-free communication in motor vehicles*, International Telecommunication Union, Geneva, March 2011
8. ITU-T Recommendation P.1110, *Wideband hands-free communication in motor vehicles*, International Telecommunication Union, Geneva, Dec 2009
9. ITU-T Focus Group Distraction, http://www.itu.int/en/ITU-T/focusgroups/distrac-tion/Pages/default.aspx

10. ITU-T Recommendation G.114, *One-way transmission time*, International Telecommunication Union, Geneva, May 2003
11. ITU-T Recommendation P.79, *Calculation of loudness ratings for telephone sets*, International Telecommunication Union, Geneva, Nov 2007
12. ITU-T Recommendation P. 340, *Transmission characteristics and speech quality parameters of hands-free telephones*, International Telecommunication Union, Geneva, May 2000
13. ITU-T Recommendation P.58, *Head and torso simulator for telephonometry*, International Telecommunication Union, Geneva, May 2013
14. ITU-T Recommendation P.501, *Test signals for use in telephonometry*, International Telecommunication Union, Geneva
15. ITU-T Recommendation P.50, *Artificial voices*, International Telecommunication Union, Geneva, Jan 2012
16. ITU-T Recommendation P.502, *Objective test methods for speech communication systems using complex test signals*, May 2000
17. H.W. Gierlich, in *Methods of Determining the Communicational Quality of Speech Transmission Systems*, ed. by D. Havelock, S. Kuwano, M. Vorländer. Handbook of Signal Processing in Acoustics (Springer, New York, 2008), ISBN 078-0-387-77698-9
18. F. Kettler, H.W. Gierlich, *Evaluation of hands-free Terminals* ed. by E. Hänsler, G. Schmidt, Speech and Audio Processing in Adverse Environments, (Springer, New York, 2008), ISBN 978-3-540-70601-4
19. ITU-T Recommendation P.862, *Perceptual evaluation of speech quality (PESQ): an objective method for end-to-end speech quality assessment of narrow-band telephone networks and speech codecs*, International Telecommunication Union, Geneva, Feb 2001
20. J. Berger, Instrumentelle Verfahren zur Sprachqualitätsschätzung – Modelle auditiver Tests, Ph.D. thesis, Kiel, 1998
21. ITU-T Recommendation P.863, *Perceptual perceptual objective listening quality assessment*, International Telecommunication Union, Geneva, Jan 2011
22. S. Möller, J. Berger, *Performance of P.863 on P.AAM acoustic data, ITU-T SG 12 Study Period 2009–2012*; COM 12-C276E, 2011
23. ITU-T Recommendation G.131, *Control of talker echo*, International Telecommunication Union, Geneva, Nov 2003
24. M. Lepage, F. Kettler, J. Reimes, Scalable, Perceptual based echo assessment method for aurally adequate evaluation of residual single talk echoes, IWANEC, Aachen, Sept 4–6 2012
25. J. Reimens, H.W. Gierlich, F. Kettler, S. Poschen, M. Lepage, The relative approach algorithm and its application in new perceptual models for noisy speech and echo performance, Acta. Acustica. **97**, 2 (2011)
26. ETSI EG 202 396-3, *Speech processing, transmission and quality aspects (STQ); speech quality performance in the presence of background noise; Part 3: background noise transmission – objective test method*, Feb 2011
27. ITU-T Recommendation P. 505, *One-view visualization of speech quality measurement results*, International Telecommunication Union, Geneva, Nov 2005
28. H.W. Gierlich, F. Kettler, S. Poschen, J. Reimes, A new objective model for wide- and narrowband speech quality prediction in communications including background noise, *EUSIPCO*, Lausanne, Switzerland, 2008
29. H.W. Gierlich, F. Kettler, *Wideband hands-free in cars – new challenges for system design and testing*, ed. by J. Hansen, P. Boyraz, K. Takeda, H. Abut, Digital Signal Processing for In-Vehicle Systems and Safety, (Springer, 2011), ISBN 1441996060

Chapter 6
A Wideband Automotive Hands-Free System for Mobile HD Voice Services

Marc-André Jung and Tim Fingscheidt

Abstract Wideband mobile telephony supporting a speech bandwidth from 50 to 7,000 Hz gets more and more employed. These so-called mobile HD Voice services consequently find their way into automobile applications. In this chapter we present a wideband hands-free system for automotive telephony applications with a synchronously adapted acoustic echo canceller and postfilter. It is based on a frequency domain adaptive filter approach and Kalman filter theory and makes use of a generalized Wiener postfilter for residual echo suppression and noise reduction in a consistent way. To provide a high convergence rate in case of time-variant echo paths, the echo canceller with very robust double-talk performance is supported by a fast converging shadow filter, which allows for a good tracking performance. A decimation approach is used to decrease algorithmic delay and computational complexity without loss of quality. Experimental results with car cabin impulse responses show good echo cancellation capabilities with fast convergence times along with extraordinary full-duplex performance while still keeping an almost untouched speech component in the converged state.

Keywords AEC • FDAF • Hands-free system • HD voice • Shadow filter • Wideband

6.1 Introduction

Mobile HD Voice services supporting wideband speech (50–7,000 Hz) as opposed to narrowband speech (300–3,400 Hz) allow for a high-quality and high-intelligibility telephony experience. Syllable articulation (i.e., human syllable recognition rate) increases from 90 % to about 98 %, making the use of spelling alphabets for

M.-A. Jung (✉) • T. Fingscheidt
Institute for Communications Technology, Technische Universität Braunschweig, Schleinitzstr. 22, Braunschweig D-38106, Germany
e-mail: jung@ifn.ing.tu-bs.de; fingscheidt@ifn.ing.tu-bs.de

G. Schmidt et al. (eds.), *Smart Mobile In-Vehicle Systems: Next Generation Advancements*, DOI 10.1007/978-1-4614-9120-0_6,

proper names widely obsolete. Apart from proper names, in narrowband speech transmission listeners typically are able to employ their language model in order to reconstruct missing syllables in an interpolative fashion. This, however, does not work sufficiently well in the case of foreign-language conversations. Also in situations with a high level of background noise—as it is typical for mobile telephony in general and automotive telephony in particular—the further drop in syllable articulation cannot be compensated sufficiently for. All these aspects were driving forces for the worldwide deployment of mobile wideband speech services in the past years, commonly being known as mobile HD Voice services.

High-quality hands-free capabilities are a greatly demanded feature of telecommunication systems in office, home, or car environments and—referring to the latter—are even mandatory in many countries. Several state-of-the-art algorithms have been developed to fulfill technical requirements, such as full-duplex speech transmission capability, sufficient acoustic echo cancellation even for highly time-variant echo paths, and minimal speech distortion (e.g., [1–6]). Nevertheless, those requirements often collide with practical restrictions such as low complexity and algorithmic delay [7–9].

Hands-free systems are usually designed to cope with signal degradations stemming from the acoustic environment. These degradations are typically caused by acoustic echo and additive noise, leading to reduced intelligibility and speech quality. This is specifically the case for long round-trip delays or high noise immissions, as can be often found in automotive mobile phone usage. As a countermeasure, acoustic echo cancellers (AECs) [1, 6, 10, 11] and postfilters (PFs) for residual echo suppression (RES) [12, 13] and noise reduction (NR) approaches [14–16] have been proposed, typically working at a sampling rate of $f_s = 8$ kHz (narrowband speech).

With upcoming mobile wideband speech transmission (HD Voice services) at a sampling rate of $f_s = 16$ kHz, there are a couple of obstacles to be solved when designing a hands-free system. The doubled sampling rate causes a non-negligible increase of algorithmic complexity and can also lead to other unwanted effects when porting an algorithm from narrowband to wideband [9].

Typical hands-free system representatives in the time domain are based on the normalized least mean square (NLMS) [17], affine projection (AP) [17–19], recursive least squares (RLS) [20], or Kalman algorithm [6, 21]. These approaches usually feature a simple algorithmic structure with the ability to work on a per-sample base. On the one hand, this usually leads to zero or low delay; on the other hand, modeling of longer impulse responses (IRs) can lead to exceedingly high computational complexity if the filter is adapted in every single sample. This problem can be addressed by block processing, where the filter is only adapted once per block of samples. Albeit computationally efficient, this block processing leads to algorithmic delay and a slower convergence rate. Due to the fact that most of these algorithms make the assumption of a spectrally white echo signal but speech signals usually still have some inherent correlation, adaptation can only take place in the limited direction of the error signal vector. This decreased convergence rate can partly be avoided by using some kind of decorrelation technique for the excitation signal [6]. Whereas convergence speed can be increased especially with the RLS and Kalman algorithms, tracking performance often still suffers since adaptation of a well-converged system model to IR changes only takes place in little steps [7]. Another well-known problem of time domain AEC approaches is the poor double-talk performance. Presence of near-end speech or noise leads to undesired adaptation and therefore misestimation of the true impulse response. To avoid this, a—more or less—robust double-talk detection (DTD) scheme is often applied [7], which triggers an adaptation speed reduction during double-talk.

Adaptation in a transform domain like subband or frequency domain may circumvent some of the abovementioned deficiencies. However, it should be mentioned that transformation domain processing may introduce other, possibly more perturbing, problems. Having said this, these algorithms may be a very good choice if applied appropriately. Filter adaptation in the subband domain, for example, can lead to a significantly reduced computational complexity if long impulse responses have to be modeled. This is made possible by splitting the fullband signal into several subbands by means of a filter bank. Due to this, each of the subband signals is analyzed separately, whereas subsampling can be applied, and individual filter lengths for each subband can be chosen. Furthermore, convergence speed is highly improved since each subband signal can be esteemed as spectrally white. It should be considered, however, that algorithmic delay increases. Note that also low-delay filter-bank approaches exist, e.g., [4, 22]. However, the problem of poor double-talk performance with the need of DTD often remains. Furthermore, the design of a filter-bank analysis and synthesis structure is typically realized with prototype filters [23], which might be tedious to some extent.

As an alternative to these subband algorithms, the so-called frequency domain adaptive filter (FDAF) algorithms can be used [24]. Adaptation of the impulse response model and estimation of the echo signal is performed in the frequency domain. This allows to compute frequency-dependent parameters like optimal stepsize vectors. In our case, the inversely transformed estimated echo signal is then used to filter the microphone signal in the time domain [3, 25]. Due to the inherent block processing of the fast convolution in the frequency domain, in many cases complexity can be drastically decreased. A further significant advantage is the extraordinary double-talk performance of some FDAF algorithms, which makes DTD obsolete [3, 25, 26]. Furthermore, they are able to preserve a very good quality of the speech component in the uplink (send) path. Unfortunately, having to buffer a block of samples for the discrete Fourier transform (DFT) introduces delay in the uplink signal path. As another drawback, large DFTs, as they are needed to sufficiently cover long impulse responses, also lead to comparably slow convergence times.

Since AEC filters typically achieve a yet insufficient amount of echo suppression, a subsequent postfilter is needed. This also covers nonlinear echo components and can additionally serve as NR filter [12]. Whereas time domain AEC algorithms are frequently complemented with time domain gain loss control (GLC) postfilters [5, 7, 27], transform domain AEC filters often make use of postfilters within the same domain [13, 22, 26, 28, 29]. Here, the group of GLC postfilters could be shortly described as computationally efficient with the drawback of poor double-talk performance, while the group of frequency domain or subband postfilters often show better performance—especially during double-talk—with the drawback of additional signal delay.

The focus of our work lies in a wideband hands-free system for automotive applications with relatively short impulse responses. In contrast to the mobile use case, here the demand of very low complexity is of subordinate importance. A well-balanced double-talk performance, on the contrary, is a crucial point to keep the mental distraction of the driver at a low level. Additionally, algorithmic delay should be kept low to avoid a large contribution to the round-trip delay. Due to its excellent double-talk performance with still tolerable algorithmic delay, an FDAF-based Kalman filter algorithm [3, 25] is chosen for the following investigations and implemented for wideband speech. The algorithm is supplemented with a shadow filter (SF), which leads to a drastic reduction of convergence time. Furthermore, a modified postfilter setup is suggested, which is able to significantly reduce algorithmic delay at a given echo suppression by means of decimation in the DFT domain.

In Sect. 6.2 the FDAF hands-free algorithm based on [3, 25] is presented but already adopted to wideband speech. Section 6.3 presents the latency reduction by decimation in the DFT domain as well as the shadow-filter-enhanced FDAF algorithm. In Sect. 6.4 experimental results of single- and double-talk simulations are given. Echo suppression, convergence behavior, algorithmic delay, and quality of the speech component are discussed.

6.2 State-of-the-Art FDAF

We have motivated before that hands-free systems with adaptation of the filter coefficients in the frequency domain are generally a good choice if low computational complexity for long impulse responses, good double-talk performance, and little degradation of the near-end speech signal component are desired. The adaptive filter is placed in parallel to the electroacoustic echo path or loudspeaker-enclosure-microphone (LEM) system, trying to estimate a replica echo signal. In case of the FDAF algorithm, the adaptation of the filter coefficients and the computation of the estimated echo signal are performed in the frequency domain.

As depicted in Fig. 6.1, in a digital model of the LEM system, the echo signal $d(n)$ with sample index n is the result of the convolution of the far-end signal $x(n)$ with the LEM impulse response. The microphone signal is then given by $\boldsymbol{y} = [y(n-R+1), \ldots, y(n)]^T$, with R being the frame shift, also called block length, $[\cdot]^T$ being the transpose, and $y(n) = s(n) + n(n) + d(n)$, whereas $s(n)$ is the near-end speech signal and $n(n)$ is the noise component.

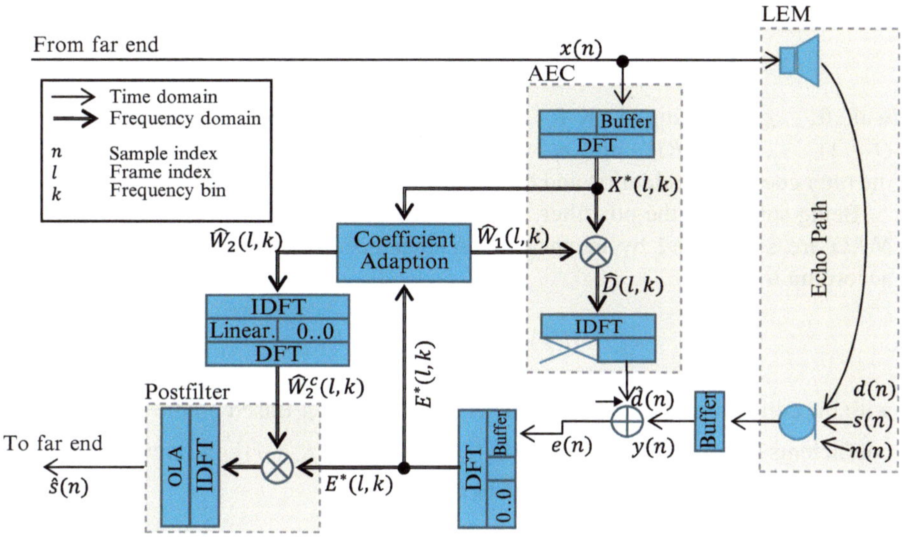

Fig. 6.1 State-of-the-art FDAF-based hands-free system

Then the loudspeaker signal $x(n)$ is transformed into the DFT domain by

$$\begin{aligned} \boldsymbol{X}_l &= [X(l,0), \ldots X(l,k), \ldots X(l,K-1)]^T \\ &= DFT\left\{[x(n-K+1), \ldots, x(n-R), x(n-R+1), \ldots, x(n)]^T\right\}, \end{aligned}$$

with frame index l and frequency bin index k. By making use of the FDAF approach based on Kalman filter theory, the DFT domain adaptive filter coefficients $\hat{W}_1(l,k)$ are estimated [26, 28]. An estimate of the frequency domain replica echo signal is then computed by

$$\hat{D}(l,k) = \hat{W}_1(l,k) \cdot X^*(l,k) \tag{6.1}$$

for $k = 0, \ldots, K-1$, with $(\cdot)^*$ being the conjugate complex operator. Its inverse DFT (IDFT) delivers $\left[\ldots, \hat{\boldsymbol{d}}_l^T\right]^T = IDFT\{\hat{\boldsymbol{D}}_l\}$, with $\hat{\boldsymbol{D}}_l = \left[\hat{D}(l,0), \ldots, \hat{D}(l,K-1)\right]^T$ and $\hat{\boldsymbol{d}}_l = \left[\hat{d}(n-R+1), \ldots, \hat{d}(n)\right]^T$, which is then used to compute R samples of an error signal

$$e(n) = y(n) - \hat{d}(n). \tag{6.2}$$

The residual echo $r(n) = d(n) - \hat{d}(n)$ is contained in the error signal as $e(n) = r(n) + s(n) + n(n)$. The DFT error signal

$$\mathbf{E}_l = DFT\left\{\left[\mathbf{0}^T_{K-R-O}, \left(\mathbf{e}^-_{l-1}\right)^T, \mathbf{e}^T_l\right]^T\right\} \tag{6.3}$$

with $\mathbf{0}_{K-R-O}$ being a (K-R-O)-dimensional zero vector, $\mathbf{e}^-_{l-1} = [e(n-R-O+1), \ldots, e(n-R)]^T$, and $\mathbf{e}_l = [e(n-R+1), \ldots e(n)]^T$ is made available for the filter coefficient adaption and for postprocessing, and O being the overlap length.

Being subject to the postfilter, the inherent residual echo $R(l,k)$ and noise signal $N(l,k)$ are suppressed by means of a Wiener postfilter in the frequency domain according to

$$\hat{S}(l,k) = \hat{W}^c_2(l,k) \cdot E^*(l,k), \tag{6.4}$$

with the constrained postfilter coefficients $\hat{W}^c_2(l,k)$ [1]. Based on the unconstrained coefficients $\hat{\mathbf{W}}_{2,l} = \left[\hat{W}_2(l,0), \ldots, \hat{W}_2(l,k), \ldots, \hat{W}_2(l,K-1)\right]^T$, a linear constraint is obtained using $\hat{\mathbf{w}}_{2,l} = IDFT\left\{\hat{\mathbf{W}}_{2,l}\right\}$ to assemble

$$\begin{aligned}\hat{w}^c_{2,l} = {} & \left[\hat{w}_{2,l}\left(n = K - N_p/2\right), \ldots, \hat{w}_{2,l}(n = K-1),\right. \\ & \left.\hat{w}_{2,l}(n=0), \ldots, \hat{w}_{2,l}\left(n = N_p/2 - 1\right), \mathbf{0}^T_{K-N_p}\right]^T,\end{aligned} \tag{6.5}$$

which contains the linear phase postfilter impulse response of length $N_p \leq K - R - O$. The constrained K-point DFT domain postfilter coefficients are then computed by $\hat{\mathbf{W}}^c_{2,l} = DFT\left\{\hat{\mathbf{w}}^c_{2,l}\right\}$.

As shown in the coefficient adaption block in Fig. 6.1, the spectral filter coefficients for the echo canceller and Wiener postfilter, $\hat{W}_1(l,k)$ and $\hat{W}_2(l,k)$, are synchronously estimated. This is done by introducing a Markov assumption for the time-varying echo path and exploiting Kalman filter theory [3, 26].

After postfiltering, inverse DFT, and subsequent overlap-add (OLA), the enhanced speech signal $\hat{s}(n)$ is transmitted to the far-end communication partner.

In this setup, an algorithmic delay of $N_p/2 - (R + O) + R = N_p/2 - O$ samples is introduced with $N_p/2$ accounting for the linear phase constrained postfilter, $(R + O)$ being the number of nonzero samples in $IDFT\{\mathbf{E}_l\}$ and R being the frame buffer for block processing.

6.3 New Latency-Reduced FDAF with Shadow Filter

6.3.1 *Latency Reduction by Postfilter Decimation*

Apart from the previously mentioned frame buffering, the linearly constrained postfilter is the only contributor to algorithmic delay. As it can be seen by comparing Figs. 6.1 and 6.2, we now introduce decimation in the DFT domain to reduce the

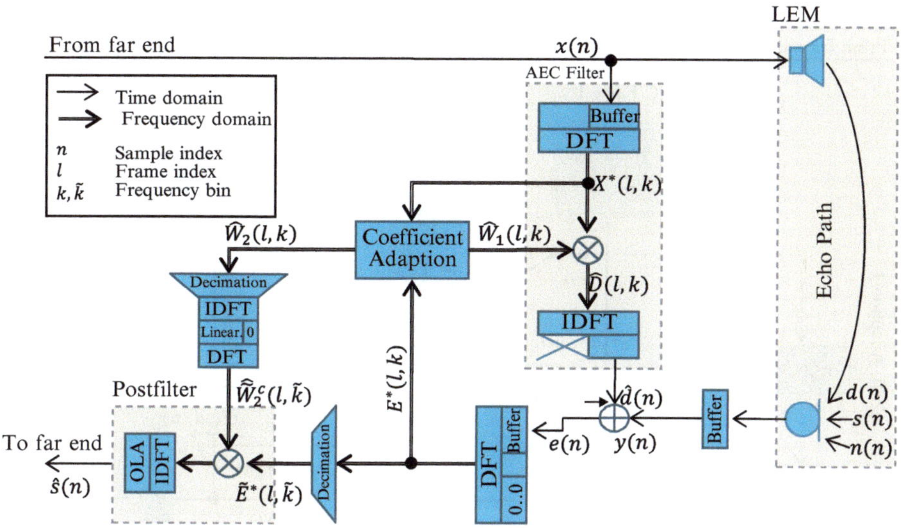

Fig. 6.2 New FDAF-based hands-free system with latency reduction by decimation in the DFT domain

number of DFT bins, which in turn reduces algorithmic delay and computational complexity. In our case, decimation is performed according to

$$\hat{\widetilde{W}}_{2,l}\left(\widetilde{k}\right) = \begin{cases} \hat{W}_{2,l}(k=0), & \text{for } \widetilde{k}=0, \\ \overline{W}_{2,l}\left(\widetilde{k}\right), & \text{for } 1 \le \widetilde{k} \le \widetilde{K}/2-1, \\ \hat{W}_{2,l}(k=K/2), & \text{for } \widetilde{k}=\widetilde{K}/2, \end{cases} \tag{6.6}$$

with $\overline{W}_{2,l}\left(\widetilde{k}\right) = \frac{1}{3}\hat{W}_{2,l}\left(k=2\widetilde{k}-1\right) + \frac{1}{3}\hat{W}_{2,l}\left(k=2\widetilde{k}\right) + \frac{1}{3}\hat{W}_{2,l}\left(k=2\widetilde{k}+1\right)$, and $\hat{\widetilde{W}}_{2,l}\left(\widetilde{k}\right)$ for $\widetilde{k} > \widetilde{K}/2$ being defined via the conjugate complex property. Additionally, the decimated DFT error signal $\widetilde{E}\left(l,\widetilde{k}\right)$ is computed in analogy to $\hat{\widetilde{W}}_{2,l}\left(\widetilde{k}\right)$. This decimation in the DFT domain therefore leads to a reduced algorithmic delay contribution of the constrained postfilter of $\widetilde{N}_p/2 - (R+O)$ with $\widetilde{N}_p \le \widetilde{K} - R - O$ at even lower computational complexity. Furthermore, due to the inherent spectral smoothing of the postfilter coefficients $\hat{\widetilde{W}}^{c}_{2,l}$, speech quality can be improved by reducing spectral artifacts.

6.3.2 Shadow-Filter Approach

The FDAF algorithm shows an excellent double-talk performance even in noisy conditions, without the requirement of an explicit DTD. As a drawback, however, at

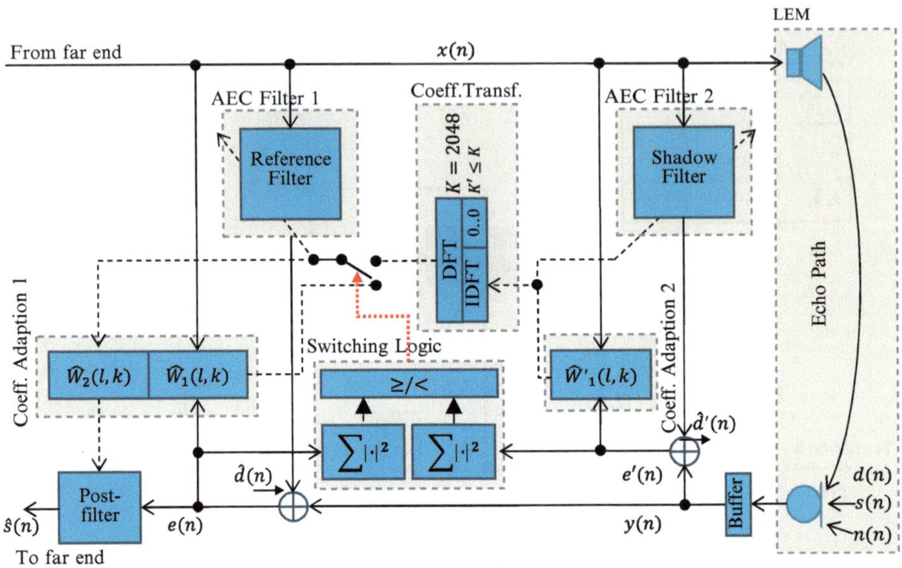

Fig. 6.3 FDAF shadow-filter approach

a sampling rate of 16 kHz, it sometimes cannot achieve a sufficient convergence speed, especially in the case of long filter lengths or highly time-variant echo paths. Luckily, echo path variations in the car cabin are typically not as dynamic as, e.g., in a mobile phone application. Nevertheless, possible solutions for a required faster convergence are the reduction of the filter length (which leads to additional residual echo if the LEM impulse response is not completely covered by the filter) or a faster adaptation rate (which may decrease the double-talk performance). As an alternative, we propose here a shadow-filter approach to overcome this drawback of the FDAF approach [7].

As depicted in Fig. 6.3, we enhance a slowly but accurately converging reference filter (RF) in the foreground with a rapidly converging shadow filter (SF) in the background. The faster convergence of the shadow filter can be achieved by using a shorter filter length and/or different parameters, e.g., for the Markov model of the time-varying echo path [3]. The thereby accelerated shadow filter is better able to follow faster changes of the impulse response and thus leading to a lower error signal energy during highly time-variant time periods. However, care has to be taken to assure robustness against near-end disturbances, since double-talk may be erroneously taken for an IR change.

Changes of the echo path are detected as follows: If the error signal energy of the reference filter $\mathbf{e}_l^T \cdot \mathbf{e}_l$ is α-times bigger than that of the shadow filter for $1 + L^-$ consecutive frames, a change of the echo path is assumed, and the switching logic, shown in the center of Fig. 6.3, triggers an exchange of filter coefficients in the reference filter (symbolized by the dotted arrow pointing to the switch). In this case, the shadow-filter coefficients $\hat{W}'_1(l,k)$ are expected to better represent the LEM IR. On the other hand, in a time-invariant/slowly changing echo path case, the error signal energy of the reference filter will likely be smaller than that of the shadow filter, and the native reference filter coefficients are used:

$$\hat{W}_{1,l} = \begin{cases} DFT\left\{\left[IDFT'\left\{\hat{W}'^{T}_{1,l}\right\}, 0^{T}_{K-K'}\right]\right\}^{T}, & \text{if } e^{T}_{\lambda}\cdot e_{\lambda} > \alpha\, e'^{T}_{\lambda}\cdot e'_{\lambda} \\ & \forall\lambda = l, l-1, \ldots, l-L^{-}; \\ \hat{W}_{1,l}, & \text{else}; \end{cases} \tag{6.7}$$

with $IDFT'\{\cdot\}$ having a reduced length $K' < K$ which stems from the shadow filter.

Please note that the error signal $e'(n)$ of the shadow filter is only deployed to detect changes of the echo path and to adapt the shadow-filter coefficients. In the end, only the error signal of the reference filter $e(n)$, which is either using its native coefficients or a transformed version of the shadow-filter coefficients, is passed over to the postfilter for later transmission.

In so doing, an immediate improvement of the model mismatch can be achieved, leading to a faster convergence. Since the number of DFT coefficients for the reference and shadow filter differs (in our case $K' = K/2$), a transformation of the coefficients has to be performed (shown as "coefficient transformation" block in Fig. 6.3).

6.3.3 Combined Postfilter Decimation and Shadow Filter

In combining both the postfilter decimation of Sect. 6.3.1 and the shadow filter of Sect. 6.3.2 in a new joint approach, advantages of both strategies can be exploited. As proposed before, different parameters were used for the reference filter and the shadow filter to assure good convergence behavior and tracking speed. The whole parameter setting for this joint approach is shown in Table 6.1. This includes the forgetting factor A of the first-order Markov model [3, 6], the AEC filter length $N_w = K - R$ [3, Eq. (20)], the impulse response length of the decimated postfilter that can be chosen to some $\widetilde{N}_p \leq N_p - \widetilde{K}$, and the error power spectral density (PSD) smoothing factor $\lambda_{\phi ee}$. Additionally, a decimation factor of $K/\widetilde{K} = 2$ is chosen. Section 6.4 presents the simulation results of this joint approach.

6.4 Simulations

Our proposed approach has been evaluated by simulation of an LEM system in a car cabin (Volkswagen Touran). Two impulse responses have been measured, originating from both front-door loudspeakers to the car's hands-free microphone

Table 6.1 Parameter settings for the new FDAF approach with decimation, reference filter (RF), and shadow filter (SF)

Description	Value RF	Value SF	Description	Value RF	Value SF
DFT length	$K = 2048$	$K' = 1024$	PF length	$N_p = 1824$	n/a
Dec. DFT length	$\tilde{K} = 1024$	n/a	Dec. PF length	$\tilde{N}_p = 800$	n/a
Frameshift	$R = 160$	$R' = 160$	SF loopback	n/a	$L^- = 6$
OLA length	$O = 64$	$O' = 64$	SF overestimation	n/a	$\alpha = 3$
Forgetting factor	$A = 0.9995$	$A' = 0.99$	Error PSD smoothing	$\lambda_{\phi ee} = 0.8$	$\lambda'_{\phi ee} = 0.999$
AEC filter length	$N_w = 1888$	$N'_w = 864$			

in the ceiling above the central console. Both measurements were performed in the quiet car with reverberation times of $t_{60} = 35$ ms each. Whereas the front left and the rear right seats were occupied by a quiet passenger during both measurements, the front passenger switched position a bit for the second measurement, but keeping a typical driving position in both cases.

These two impulse responses were used to compute the echo signal $d(n)$ of the far-end speaker, shown as waveform in the upper part of Figs. 6.4 and 6.5, by convolution with the far-end signal $x(n)$. The used impulse response was switched from the first measurement to the second after 25 s.

The near-end speech signal $s(n)$ waveform in double-talk is shown in front of the echo signal waveform in Fig. 6.5. Speech signals are concatenated samples of the NTT wideband speech database. Whereas the male far-end speaker is continuously active during both the single- and double-talk scenarios (Figs. 6.4 and 6.5), the female near-end speaker is only intermittently active during the double-talk scenario (Sect. 6.5). All simulations are performed at an input signal-to-echo ratio of SER = 0 dB.

For evaluation of the performance of the underlying hands-free system, different instrumental measures have been used. For assessment of the system's ability to suppress the echo signal, echo return loss enhancement (*ERLE*) plots for the single- and double-talk scenario are given in Figs. 6.4 and 6.5 as lines above the waveforms. *ERLE* is defined and recursively estimated as follows:

$$\begin{aligned} ERLE(n) &= \frac{E\{d^2(n)\}}{E\left\{\left(d(n) - \hat{d}(n)\right)^2\right\}} \\ &\approx \frac{(1-\beta)d^2(n) + \beta d^2(n-1)}{(1-\beta)\left(d(n) - \hat{d}(n)\right)^2 + \beta\left(d(n-1) - \hat{d}(n-1)\right)^2} \end{aligned}$$

with smoothing factor $\beta = 0.9996$.

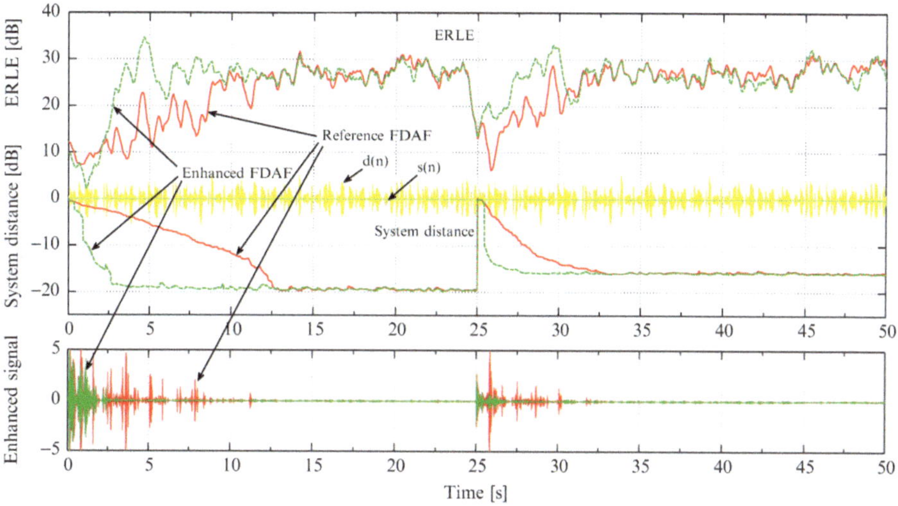

Fig. 6.4 Reference FDAF algorithm (*solid lines*) and enhanced FDAF with postfilter decimation and shadow-filter approach (*dashed lines*) in *single-talk*. The IR changes at 25 s. *Above*: ERLE (*top*) and system distance (*bottom*) for single-talk signal $d(n)$. *Below*: Enhanced microphone signals

To evaluate the convergence and tracking performance in single- and double-talk, also the normalized system distance is shown in Figs. 6.4 and 6.5, defined as

$$\frac{||h_\Delta{}^2||}{||h_i{}^2||} = \frac{||h_i - \hat{w}_{1,l}{}^2||}{||h_i{}^2||} \tag{6.8}$$

with h_i, $i = 1, 2$, being one of the two measured impulse responses, and $\hat{w}_{1,l} = IDFT\{\hat{W}_{1,l}\}$.

For both the *ERLE* and system distance plots, solid lines correspond to the reference setup and dashed lines correspond to the enhanced system, making use of shadow filtering and decimation. In the lower part of Figs. 6.4 and 6.5, the enhanced signals $\hat{s}(n)$ are shown.

For evaluation of the degradation of the wideband (uplink) speech component, the perceptual evaluation of speech (PESQ) measure according to ITU-T recommendation P.862.2 [30] is used, yielding objective listening-quality mean opinion scores (MOS_{LQO}). The so-called MOS_{LQO}^{PF} score is used here to evaluate only the quality degradation of the speech *component* by *postfiltering*, marked by the superscript “PF”.

6.4.1 Far-End Single-Talk Scenario

As it can be easily seen in the upper part of Fig. 6.4, the continuous far-end speech input leads to a final normalized system distance of −19 dB and −16 dB for the two

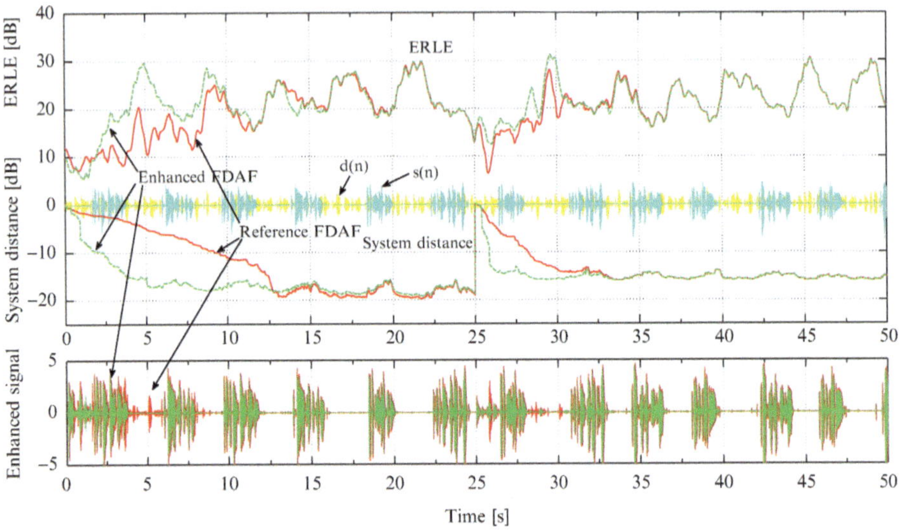

Fig. 6.5 Reference FDAF algorithm (*solid lines*) and enhanced FDAF with postfilter decimation and shadow-filter approach (*dashed lines*) in *double-talk*. The IR changes at 25 s. *Above*: ERLE (*top*) and system distance (*bottom*) for double-talk signal $d(n) + s(n)$. *Below*: Enhanced microphone signals

measured IRs, respectively. Due to the quite long adaptive filter length of $N_w = K - R = 2048 - 160 = 1888$ taps, adaptation of the reference filter (shown as solid lines in Fig. 6.4) is quite slow. A system distance of −10 dB is reached 8.5 s after initialization and 3 s after the IR switch. Since this convergence and tracking time is simply not sufficient, shadow filtering has been used to achieve faster filter adaptation. In addition, decimation is used for the postfilter. The *ERLE* and system distance plots of this enhanced setup are shown as dashed lines in the upper part of Fig. 6.4. These measures lead to significantly reduced convergence times after initialization and for tracking of abrupt IR changes. In both cases, convergence time is reduced to less than 1 s to reach −10 dB system distance.

Table 6.2 shows the mean *ERLE*, mean system distance, and algorithmic delay for four different approaches. The mean values have been computed for the signals as they are shown in Fig. 6.4. However, simulations based on different NTT datasets show comparative results. The reference algorithm, as described in Sect. 6.2, is shown in the first row. Due to the quite long convergence time of this approach, the mean *ERLE* and system distance values in the evaluated period of time significantly differ from the mean values of the fully converged filter. The algorithmic delay of $(N_p/2 - (R + O) + R) \cdot 1/16$ kHz = 53 ms of this approach is mainly accounted to the delay introduced by the Wiener postfilter (6.4) which amounts to $(N_p/2 - (R + O)) \cdot 1/16$ kHz = 43 ms. The remaining $R \cdot 1/16$ kHz = 10 ms delay is caused by the buffering, which is necessary for the block processing in the frequency domain (see Fig. 6.1).

Table 6.2 Performance evaluation in single-talk

	$\overline{ERLE}$	$\overline{SYSDIS}$	Delay	MOS_{LQO}^{PF}
Reference	23.6 dB	−13.7 dB	53 ms	n/a
Ref. + SF.	25.5 dB	−16.6 dB	53 ms	n/a
Ref. + Dec.	23.7 dB	−13.7 dB	21 ms	n/a
Ref. + Dec. + SF.	25.7 dB	−16.8 dB	21 ms	n/a

The introduction of shadow filtering, as it is depicted in Fig. 6.3, leads to better convergence behavior. This can also be seen by looking at the improved mean values for *ERLE* and system distance in the second row of Table 6.2. In this special case the mean *ERLE* could be increased by around 2 dB, and the normalized system distance could be decreased by nearly −3 dB. The algorithmic delay remains unchanged.

Further improvements can be achieved by decimation of the DFT coefficients for the spectral gain and input signal of the postfilter, as it is shown in Fig. 6.2. In so doing, three improvements can be achieved simultaneously: First, due to the halved DFT and IDFT lengths in the postfilter, the computational complexity is somewhat reduced; the decimation in the frequency domain acts as smoothing of the postfilter weights and the input speech vector; this is leading to a better residual echo suppression; and as third factor, quality degradation of the speech component can be reduced, as will be shown in Sect. 6.4.2. However, decimation should only be introduced carefully. The effect of this decimation together with a shadow filter can be seen by looking at the results in the fourth row of Table 6.2. *ERLE*, and system distance can be further improved by 0.2 dB and −0.2 dB, compared to the shadow-filter-only approach, shown in the second row of Table 6.2. Whereas these improvements are rather small, the important effect of a much smaller algorithmic delay of 21 ms compared to the former 53 ms is achieved by the considerably smaller length of the postfilter. By only applying postfilter decimation, as shown in the third row, algorithmic delay remains low at 21 ms, with *ERLE* and system distance values being comparable to the reference.

6.4.2 Double-Talk Scenario

Some effects become even more clear when regarding a double-talk scenario, as it is shown in Fig. 6.5. The presence of near-end speech or noise is posing an interference to the adaptive filter, hence leading to slower convergence or misadaptation to the interfering signal. However, looking at Fig. 6.5, the double-talk performance of the FDAF algorithm can still be considered as excellent. The *ERLE* values drop by around −8 dB during double-talk but still keep a minimum value of around 20 dB in the converged state. Furthermore, due to this high robustness, convergence times more or less stay the same as it can be seen by looking at the system distance plot of the reference filter in Fig. 6.5.

Table 6.3 Performance evaluation in double-talk

	$\overline{\text{ERLE}}$	$\overline{\text{SYSDIS}}$	Delay	$\text{MOS}_{\text{LQO}}^{\text{PF}}$
Reference	20.1 dB	−13.0 dB	53 ms	3.3
Ref. + SF.	20.9 dB	−15.0 dB	53 ms	3.3
Ref. + Dec.	20.1 dB	−13.0 dB	21 ms	3.6
Ref. + Dec. + SF.	21.3 dB	−15.3 dB	21 ms	3.6

Looking at the upper part of Fig. 6.5 (*ERLE*), it can be seen that the shadow-filter-enhanced approach is still performing well in a double-talk scenario. In this case, convergence time to reach a system distance of −10 dB could be reduced to about 2 s. For both scenarios, this convergence time is of course dependent on the speech signals as well as on proper tuning of the shadow-filter parameters.

Comparing the mean *ERLE* and system distance values from the single-talk scenario (Table 6.2) to the double-talk scenario (Table 6.3), an expected, yet moderate, performance degradation is observed. For the reference approach, shown in the first row, the mean *ERLE* value drops to 20.1 dB, and the system distance slightly increases to −13.0 dB. Of course, algorithmic delay stays at 53 ms. As introduced before, perceptual quality degradation by the postfilter is evaluated by the MOS_{LQO}^{PF}. In this case, a fair to good score of 3.3 is achieved.

By making use of the shadow filter, both the mean *ERLE* and mean system distance values could be improved. $\overline{ERLE}$ increases to 20.9 dB, whereas $\overline{SYSDIS}$ improves to −15.0 dB. Algorithmic delay as well as MOS_{LQO}^{PF} remains constant when shadow filtering is applied.

Additional application of the decimation approach is again able to further improve the performance, as the results in the fourth row show. Here, $\overline{ERLE}$ increases to 21.3 dB, whereas $\overline{SYSDIS}$ slightly improves to −15.3 dB. This approach additionally offers two further advantages: As already shown in Table 6.2, algorithmic delay is reduced to 21 ms, and it also leads to an improvement of the perceptual quality of the speech component. This can be seen from the surprising fact that the MOS_{LQO}^{PF} score improves from 3.3 (without decimation) to 3.6 (with decimation). Again, postfilter decimation alone, as shown in the third row, does not show different *ERLE* or system distance results compared to the reference, whereas algorithmic delay and the MOS_{LQO}^{PF} score are improved.

6.5 Conclusions

We have presented a wideband automotive hands-free system for mobile HD Voice services. It is based on a shadow-filter approach for an FDAF-based acoustic echo canceller, which is significantly improving the convergence speed and tracking performance. Our approach excels in double-talk performance, revealing a high quality of the speech component in uplink direction. By decimation of the frequency domain postfilter coefficients, the computational complexity, algorithmic delay, and perceptual speech quality could be improved. Experimental results show a good performance in a simulated car environment.

References

1. J.J. Shynk, Frequency-domain and multirate adaptive filtering. IEEE Signal Process. Mag. **9**(1), 14–37 (1992)
2. C. Breining, P. Dreiseitel, E. Hänsler, A. Mader, B. Nitsch, H. Puder, T. Schertler, G. Schmidt, J. Tilp, Acoustic echo control. An application of very-high-order adaptive filters. IEEE Signal Process. Mag. **16**(4), 42–69 (1999)
3. G.W. Enzner, P. Vary, Frequency-domain adaptive Kalman Filter for acoustic echo control in hands-free telephones. Signal Process. Elsevier **86**, 1140–1156 (2006)
4. H.W. Löllmann, P. Vary, Uniform and low delay filter-banks for speech enhancement. EURASIP Speech Commun. **49**(7–8), 574–587 (2007)
5. H. Puder, P. Dreiseitel, Implementation of a hands-free car phone with echo cancellation and noise-dependent loss control, in *Proceedings of IEEE International Conference Acoustics, Speech, and Signal Processing* (*ICASSP*), vol. 6 (2000), pp. 3622–3625
6. S. Haykin, *Adaptive Filter Theory*. Prentice Hall, 4 edn., (2002)
7. E. Hänsler, G. Schmidt, *Acoustic echo and noise control: a practical approach* (Wiley, Hoboken, NJ, 2004)
8. N.K. Jablon, On the complexity of frequency-domain adaptive filtering. IEEE Trans. Signal Process. **39**(10), 2331–2334 (1991)
9. C. Beaugeant, M. Schönle, I. Varga, Challenges of 16 kHz in acoustic pre- and post-processing for terminals. IEEE Commun. Mag. **44**(5), 98–104 (2006)
10. B. Widrow, P. N. Stearns, *Adaptive Signal Processing*. Prentice Hall, 1 edn.,(1985)
11. A. H. Sayed, *Fundamentals of Adaptive Filtering*. Wiley-IEEE Press, 1 edn., (2003)
12. S. Gustafsson, R. Martin, P. Vary, Combined acoustic echo control and noise reduction for hands-free telephony. Signal Process. **64**(1), 21–32 (1998)
13. C. Yemdji, M. Idrissa, N. Evans, C. Beaugeant, Efficient low delay filtering for residual echo suppression, in *European Signal Processing Conference* (*EUSIPCO*), (2010)
14. Y. Ephraim, D. Malah, Speech enhancement using optimal non-linear spectral amplitude estimation, in *Proceedings of IEEE International Conference Acoustics, Speech, and Signal Processing* (*ICASSP*), vol. 8 (1983), pp. 1118–1121
15. T. Lotter, P. Vary, Speech enhancement by MAP spectral amplitude estimation using a super-Gaussian speech model. EURASIP J. Appl. Signal Process. **7**, 1110–1126 (2005)
16. P. Scalart, J. Filho, Speech enhancement based on a priori signal to noise estimation, in *Proceedings of IEEE International Conference Acoustics, Speech, and Signal Processing* (*ICASSP*), vol. 7 (1996), pp. 629–632
17. H.-C. Shin, A. Sayed, Variable step-size NLMS and affine projection algorithms. IEEE Signal Process. Lett. **11**, 132–135 (2004)
18. K. Ozeki, T. Umeda, An adaptive filtering algorithm using an orthogonal projection to an affine subspace and its properties. Electron. Commun. Jpn. **67**(5), 19–27 (1984)
19. S. Gay, S. Travathia, The fast affine projection algorithm, in *Proceedings IEEE International Conference Acoustics, Speech, and Signal Processing* (*ICASSP*), vol. 3 (1995), pp. 3023–3027
20. J. Cioffi, T. Kailath, Fast, recursive-least-squares transversal filters for adaptive filtering. IEEE Trans. Acoustics Speech Signal Process. **32**(2), 304–337 (1984)
21. R.E. Kalman, A new approach to linear filtering and prediction problems. Trans. ASME J. Basic Eng. **82**, 35–45 (1960)

22. K. Steinert, M. Schönle, C. Beaugeant, T. Fingscheidt, Hands-free system with low-delay subband acoustic echo control and noise reduction, in *Proceedings of IEEE International Conference Acoustics, Speech, and Signal Processing (ICASSP)*, (2008) pp. 1521–1524
23. R. Crochiere, L.R. Rabiner, *Multirate digital signal processing* (Prentice Hall, Englewood Cliffs, NJ, 1983)
24. J. Shynk, R. Gooch, Frequency-domain adaptive pole-zero filtering. Proc. IEEE **73**, 1526–1528 (1985)
25. G. W. Enzner, *A model-based optimum filtering approach to acoustic echo control: theory and practice*. PhD thesis, Rheinisch-Westfälische Technische HochschuleAachen, 2006
26. G. W. Enzner, P. Vary, Robust and elegant, purely statistical adaptation of acoustic echo canceler and postfilter, in *International Workshop on Acoustic Echo and Noise Control (IWAENC)*, (Kyoto, 2003), pp. 43–46
27. M. Ihle, K. Kroschel, Integration of noise reduction and echo attenuation for handset-free communication, in *International Workshop on Acoustic Echo and Noise Control (IWAENC)*, (1997) pp. 69–72
28. E. Hänsler, U. Schmidt, Hands-free telephones—joint control of echo cancellation and postfiltering. IEEE Signal Process. Mag. **80**, 2295–2305 (2000)
29. C. Beaugeant, V. Turbin, P. Scalart, A. Gilloire, New optimal filtering approaches for hands-free telecommunication terminals. Signal Process. **64**(1), 33–47 (1998)
30. ITU-T, Rec. P.862.2: *Wideband Extension to Recommendation P.862 for the Assessment of Wideband Telephone Networks and Speech Codecs*, (2007)

Chapter 7
In-Car Communication

Christian Lüke, Gerhard Schmidt, Anne Theiß, and Jochen Withopf

Abstract Communicating inside a car can be difficult because there is usually a high level of background noise and also the talking and the listening passengers do not necessarily face each other as they would do in a natural conversation. In-car communication (ICC) systems are a solution to this problem. They record the talkers' speech signal by means of microphones and reproduce it over loudspeakers that are located close to the listening passengers. However, such systems operate in a closed electroacoustic loop which significantly limits the gain that can be introduced by the system. In order to improve this gain margin and to achieve additional signal enhancement, several signal processing techniques are applied in ICC systems. Special care has to be taken about the signal delay: If it is too large, the reverberation inside the car is increased considerably and the speech reproduced over the loudspeakers might be perceived as an echo by the speaking passengers. In this chapter, an overview of the signal processing components of an ICC system is given. The necessary signal processing steps are explained and approaches to implement them are shown, especially with a focus on low processing delays.

Keywords Echo cancellation • Feedback • In-car communication (ICC) • Low-delay filter banks • Noise reduction • Speech intelligibility

7.1 Introduction

The communication in cars often lacks of quality in the sense of intelligibility. Especially at higher speed, the conversation comfort is reduced due to the high background noise (engine, wind, tire noise, etc.). Also the sound absorbing materials in the car which reduce the noise inside the passenger compartment

C. Lüke (✉) • G. Schmidt • A. Theiß • J. Withopf
Christian-Albrechts-Universität zu Kiel, Digital Signal Processing and System Theory, Kaiserstr. 2, Kiel D-24143, Germany
e-mail: cl@tf.uni-kiel.de; gus@tf.uni-kiel.de; ath@tf.uni-kiel.de; jow@tf.uni-kiel.de

G. Schmidt et al. (eds.), *Smart Mobile In-Vehicle Systems: Next Generation Advancements*, DOI 10.1007/978-1-4614-9120-0_7,

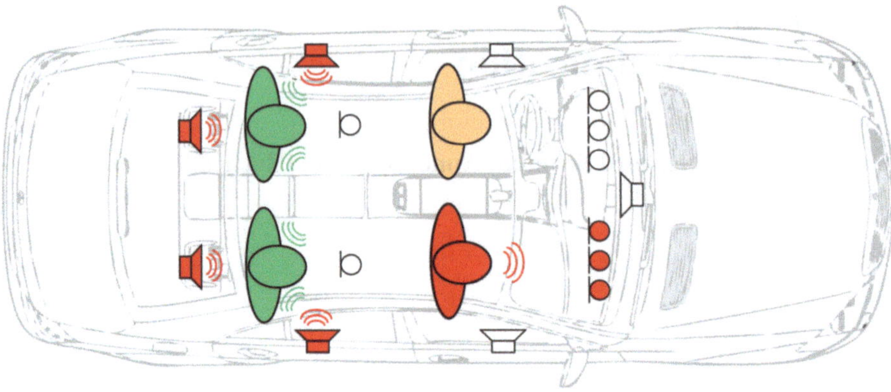

Fig. 7.1 Principle of an ICC system: the front passenger's voice is recorded by one or more microphones and played back over loudspeakers close to the listening passengers on the rear seat

degrade the speech intelligibility. This results from the fact that early reflections of the speech components are attenuated. These reflections usually contribute in a positive sense to the speech intelligibility [1]. In large vehicles, for instance minivans and buses, there is also a considerable attenuation of the acoustic signals due to the distance between the talking and the listening passenger. The usual reaction is that the rear passengers speak louder and lean forward to the front passengers. The problem of reduced intelligibility increases further in the communication from a front passenger to a rear passenger as the front passenger talks towards the windshield (see Fig. 7.1). The front passenger usually turns over which is uncomfortable for a longer time and, in addition, a security risk if the driver does so.

To overcome these problems, in-car communication (ICC) systems record the talking passengers and distribute the seat-dedicated microphone signals to the loudspeakers located close to the listening passengers [2, 3]. However, this technical support of the conversation contains some challenges due to the interfering signals (noise, music, etc.) and the closed-loop operation. Various signal processing techniques are required to reduce feedback, echo, and noise as well as to prevent system instability.

If the system delay exceeds 10–15 ms, passengers start to perceive the additional playback as a separate source [4, 5]. The system delay consists of the delay caused by the analog-to-digital and digital-to-analog converters, amplifiers, block-based signal transport on the car's signal processing hardware, acoustical paths, and also the signal processing. Subsequently, all algorithms should be designed to cause as little delay as possible. However, selected loudspeaker signals might also be delayed on purpose in order to overcome a localization mismatch between the acoustically perceived talker location and the actual one. Another aspect in rating on the quality of an ICC system is how much the talker is disturbed by hearing the own voice played back over the loudspeakers. Subjective tests with the presented ICC system showed that the talker tolerates a higher system gain for a system delay

Fig. 7.2 Test car equipped with the presented ICC system

between 12 and 15 ms. Thus, the boundary conditions set by the perception of separate sources and self-masking are close together.

The ICC system that is presented in this chapter has been implemented in the Kiel Real-Time Audio Toolkit (KiRAT) in the programming language C. For testing the algorithms in a car, the software runs on a PC platform with audio connections over low-delay ASIO soundcards. The delay (without that originating from signal processing) of this configuration is approximately 5.7 ms. Microphones placed at different positions are connected to the system as it can be seen in Fig. 7.2 where the microphones are highlighted by ellipses.

Section 7.2 gives an overview of the ICC system and briefly explains the contained components and how they interact. More details about the algorithms employed in certain modules are given in Sects. 7.3–7.8. Examples are shown after the algorithms are introduced in order to demonstrate the performance. Finally, conclusions are drawn in Sect. 7.9.

7.2 Overview

Figure 7.3 shows an overview of the signal processing in an ICC system containing the essential components. First, preprocessing is applied to each microphone signal. This contains a signal analysis, where, e.g., clipping or complete blackout of a microphone is detected. In automotive environments, usually the background noise is dominating the speech components at low frequencies. For this reason, there is a high-pass filter to remove these frequencies of bad signal quality. The high-pass

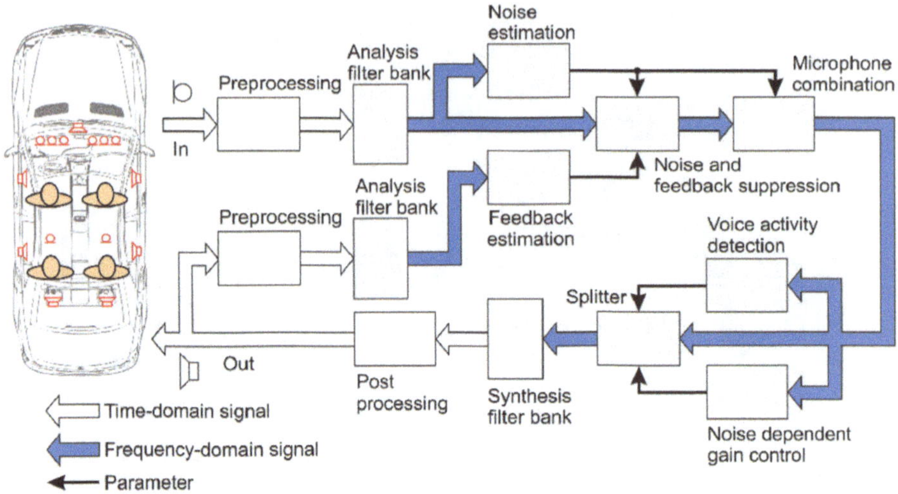

Fig. 7.3 Overview of the ICC system

filter that is used in the presented system has Butterworth characteristic and is of second order (two poles and two zeros). The 3 dB cut-off frequency is set to 200 Hz, but this value depends on the user preferences as well as on the properties of the vehicle. Most of the remaining signal processing takes place in the frequency domain which allows for reduced computational complexity. The next block is therefore an analysis filter bank (see Sect. 7.3 for details) that computes a subband-signal representation.

All signal spectra of microphones that are assigned to talking passengers are then enhanced in terms of their signal quality. This mainly consists of noise- and feedback reduction by a Wiener-type filter as explained in Sect. 7.6. For this filter, noise and feedback estimates have to be computed as presented in Sects. 7.4 and 7.5, respectively.

The remaining part of the signal processing is concerned with the distribution of signals from the input microphones to the output loudspeakers and adjusting the signals for good playback quality. If multiple microphones are available for one talking passenger, first one signal per talker has to be extracted. This can be done by combining the signals, e.g., by beamforming where knowledge about the position of the talkers can be exploited. Another method is to detect which microphone offers the best signal quality in terms of SNR. Any method used here should work adaptively because the noise level might change, e.g., when a window is opened or the ventilation is turned on. Based on the output signal of this signal-combination module, a voice activity detection (VAD) as described in Sect. 7.7 is necessary to determine the active talking passenger for correctly managing the subsequent steps of the signal distribution.

With the information of the VAD, the signals of nonactive talkers are attenuated by a unit called loss control. Then, the talker signals are mapped to the listening

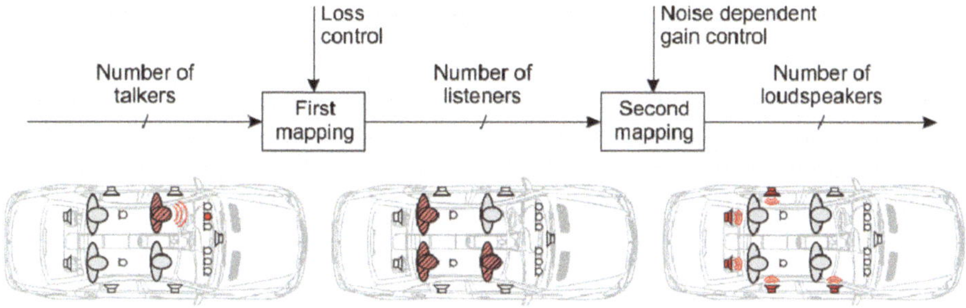

Fig. 7.4 The two-stage mapping of talker dedicated signals to the available loudspeakers

passengers and further mapped to the loudspeakers that are available for a specific listener. In this last mapping, the gain of each signal is adjusted according to the background noise level. While no support of the system is usually needed during standstill, more gain is required with increasing speed. Because the noise might vary considerably between the seats of a car, each listener can be assigned one or more microphones that are used to estimate the noise level at his position. This noise estimate is then converted into a gain factor by the noise-dependent gain control (NDGC, see Sect. 7.8). This gain factor is computed individually for each loudspeaker of a listener because, due to their position, some loudspeakers are more critical in terms of feedback. The principle of the two-stage signal mapping is shown in Fig. 7.4.

The loudspeaker-dedicated signals can be enhanced by different processing units before playback. Two different equalizers are implemented to improve the sound impression, but also to optimize the feedback properties of the system by attenuating those frequencies that exhibit the largest coupling to the microphones. The first one operates in the frequency domain and provides zero-phase equalization with low computational complexity.[1] After this frequency-domain equalizer, the signals are transformed back to the time-domain by a synthesis filter bank. A so-called peak-filter equalizer [6] can be used to realize narrow band corrections of the frequency response. Setting such narrow notches or peaks would not be possible with the frequency-domain equalizer. Other components contained in the post-processing are a gain- and delay-element that can be used to adjust the spatial hearing impression and a limiter to prevent clipping of the digital/analog converters. Because the estimation of the feedback component needs information about the loudspeaker signals and operates in the subband-domain, another analysis filter bank that also contains the preprocessing which is applied to the input microphones is computed.

[1] If the delay and the computational load of the analysis and synthesis filter banks are neglected.

7.3 Analysis and Synthesis Filter Bank

Filter banks provide a conversion between time and frequency domain. Both parts, the analysis and the synthesis filter bank, need to be matched for proper operation. Their performance can be improved by applying pre- and de-emphasis filters before the analysis and after the synthesis stage.

7.3.1 Complex Modulated Filter Banks

Most often, filter banks are viewed as a set of N parallel bandpass filters. After the input signal $x(n)$ has been passed through these filters, downsampling by factor $R \leq N$ can be applied to the band limited signals to obtain the subband signals[2] $X(\mu,k)$ for frequency bin μ and frame $k = n/R$. If the bandpass filters $v_\mu(n)$ are derived from prototype lowpass filters $v(n)$ by complex modulation

$$v_\mu(n) = v(n)\, e^{-j\frac{2\pi}{N}\mu n}, \tag{7.1}$$

this structure can be implemented efficiently as a discrete Fourier transform (DFT) filter bank by using the fast Fourier transform (FFT) when N is chosen as a power of two. This implementation can also be seen as an STFT with a sliding window that is evaluated every R samples:

$$\begin{aligned} X(\mu,k) &= \mathrm{DFT}_N\{\, v(l)\, x(kR+l)\} \\ &= \sum_{l=0}^{N-1} v(l)\, x(kR+l)\, e^{-j\frac{2\pi}{N}\mu l}. \end{aligned} \tag{7.2}$$

Here, $l = 0, \ldots, N-1$ is the local time index within the analyzed signal part. Only the first $N/2 + 1$ frequency bins have to be stored and processed for real-valued input signals, while the remaining ones are the complex conjugate and can be recreated before the synthesis operation.

Signal synthesis is done in a straightforward manner by first computing the inverse DFT of each processed spectrum $Y(\mu,k)$ and overlapping the time-domain signals after weighting with the synthesis window $w(n)$. This window $w(n)$ corresponds to an anti-imaging filter that has to be applied after upsampling the subband signals in the concept of parallel bandpass filters in order to suppress repetitions of the signal spectrum. From a time-domain point of view, it interpolates between the samples after filling the downsampled signals with zeros. The effective window

[2] The subbands can also be seen as time-aligned spectra if all filter bank channels are considered at a certain time instance.

length is M, meaning that $w(n) = 0, \forall\, n \notin \{0, \ldots, M-1\}$. For producing a single output sample, this *overlap-add* (OLA) procedure can be expressed by

$$\begin{aligned} y(kR+r) &= \sum_{\kappa}^{K-1} w(\kappa R + r)\, N\, \mathrm{IDFT}_N^{(kR+r)}\{Y(\mu, k+K-\kappa)\} \\ &= \sum_{\kappa}^{K-1} w(\kappa R + r) \sum_{\mu=0}^{N-1} Y(\mu, k+K-\kappa)\, e^{j\frac{2\pi}{N}(\kappa R + r)}, \end{aligned} \tag{7.3}$$

where $K = M/R$. The local time index $r = 0, \ldots, R-1$ describes the position of the output sample within the current output frame. The notation $\mathrm{IDFT}_N^{(kR+r)}\{Y(\mu,k)\}$ denotes taking sample $kR + r$ of the inverse discrete Fourier Transform (IDFT) of order N.

7.3.1.1 Delay Considerations

Since most automotive sound processing hardware processes input samples blockwise, output samples will be aggregated until $r = R-1$ before writing them to the output buffer. This means that a delay of K frames is introduced by the synthesis through (7.3). The delay for collecting R input samples in the analysis (7.2) is then included. Note that (7.2) covers only the case where the window length is equal to the DFT order N. For polyphase implementations where the window length is a multiple of the DFT order, additional delay could be introduced depending on the shape of $v(n)$.

For DFT filter banks, most often the synthesis window length is chosen as the DFT order M for obtaining good anti-imaging properties by having a large filter order. However, this causes a large delay according to (7.3). An *overlap-save* (OLS) filter bank results from setting $M = R$. In order to avoid drawbacks from the OLS approach (e.g., the need for projection filters) and still reducing the delay, we propose a filter length of $R < M < N$ resulting in somewhat degraded anti-imaging properties.

7.3.2 *Approaches to Filter Design*

A standard approach in the design of prototype lowpass filters is to use raised-cosine (e.g., Hann or Hamming) windows. For downsampling ratios $N/R = 2^{\beta}$, $\beta \in \mathbb{N}$, and $M = N$ they fulfill the condition

$$\sum_{k=0}^{K-1} v(n+kR)\, w(n+kR) = g \tag{7.4}$$

for perfect reconstruction, meaning that the filter bank without processing of the subband signals introduces only a delay and a gain g. Several methods have been proposed for iteratively optimizing these filters. A mathematical formulation of the so-called *in-band aliasing* and the *total aliasing* is given in [7]. Minimization of these error criteria is done subject to the constraint of near-perfect reconstruction, meaning that amplitude distortions are allowed to a certain extend. Similar approaches are reported in [8] and [9] which mainly differ in details about the error function for optimization. Different filter lengths $M \neq N$ are generally allowed, but only taken into account when the filter group delay is included in the error function. This is not the case in the approach described in [10], which has an emphasis on designing pairs of analysis/synthesis windows that can be switched to trade-off time and frequency resolution depending on the input signal. Once again, different criteria related to the filter stopband attenuation are optimized in an iterative way, limiting the usage of this method to offline filter design. A different approach for $M = N$ is followed in [11] where a raised-cosine sequence serves as a prototype.

7.3.3 Non-iterative Filter Design

Given an analysis window $v(n)$, a matching synthesis window ensuring perfect reconstruction for arbitrary values of N and R can be obtained by the transformation

$$\boldsymbol{w} = W\{\boldsymbol{v}\} = \left[\boldsymbol{V}^{\mathrm{T}}\boldsymbol{V}\right]^{-1} \boldsymbol{V}^{\mathrm{T}}\, \boldsymbol{c}. \tag{7.5}$$

The $N \times 1$ vectors $\boldsymbol{v} = [v(0), v(1), \ldots, v(N-1)]^{\mathrm{T}}$ and $\boldsymbol{w}$ contain the filter coefficients, $\boldsymbol{c}$ is a vector of R ones, and the matrix $\boldsymbol{V} = [\boldsymbol{V}_0, \boldsymbol{V}_1, \ldots, \boldsymbol{V}_K]$ contains the matrices

$$\boldsymbol{V}_k = \begin{bmatrix} v(kR) & 0 & & \cdots & 0 \\ 0 & v(kR+1) & & \cdots & 0 \\ \vdots & \vdots & \ddots & & \vdots \\ 0 & 0 & \cdots & & v(kR+R-1) \end{bmatrix} \tag{7.6}$$

with coefficients $v(n)$ on the diagonals [12]. This solution with the pseudo-inverse matrix yields the matching coefficients with minimum norm but, however, there are other solutions to this problem. If the method according to (7.5) is applied for shortened windows with $M < N$ an additional smoothing window $\boldsymbol{h} = [h(0), \ldots, h(N-1)]^{\mathrm{T}}$ should be introduced to avoid discontinuous results:

$$w(n) = \begin{cases} \dfrac{h(n)\,[W\{\boldsymbol{h}\}]^{(n)}}{\max\{v(n),\ \varepsilon\}} & \text{, for } n = 0, \ldots, M-1, \\ 0 & \text{, else.} \end{cases} \tag{7.7}$$

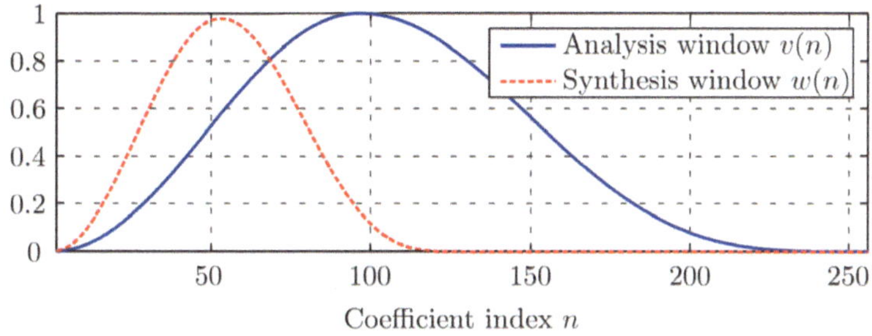

Fig. 7.5 Analysis and synthesis windows

The notation $[W\{\boldsymbol{h}\}]^{(n)}$ denotes taking the nth element of the transformed vector. The maximum operation with the variable ε avoids divisions by zero or very small values which can occur at the edges of $v(n)$. Note that for values $\varepsilon < 0$ condition (7.4) is violated. However, near-perfect reconstruction can be achieved which is sufficient in many applications [10].

The remaining free parameters in this design scheme of (7.7) are the analysis window $v(n)$ and the intermediate smoothing window $h(n)$. Both have been derived by setting different parameters in the prototype window

$$g(n) = \begin{cases} \left[\cos^2\left(\dfrac{\pi n}{2N_1} + \dfrac{\pi}{2}\right)\right]^{\alpha_1} & \text{, for } 0 \leq n < N_1, \\ \left[\cos^2\left(\dfrac{\pi(n - N_1 + N_2)}{2N_2} + \dfrac{\pi}{2}\right)\right]^{\alpha_2} & \text{, for } N_1 \leq n < N. \end{cases} \tag{7.8}$$

that consists of two raised-cosine parts. The length $N_1 = N - N_2$ can be set to produce asymmetric windows, the powers α_1 and α_2 allow stronger tapering towards the left and right end. For a design with $M = N$ we propose to set $N_1 = (N - M)/2$, $\alpha_1 = 1$, and $\alpha_2 = 2$ for the analysis window. For the smoothing window $h(n)$ we propose $N_1 = N_2 = M/2$, $\alpha_1 = 1$, and $\alpha_2 = K/10$.

7.3.3.1 Design Example

Figure 7.5 shows a window pair that has been designed for a low-delay speech enhancement system with the parameters $N = 256$, $M = N/2$, and $R = 32$. The sampling rate is $f_s = 22050$ Hz, so the resulting filter bank delay is $M/f_s = 5.8$ *ms*. Tests with a subband echo cancellation it within this systems showed echo reduction of about 35 dB and thus sufficient suppression of aliasing distortions. Compared to a filter bank with Hann windows of these lengths, perfect reconstruction is achieved and the echo cancellation could be improved only by approximately 5 dB.

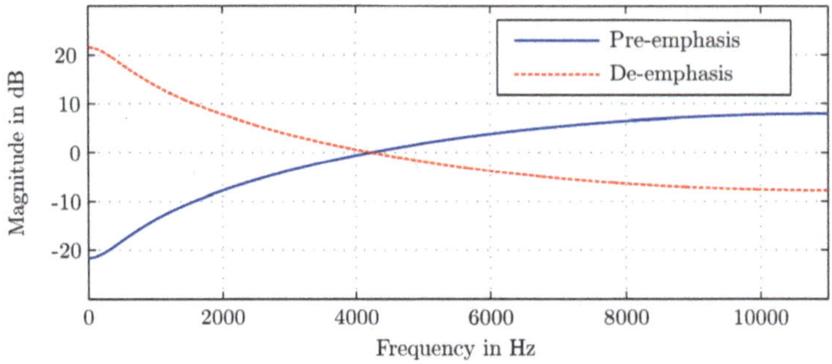

Fig. 7.6 Pre- and de-emphasis filters of second order designed as prediction error filters

7.3.4 *Pre- and De-emphasis Filters*

Due to the limited amount of subbands, the resolution of a filter bank is limited. Even with a proper design of the analysis window, aliasing in the frequency domain cannot be avoided totally. Therefore, a pre-emphasis filter is used to whiten the signal and thus achieve an approximately constant power of the aliasing distortion over the subbands.

Since speech signals cannot be assumed to be stationary, the desired de-correlation of the time-domain signals cannot be achieved exactly with a fixed pre-emphasis filter. However, it can be used to remove the high-frequency roll-off that is common to all speech signals. This means that low filter orders are sufficient. After the synthesis filter bank, a de-emphasis filter has to be applied in order to revert the filtering introduced in the pre-emphasis stage. One method is to design a prediction error filter for the pre-emphasis, as these filters are always minimum phase and thus straightforward to invert [13]. An example for pre- and de-emphasis filters designed with this method is shown in Fig. 7.6.

7.4 Feedback Estimation

In order to obtain sufficient system gain, it is necessary to investigate the electro-acoustic feedback loop. One possibility to attack the feedback problem is to estimate the feedback component for every microphone and suppress it with a frequency-dependent attenuation factor as described in Sect. 7.6.

The model for estimating the power spectral density (PSD) of the feedback from microphone m to loudspeaker l in frame k and subband μ can be described as [14, 15]

$$\hat{S}_{ff}^{(lm)}(\mu,k) = \alpha_{lm}(\mu)\,\hat{S}_{ff}^{(lm)}(\mu,k-1) + \beta_{lm}(\mu)\,\hat{S}_{yy}^{(l)}(\mu,k-d_{lm}), \tag{7.9}$$

where the quantities are as follows:

$\hat{S}_{ff}^{(lm)}(\mu,k)$	Estimated feedback PSD
$\hat{S}_{yy}^{(l)}(\mu,k)$	PSD of loudspeaker l
$\beta_{lm}(\mu)$	Room coupling factor
$\alpha_{lm}(\mu)$	Attenuation factor
d_{lm}	Signal delay in frames

The loudspeaker PSD $\hat{S}_{yy}^{(l)}(\mu,k)$ can be estimated from the loudspeaker signal by computing the squared magnitude

$$\hat{S}_{yy}^{(l)}(\mu,k) = \left|Y^{(l)}(\mu,k)\right|^2. \tag{7.10}$$

According to this first-order infinite impulse response (IIR) model, the feedback component is comprised of the previous estimate, weighted by the attenuation constant $\alpha_{lm}(\mu)$ which describes how fast the feedback decays in subband μ. This system is driven by the loudspeaker output signal, delayed by the length of the acoustic path d_{lm} between loudspeaker l and microphone m and weighted by the coupling factor $\beta_{lm}(\mu)$.

The complete feedback PSD $\hat{S}_{ff}^{(lm)}(\mu,k)$ at microphone m can be estimated by summing over all contributions of the N_{lsp} loudspeakers:

$$\hat{S}_{ff}^{(m)}(\mu,k) = \sum_{l=0}^{N_{\text{lsp}}-1} \hat{S}_{ff}^{(lm)}(\mu,k). \tag{7.11}$$

All model parameters of (7.9) can be estimated from the impulse responses which describe the feedback paths [16]. The attenuation factor $\alpha_{lm}(\mu)$ can also be converted to the more familiar reverberation time T_{60} in seconds by

$$T_{60} = -\frac{60}{20\log_{10}(\alpha_{lm}(\mu))\,\frac{f_s}{R}}. \tag{7.12}$$

The reverberation time is the time it takes an impulse response to decay by 60 dB. For cars, T_{60} is usually around 50 ms and the coupling $\beta_{lm}(\mu)$ for typical loudspeaker and microphone positions between 0 and −60 dB. Especially the coupling depends heavily on the frequency and is usually larger for low frequencies. Figure 7.7 shows values for the reverberation time T_{60} and the coupling $\beta_{lm}(\mu)$ that have been measured inside a car for one feedback path. These parameters could also be updated and adapted to changing environments during operation by estimating the impulse responses online. This is of particular interest if the ICC system is also equipped with echo cancellation where the needed measurements are already available.

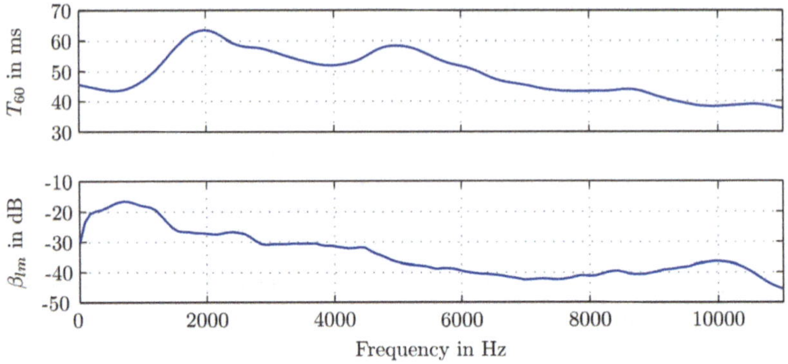

Fig. 7.7 Reverberation time T_{60} and the coupling $\beta_{lm}(\mu)$

7.5 Noise Estimation

It cannot be avoided that, besides the desired speech signal, the microphones also pick up background noise. If this background noise would be played back over the loudspeakers, the overall noise level in the car would increase which is of course undesirable. The noise reduction algorithm described in Sect. 7.6 needs an estimate of the background noise PSD $\hat{S}_{bb}(\mu, k)$ which can be obtained for (nearly) stationary noise processes in various ways. Here, we propose a rather simple scheme. First, the squared magnitude of the input spectrum $X(\mu,k)$ is smoothed over time with a first-order IIR filter:

$$\left|\overline{X}(\mu, k)\right|^2 = \beta_{\text{sm}} \left|X(\mu, k)\right|^2 + (1 - \beta_{\text{sm}}) \left|\overline{X}(\mu, k - 1)\right|^2. \tag{7.13}$$

The smoothing time constant β_{sm} describes how fast the smoothed squared magnitude $\left|\overline{X}(\mu, k)\right|^2$ may vary over time. Since its value depends on the sampling rate f_s and the frameshift R, it is convenient to define it in the physical unit of dB/s by the conversion[3]

$$\widetilde{\beta}_{\text{sm}} = 20\, log_{10}(1 - \beta_{\text{sm}}) \frac{f_s}{R}. \tag{7.14}$$

A time constant of, e.g., $\widetilde{\beta}_{\text{sm}} = 300\,\text{dB/s}$ helps to remove so-called outliers efficiently.

[3] From now on, the tilde is used to annotate these "user-friendly" variables.

The smoothed short-term power estimate $\left|\overline{X}(\mu,k)\right|^2$ is then compared to the previous estimate of the noise PSD $\hat{S}_{bb}(\mu,k)$ to update the estimated value:

$$\hat{S}_{bb}(\mu,k) = \begin{cases} \gamma_{\text{inc}}\,\hat{S}_{bb}(\mu,k-1), & \textit{if } \left|\overline{X}(\mu,k)\right|^2 > \hat{S}_{bb}(\mu,k-1), \\ \gamma_{\text{dec}}\,\hat{S}_{bb}(\mu,k-1), & \text{else.} \end{cases} \tag{7.15}$$

The increment and decrement time constants could be chosen, e.g., like[4] $\widetilde{\gamma}_{\text{inc}} = 3\ \text{dB/s}$ and $\widetilde{\gamma}_{\text{dec}} = -10\ \text{dB/s}$. If γ_{inc} is chosen much higher, the noise estimate will increase too fast during speech periods, if it is set too small, the noise estimator cannot follow changes in the noise power fast enough. Usually, the decrement is set to a "faster" value than the increment. The noise estimator is initialized to a rather high value because the estimate drops faster and thus reaches the correct value earlier after the estimation procedure is started.

7.5.1 Improved Performance in Nonstationary Noise

The computationally "cheap" noise estimation scheme described in Sect. 7.5 yields reliable and robust results for stationary or slowly varying background noise. However, it fails in the estimation of nonstationary noise components which can originate, e.g., from open windows, fans, or air conditions blowing on the sensors, or when passengers touch a microphone.[5] In these cases, the signal is typically misinterpreted as speech by algorithms that rely on proper estimates such as a noise reduction or VAD. One solution to this problem is to employ more sophisticated noise estimation algorithms, e.g., extended minimum statistics [17].

Another option is to detect the presence of nonstationary noise for each microphone. Different single channel approaches have been reported for this [18–20], but the performance of these algorithms can be improved considerably if multiple microphones are available and thus the spatial characteristics of desired and undesired signal components can be exploited [21, 22]. The method that we suggest assumes that not all microphones are disturbed by nonstationary noise at the same time. This is usually fulfilled for typical use cases when the microphones are touched by the passengers. Wind buffets will reach the microphones delayed individually because their propagation velocity is low compared to the speed of sound. Based on this assumption, a reference spectrum for the undisturbed signal is derived and the presence of nonstationary noise can be detected. In this case, the subsequent ICC algorithms that assume stationary signals or linear systems might be paused or modified. Details can be found in [23].

---

[4] The user-friendly variables are obtained by the conversion similar to (7.12).

[5] This is likely to happen if microphones are integrated into the seat belts.

7.6 Noise and Feedback Reduction

For suppression of the undesired background noise and feedback components, the microphone spectrum $X(\mu,k)$ is multiplied with a frequency-dependent attenuation factor $G(\mu,k)$ to form the enhanced spectrum

$$X_{\mathrm{enh}}(\mu,k) = X(\mu,k)\,G(\mu,k). \tag{7.16}$$

The attenuation coefficients are found by a modified Wiener characteristic

$$G(\mu,k) = max\left\{G_{\min}, 1 - \frac{\beta_b\,\hat{S}_{bb}(\mu,k) + \beta_f\,\hat{S}_{ff}(\mu,k)}{\hat{S}_{xx}(\mu,k)}\right\}. \tag{7.17}$$

where $\hat{S}_{bb}(\mu,k)$ and $\hat{S}_{ff}(\mu,k)$ are estimates for the background noise and feedback PSDs, respectively. $\hat{S}_{xx}(\mu,k)$ is the microphone signal PSD of the current frame k and can be estimated as squared magnitude $|\overline{X}(\mu,k)|^2$ of the microphone spectrum:

$$\hat{S}_{xx}(\mu,k) = |X(\mu,k)|^2. \tag{7.18}$$

The overestimation factors β_b and β_f are used to correct or to intentionally introduce a bias in the estimates. Values greater than one make the filter more "aggressive," i.e., the filter attenuates more often. Subsequently, a compromise between suppression of unwanted signal components and speech distortion introduced by extensive filtering has to be found. An overview over the noise and feedback reduction for a single microphone channel including the estimators is shown in Fig. 7.8.

When the filter randomly attenuates for a short time and only at some subbands, this can be heard as so-called *musical tones*. They can be avoided (or masked) if some residual noise is allowed by introducing the maximum attenuation $G_{\min}$ which is typically set to values $-15\text{ dB} < G_{\min} < -9\text{ dB}$.

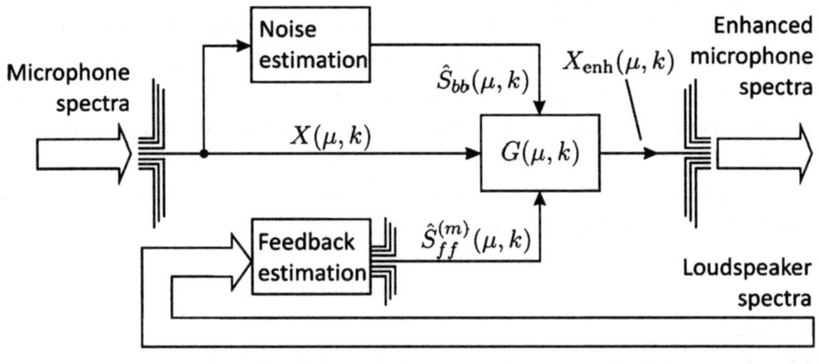

Fig. 7.8 Structure of noise and feedback reduction with the necessary estimation schemes for a single microphone channel

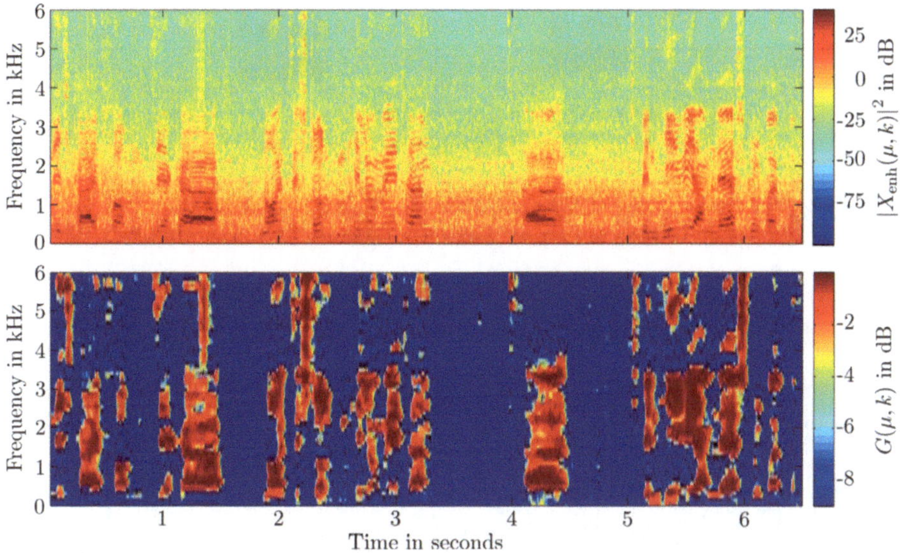

Fig. 7.9 Example for the noise reduction: Noise reduced signal (*upper plot*) and noise reduction coefficients (*lower plot*)

Figure 7.9 shows an example for noise reduction only (i.e., $\beta_f = 0$): The upper plot shows the spectrogram of a signal recorded in a car moving at a speed of 100 km/h after the noise reduction coefficients, that are shown in the plot below, have been applied according to (7.14). The plot of the attenuation coefficients clearly shows where the speech components are located.

An example for feedback reduction only ($\beta_{fb} = 0$) is shown in Fig. 7.10. The ICC system was operating at a maximum gain and the feedback reduction is turned off around 3 and around 6.5 s. The upper plot shows the output signal of a loudspeaker in the time domain. It can clearly be seen that the signal energy increases considerably in these time intervals. The spectrogram below reveals that the system starts oscillating at a frequency of approximately 500 Hz. In the lower plot, the attenuation coefficients are depicted. Again, the two periods when the feedback attenuation is switched off can be readily identified. The howling stops almost immediately after the feedback reduction is switched on again.

7.7 Voice Activity Detection

In the following, a voice activity detection (VAD) scheme that takes the difficulties of an automotive environment into account is presented. For this VAD, a noise estimation has to be computed for the talkers' signals. This is done in N_{VAD} frequency bands whose lower and upper cut-off frequencies can be set arbitrarily.

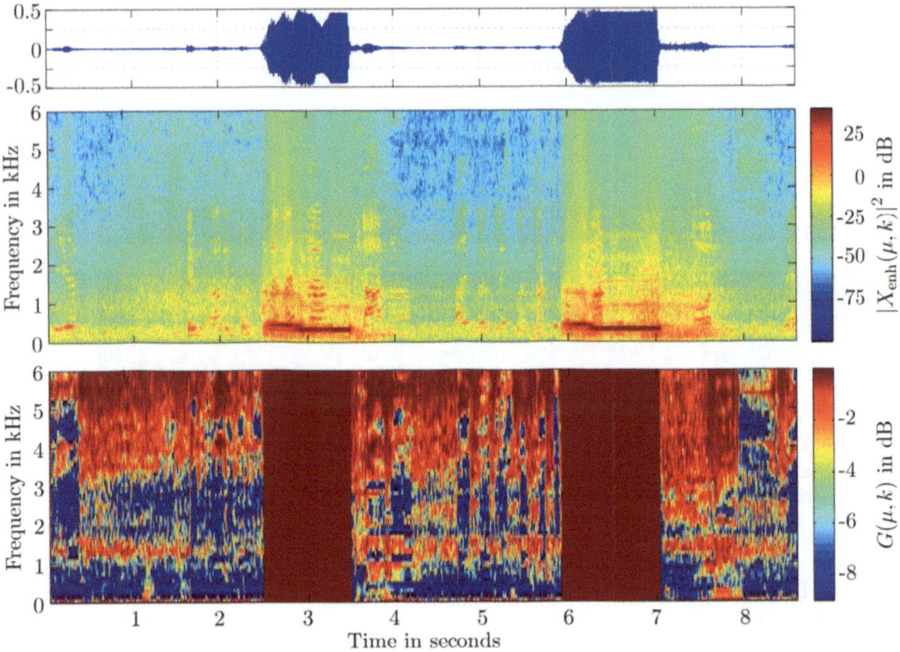

Fig. 7.10 Example for the feedback reduction: Loudspeaker output signal (*upper and middle plot*) and feedback reduction coefficients (*lower plot*). The feedback reduction is switched off between 2.5 and 3.5 and between 6 and 7 s

It is also possible to exclude certain frequency ranges, e.g., if they are known to be heavily corrupted by noise.

For the decision of voice activity, two conditions are tested for each noise estimation band:

- Does a talker achieve a minimum SNR?
- Does the large SNR originate from a neighboring talker?

If a condition is met for talker p, this is rewarded by the increase of a counter by

$$c^{(p)}(k) = min\Big\{1, c^{(p)}(k-1) + \Delta_{\mathrm{inc}}\Big\}. \tag{7.19}$$

If a condition is missed, it is penalized in a similar manner:

$$c^{(p)}(k) = \max\Big\{0, c^{(p)}(k-1) - \Delta_{\mathrm{dec}}\Big\}. \tag{7.20}$$

Due to the maximum and minimum operations, the counter is limited to the interval $c^{(p)}(k) \in [0,1]$. The counter changes should be normalized to the number of noise estimation bands, e.g., $\Delta_{\mathrm{inc}} = 1/\mathrm{N}_{\mathrm{VAD}}$.

Equation (7.19) is used if at least a minimum SNR is achieved, i.e., if

$$\hat{S}_{xx}^{(p)}(i,k) > \hat{S}_{bb}^{(p)}(i,k)\, SNR_{\min}. \tag{7.21}$$

If this is true for the noise estimation band i, the counter $c^{(p)}(k)$ is increased according to (7.19) and the second condition—if the high SNR for talker p actually originates from talker q ' s speech—is tested. A good estimator for the signal PSD needed in (7.21) is the short-term power $\left|\overline{X}^{(p)}(i,k)\right|^2$, which is available as a byproduct of the noise estimation procedure of Sect. 7.5.

Before comparing the signal PSDs $\hat{S}_{xx}^{(p)}(i,k)$ of all talking passengers, they are normalized to the background noise level in order to remove differences in the signal power that stem from inaccuracies in the hardware, e.g., different gain settings in the microphone preamplifiers. Therefore, first the mean noise level over all N_{talk} talking passengers is calculated in all noise estimation bands i:

$$\overline{S}_{bb}(i,k) = \frac{1}{N_{\text{talk}}} \sum_{p=0}^{N_{\text{talk}}} \hat{S}_{bb}^{(p)}(i,k). \tag{7.22}$$

This mean noise level is then used to find the normalization factor

$$\alpha_{\text{norm}}^{(p)}(i,k) = max\left\{N_{\min}, \min\left\{N_{\max}, \frac{\overline{S}_{bb}(i,k)}{\hat{S}_{bb}^{(p)}(i,k)}\right\}\right\}, \tag{7.23}$$

where $N_{\min}$ and $N_{\max}$ are the lower and upper boundaries of $\alpha_{\text{norm}}^{(p)}(i,k)$, respectively.

The second condition tests if the signal to interference ratio (SIR) between talker p and talker q (considered to be an interferer) is greater than a threshold:

$$\alpha_{\text{norm}}^{(p)}(i,k)\, \hat{S}_{bb}^{(p)}(i,k) > \alpha_{\text{norm}}^{(q)}(i,k)\, \hat{S}_{bb}^{(q)}(i,k)\, SIR_{\min}. \tag{7.24}$$

If the inequality (7.24) does not hold, this is penalized by decreasing the counter of talker p by applying (7.20).

After all noise estimation bands have been evaluated for updating the counters of all talkers, the score is compared to a threshold $VAD_{\min}$ to decide whether talker p is active or not

$$VAD^{(p)}(k) = \begin{cases} 1, & \text{if } c^{(p)}(k) < VAD_{\min}, \\ 0, & \text{else,} \end{cases} \tag{7.25}$$

where $VAD^{(p)}(k) = 1$ denotes speech activity. By deciding in this fashion, it is possible to classify multiple talkers as active.

7.8 Noise-Dependent Gain Control (NDGC)

The NDGC adjusts the playback volume to the noise level inside the vehicle. This is done for each listener and loudspeaker individually in order to exploit the gain-before-feedback margin as much as possible.

7.8.1 Basic Principle

The basic principle of the NDGC is depicted in Fig. 7.11: the noise estimate $\hat{N}(k)$ is mapped onto an instantaneous gain factor $\breve{g}(k)$ using a piecewise linear characteristic made up of N_{map} pieces. In order to avoid abrupt changes in the gain factor, the actual gain

$$g(k) = \begin{cases} \eta_{\text{inc}}(k)\, g(k-1), & \text{if } \breve{g}(k) > g(k-1), \\ \eta_{\text{dec}}(k)\, g(k-1), & \text{else,} \end{cases} \tag{7.26}$$

is computed by incrementing or decrementing the previous value. The corresponding time constants $\eta_{\text{inc}}(k)$ and $\eta_{\text{dec}}(k)$ can be defined in dependence on the current gain value $g(k)$. This is useful, e.g., when the microphones should be muted during standstill. A faster increase for the low-gain case would then allow reaching an appropriate system gain within a reasonable time when the noise level increases.

7.8.2 Loudspeaker- and Frequency-Dependent NDGC

The NDGC concept explained so far can be extended to a loudspeaker- and frequency-dependent design which allows better adaption to the conditions of a

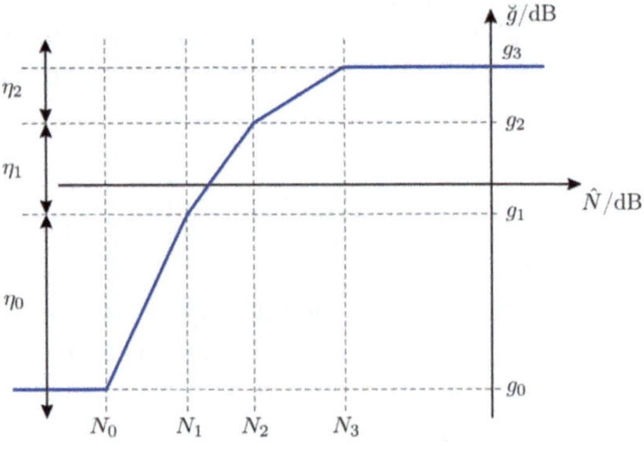

Fig. 7.11 Mapping of noise estimates to gain values

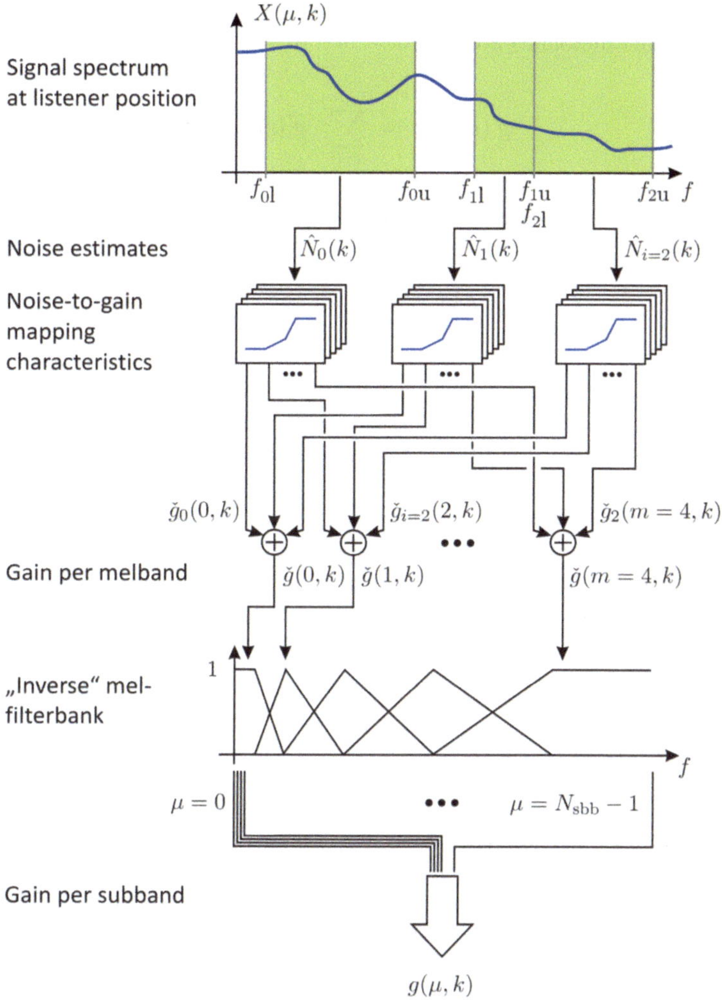

Fig. 7.12 Combination of several NDGC characteristics

given vehicle. Figure 7.12 shows how the gain vector $g(\mu,k)$ for a certain loudspeaker of a listener is computed. Several noise estimates $\hat{N}_i(k)$ can be obtained in N_{acc} noise estimation bands. In the example of Fig. 7.12, $N_{\mathrm{acc}} = 3$ noise estimation bands are used. These can be specified by their lower and upper cut-off frequencies $f_{i,\mathrm{l}}$ and $f_{i,\mathrm{u}}$ might be overlapping or with gaps in between to exclude certain frequency bands totally.

Each of the noise estimates is input to a set of N_{mel} mapping characteristics of the type of Fig. 7.11 to obtain preliminary gain values $g\,'_i(m,k)$, where $i \in [0, \ldots, N_{\mathrm{acc}} - 1]$ and $i \in [0, \ldots, N_{\mathrm{mel}} - 1]$. In Fig. 7.12, $N_{\mathrm{mel}} = 5$ melbands have been chosen.[6] To obtain one gain factor for each melband, the preliminary gains of the same melbands are added:

$$g'(m,k) \; = \sum_{i=0}^{N_{\mathrm{acc}}-1} g_i'(m,k). \tag{7.27}$$

These factors g' (m,k) are assigned to the subbands by

$$g(\mu,k) \; = \sum_{i=0}^{N_{\mathrm{mel}}-1} \alpha_{m\mu}\, g'(m,k), \tag{7.28}$$

where $\alpha_{m\mu}$ are overlapping triangular weighting functions for the extrapolation from melbands to subbands as schematically sketched in Fig. 7.12. The widths of the triangles are chosen according to the mel scale, i.e., they are increasing towards higher frequencies.

This scheme has been successfully used in practice with $N_{\mathrm{acc}} = 1$ and $N_{\mathrm{mel}} = 2$. Since the maximum possible gain in the test car was about 4 dB higher at low frequencies, some extra boosting could have been applied there when very high system gain was required. Further degrees of freedom could be added to fine-tune the system.

7.9 Conclusions

In this chapter we presented an ICC system for increasing the quality of a conversation inside a car. The individual algorithmic components have been presented in an overview followed by a more detailed description of most of the signal processing modules. Examples for suitable parameterizations of these algorithms have been given and also some processed data has been presented to demonstrate the functioning of the algorithms. All results have been obtained from an implementation of the ICC system within the KiRAT framework.

For testing, a car has been equipped with our ICC system consisting of low-latency audio equipment, a powerful automotive signal processing hardware (both PC- and DSP based) for signal processing based on the presented algorithms and amplifiers for driving the loudspeakers. The signal processing hardware is placed in the trunk of the car as it can be seen in Fig. 7.13.

Informal tests showed that the ICC system increases speech intelligibility and communication comfort at medium and high driving speed. The feedback reduction helps to improve the gain-before-feedback margin significantly. When the system operates at maximum gain and the feedback reduction is switched off, howling

[6] The concept of mel-filtering is, e.g., commonly used in the feature extraction for speaker and speech recognition, see [24].

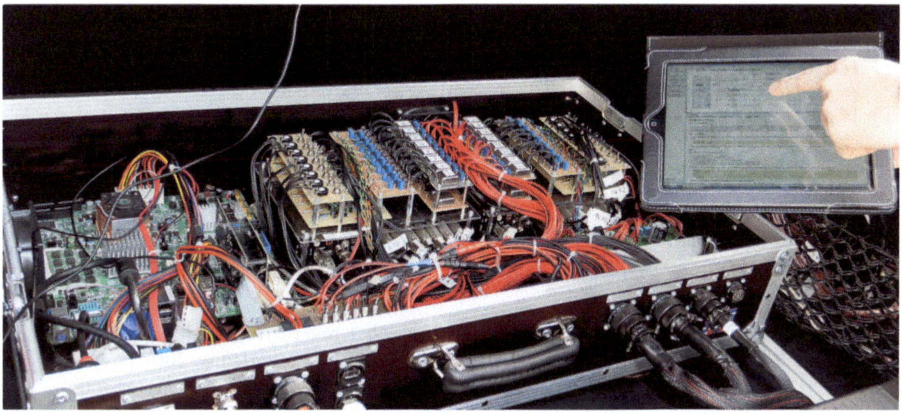

Fig. 7.13 Signal processing hardware in the trunk of the test car

occurs almost instantly. But even before the system starts oscillating, the signal quality is degraded due to an increase of reverberation caused by the feedback. The concept of the frequency- and loudspeaker-dependent NDGC helps to adapt the system to a given vehicle and to exploit the gain resources as well as possible.

At very high noise levels, even more gain than the system can currently provide might be desired. One way to improve the gain margin is to apply a feedback cancellation which works similar to the echo cancellation algorithms known from hands-free telephony. However, in the ICC scenario difficulties arise in continuously estimating the required impulse responses.

References

1. J.S. Bradley, H. Sato, M. Pickard, On the importance of early reflections for speech in rooms. JASA. **113**(6), 3233–3244 (2003)
2. T. Haulick, G. Schmidt, Signal processing for in-car communication systems. Signal Process. **86**(6), 1307–1326 (2006)
3. A. Ortega, E. Lleida, E. Masgrau, Acoustic echo control and noise reduction for cabin car communication. Proc. EUROSPEECH **3**, 1585–1588 (2001)
4. H. Haas, The influence of a single echo on the audibility of speech. J. Audio Eng. Soc. **20**, 145–159 (1972)
5. E. Meyer, G.R. Schodder, Über den Einfluß von Schallrückwürfen auf Richtungslokalisation und Lautstärke bei Sprache, *Nachrichten der Akademie der Wissenschaften in Göttingen*. Math-Phys. Kl. **6**, 31–42 (1952) (in German)
6. U. Zölzer, *DAFX: digital audio effects* (Wiley, New York, 2002)
7. H.H. Dam, S. Nordholm, A. Cantoni, J.M. de Haan, Iterative method for the design of DFT filter bank. IEEE Trans. Circ. Syst. II: Express Briefs **51**(11), 581–586 (2004)
8. C. Stöcker, T. Kurbiel, D. Alfsmann, H. Göckler, A novel approach to the design of oversampling complex-modulated digital filter banks. *J. Advances Sign. Process.* Article ID 692861, (2009)
9. T. Kurbiel, H. Göckler, Iteratively reweighted design of oversampling complex-modulated filter banks for high output signal quality. *Proc. EUSIPCO*. 2171–2175 (2010), Aalborg, Denmark

10. D. Mauler, R. Martin, Optimization of switchable windows for low-delay spectral analysis-synthesis. *Proc. ICASSP*. 4718–4721 (2010)
11. N. Fliege, Closed form design of prototype filters for linear phase DFT polyphase filter banks. *Proc. ISCAS*. **1**, 651–654 (1993)
12. P. Hannon, M. Krini, G. Schmidt, A. Wolf, in *Reducing the complexity or the delay of adaptive subband filtering. Proceedings of German Conference Speech Signal Processing (ESSV)*, (Berlin, Germany, 2010)
13. E. Hänsler, G. Schmidt, *Acoustic echo and noise control, a practical approach* (Wiley-Interscience, Hoboken, NJ, 2004)
14. M. Buck, A. Wolf, Model-based dereverberation of single-channel speech signals. *Proceedings of German Annual Conference on Acoustics (DAGA), Dresden, Germany*, 2008, pp. 261–262
15. E. A. P. Habets, Single- and multi-microphone speech dereverberation using spectral enhancement. *Ph.D. thesis, Eindhoven University*, Eindhoven, The Netherlands, 2007
16. E.A.P. Habets, S. Gannot, I. Cohen, Dereverberation and residual echo suppression in noisy environments, in *Speech and audio processing in adverse environments*, ed. by E. Hänsler, G. Schmidt (Springer, New York, 2008)
17. R. Martin, Noise power spectral density estimation based on optimal smoothing and minimum statistics. IEEE Trans. Speech Audio Process. **9**(5), 504–512 (2001)
18. E. Nemer, W. Leblanc, Single-microphone wind noise reduction by adaptive postfiltering. *IEEE Workshop on Applications of Signal Processing to Audio and Acoustics*, (2009), pp. 177–180
19. M. N. Schmidt, J. Larsen, F.-T. Hsiao, Wind noise reduction using non-negative sparse coding. *IEEE Workshop on Machine Learning for Signal Processing*, (2007), pp. 431–436
20. P. Hetherington, X. Li, P. Zakarauskas, Wind noise suppression system, *Pat No. US 7,885,420 B2*, (2011)
21. K. S. Petersen, G. Bogason, U. Kjems, T. B. Elmedyb, Device and method for detecting wind noise. *Pat No. US 7,340,068 B2*, (2003)
22. G. W. Elko, J. Meyer, S. Backer, J. Peissig, Electronic pop protection for microphones. *Appl. Signal Process. Audio Acoustics*. 46 –49 (2007)
23. J. Withopf, G. Schmidt, Suppression of non-stationary distortions in automotive environments. *Proc. of ITG* (Braunschweig, 2012)
24. J. Benesty, M. Sondhi, Y. Huang, Springer handbook of speech processing, Springer, Ch. 41, ed. by D. Reynolds, W. Campbell, W. Shen, E. Singer. Automatic language recognition via spectral and token based approaches. (2008) pp. 811–824

Chapter 8
Room in a Room: A Neglected Concept for Auralization

Markus Christoph

Abstract The term *Room in a Room* describes an old technique which can be utilized for auralization purposes. Thereby the auralization procedure can be divided into an analysis- and a synthesis task, usually dependent of each other. The analysis recordings in the source room have to be made exactly at those directions where loudspeakers are physically placed in the target room. In doing so, the spatial separation, respectively, filtering will be realized by beamforming. The outputs of the different, fixed beamformers provide the signals for the loudspeakers in the target room, representing the synthesis of this auralization technique. Different methods of how such a beamformer can robustly be designed will be presented in this chapter.

Keywords Auralization • Beamforming • 3D-beamformer • Modal beamforming

8.1 Introduction

Auralization techniques have been an object of interest since a long time. They can be used to create an illusion of being acoustically in a desired room, whilst, in reality, sitting in a completely different one. If this illusion becomes authentic, such techniques can be utilized, for example, for acoustical documentation or virtual tuning purposes. Especially in automotive acoustics, such a tool is highly desired, since up to now the majority of prototype cars are still individually tuned by an acoustician, mostly by hand, which is very time consuming. At the same time, automotive companies constantly cut time frames for such tasks, since working time on the prototypes is expensive, as many people from different disciplines have to work on them. By using auralization techniques one could measure the car once

---

M. Christoph (✉)
Harman/Becker Automotive Systems GmbH, Schlesische Str. 135, Straubing, Germany
e-mail: markus.christoph@harman.com

G. Schmidt et al. (eds.), *Smart Mobile In-Vehicle Systems: Next Generation Advancements*, DOI 10.1007/978-1-4614-9120-0_8,

and work, from this moment on, solely on this data, to get the car tuned. In doing so, valuable working time on a prototype can be reduced to a minimum, whilst at the same time; tuning time can be increased, eventually leading to a better acoustical result.

8.1.1 Background

There is a whole number of possibilities how auralization can be achieved, which can roughly be divided into two categories: headphone- and loudspeaker-based solutions. As representatives of the headphone solutions, a pure binaural recording and reproduction solution, as is more deeply described in [1], can be mentioned. Thereby the idea of the system, presented in [1], is to duplicate the binaural signal, as recorded directly at the eardrums within the source room by a headphone reproduction system. Thereby the headphones have to be adequately compensated, ideally with the inverse of the headphone transfer function, which can only be realized approximately, since this transfer function is usually not minimum phase. Despite its simplicity, this method is able to deliver a very authentic room impression. The BRS (Binaural Room Scanning) principle [2] is also based on binaural recordings, which are, in contrast to the before mentioned principle, not individualized. Due to the fact that BRS utilizes general HRTFs (Head Related Transfer Function) data, a head tracking system has to be applied in order to externalize the acoustic expression, or in other words to avoid in-the-head localizations. Thereby, the head tracker is able to measure the current head rotation. This information is then used to pick the most adequate, i.e., the two, in 2D applications, neighboring HRTFs out of a library, interpolate in between them and insert the result into a convolution machine. As a consequence an impression can be achieved that the stage does not move with the head during head rotations.

The topic of this chapter belongs to the category of loudspeaker-based auralization methods. There exist a number of different methods, from which only a few, which are considered to be the most promising ones, will briefly be mentioned in the following. In [3] the author utilized the *inverse filter theory* as three dimensional (=3D) sound reproduction technique. The aim of this technique is to compensate for any disturbing effects of the target room, by utilizing matrix inversion to calculate the necessary inverse filter matrix such, that in connection with the target room, a Dirac impulse, at the desired location within the target room, will ideally result. As soon as this task is successfully accomplished, it is easy to create any desired room impression by additionally inserting the room transfer matrix of the desired room into the system, which could, above all, be efficiently integrated into the already inevitable, inverse filter matrix. Despite its mathematical correctness, this principle suffers from diverse practical problems, such that the results are only valid at the location in the target room where the measurement had been applied but not at its close proximity. Furthermore, the acoustics are adversely affected by the inverse filter, as they usually show a great deal of *pre-ringing*.

The term *Ambisonics*, as described, e.g., in [4], stands for an analysis and synthesis method able to measure and reproduce spherical harmonics of a sound field. Thereby, the main task consists of creating a gain or filter matrix, jointly maximizing the energy as well as the particle velocity vector. Since Ambisonic is based on free-field conditions, it can only auralize the desired wave field as expected, if the target room is free of reverberation and/or early reflections, which is the case if the target room is an anechoic chamber. This problem is tackled by the *DirAC* (Directional Audio Coding) system, as introduced in [5], which can be looked upon as an extension of the previously mentioned Ambisonic system. Here, in addition to the extraction of the azimuth and elevation information from the B-Format signal, which act as input, as is the case in the Ambisonic system, spatial parameters, like the diffuseness are extracted during analysis and used in synthesis, with the objective of creating a realistic spatial impression. This of course can only approximate spatial impressions. If one wants to replicate the "real" sound field by loudspeakers, probably the most accurate way is to utilize Wave Field Synthesis (WFS), as disclosed, e.g., in [6], or High Order Ambisonics (HOA) as introduced, e.g., in [7], despite the fact that here one has to deal with still unsolved problems too, such as spatial aliasing effects. Another negative aspect in this concern is the enormous effort necessary to successfully run a WFS or HOA system, since a great number of (closely spaced) loudspeakers and the accompanying signal processing are necessary to create the desired effect.

The aim of the *Room in a Room* method, as introduced in this chapter, is to create a realistic sound field in the target room, in an easy and efficient way, thereby circumventing diverse problems inherent in some of the above-mentioned principles.

This chapter is structured as follows. In Sect. 8.2 the principles of the Room in a Room method is presented. Then, in Sect. 8.3 the first utilized beamforming technique is reviewed. Thereby, the underlying 3D microphone array is presented, as well as a new concept of how superdirective beamformers can effectively be combined by utilization of the presented 3D microphone array, which forms the "heart" of the whole method. Section 8.4 presents simulations of the novel beamforming technique and reconsiders its outputs. In Sect. 8.5 measurements of the novel beamformer are discussed. Finally, Sect. 8.6 summarizes this chapter.

8.2 Room in a Room

As one can perceive in Fig. 8.1, a microphone array is placed in the source room at a desired position, which acoustic should eventually be reproduced at a certain position within the target room. Thereby two types of recordings are feasible. Firstly it is always possible to directly record the desired sound, picked up by all microphones, which we will subsequently refer to as *signal-dependent recording*. This is necessary if one wants to "document" the acoustics, e.g., of an opus at a specific location within an opera. Secondly, if the sound which should be

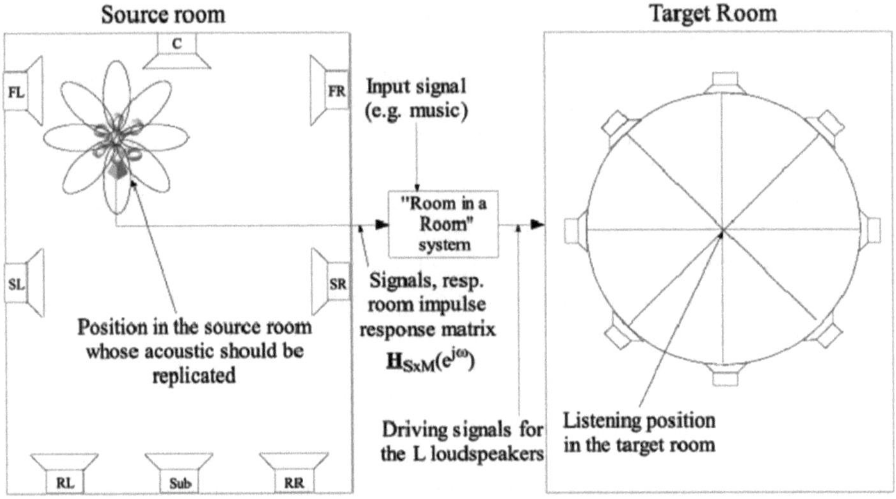

Fig. 8.1 Principle of the *Room in a Room* auralization method

reproduced stems from a sound system with a defined number and location of loudspeakers, as is the case in an automobile, it is reasonable to pick up the room impulse responses (RIR) from all S speakers to all M microphones of the microphone array in order to eventually create a signal-independent auralization system, which we will subsequently call *room-dependent recording*.

Independent of the type of recording, successively, based on the recorded data, beamforming filter will be applied, such that L beams, pointing exactly to the positions of the L loudspeakers, as located in the target room, will be created. Regarding a signal-dependent recording, further processing for the synthesis is not necessary. Figure 8.2 depicts all steps of the synthesis procedure, necessary for the room-dependent recording. In this process, based on the N input signals x, the driving signals for the S speakers (located in the target room) have to be calculated, by replication of the whole signal processing chain of the sound system, utilized in the target room. This may simply consist of a pure passive upmixing matrix $\underline{\mathbf{M}}_{N\times S}(e^{j\omega})$, broadcasting parts, combinations or originals of the N input signals to the S speakers. Afterwards, the influence of the transfer functions of the target room, stored in the RIR matrix $\underline{\mathbf{H}}_{S\times M}(e^{j\omega})$, will be considered during the course to create the M virtual microphone signals, as would be picked up by the microphone array if one were to play the desired input signal in the target room, utilizing its sound system and directly record these signals at a desired location by the M microphones of the microphone array. Finally, on grounds of these "virtual" microphone signals, L beams, pointing to the L loudspeakers, as located in the target room will be designed, whose output signals form their driving signals, the same as in the signal-dependent recording.

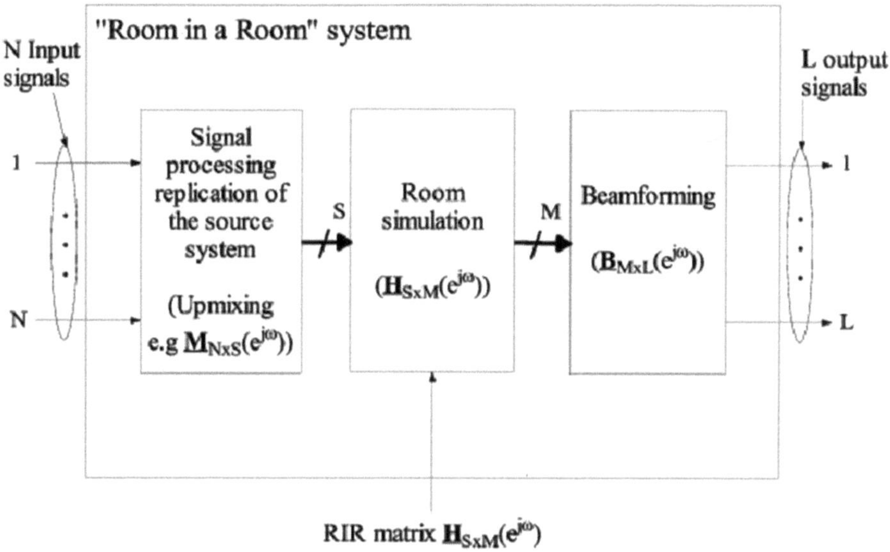

Fig. 8.2 Synthesis of room-dependent recordings

8.3 Beamforming

A microphone array consisting at least of two microphones, from which, at least one signal is filtered, with a successive filter, calculated such that a desired spatial filtering will eventually arise by combination of the processed microphone signals, is called a *beamformer*.

In the coordinate system, utilized for the design of a beamformer, as shown in Fig. 8.3, it is expected that all microphones are aligned along the x-axis. Furthermore, to design a beamformer with a beam pointing to a desired direction, e.g., as depicted by the vector $\mathbf{u}$ in Fig. 8.3, the direction of the beam will be assigned by its corresponding horizontal- (azimuth Θ) and vertical angle (elevation φ).

As revealed by Fig. 8.4, showing a beamformer, implemented in the spectral domain, the required signal processing can be divided into two parts, that is to say the so-called beamsteering on the one hand, which stands for a time delay compensation, necessary to ensure coherent, i.e., phase aligned addition of the microphone signals by $e^{j\omega\tau_i}$, and the filtering on the other hand by $\mathbf{A}(\omega)$, which performs the intrinsic spatial filtering. It should be noted that in Fig. 8.4, it is expected that free-field conditions be met, i.e., signals picked up by the microphones differ in their phasing but not in their amplitude.

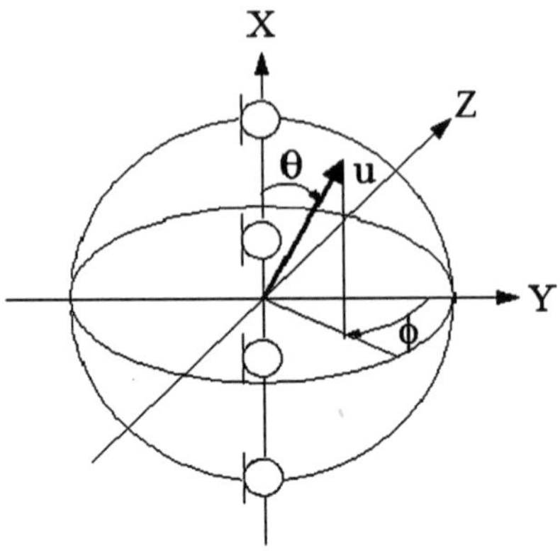

Fig. 8.3 Coordinate system utilized for the design of a beamformer

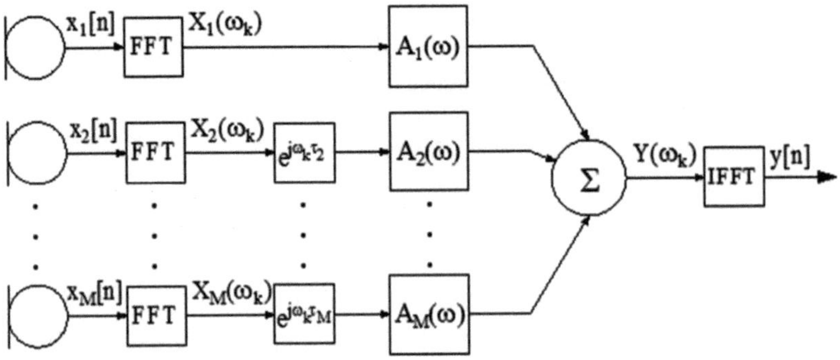

Fig. 8.4 Signal flow diagram of a beamformer, realized in the spectral domain

8.3.1 Design of Optimal Beamforming Filter

Optimal beamforming filter $\mathbf{A}(\omega)$ can generally be calculated as follow:

$$\mathbf{A}(\omega) = \frac{\underline{\varphi}_{NN}^{-1}(\omega)\mathbf{d}(\omega)}{\mathbf{d}^{\mathrm{H}}(\omega)\underline{\varphi}_{NN}^{-1}(\omega)\mathbf{d}(\omega)} \tag{8.1}$$

where

$\mathbf{A}(\omega)$=Vector holding the beamforming filter

$\mathbf{A}(\omega) = [\mathrm{A}_1(\omega), \ldots, \mathrm{A}_\mathrm{M}(\omega)]^T$, where $\mathbf{X}^\mathrm{T}$ denotes the transpose of $\mathbf{X}$,

$\underline{\varphi}_{NN}(\omega)$=Power spectral density (PSD) matrix of the background noise N,

$\mathbf{X}^\mathrm{H}$=Hermitian (conjugate transpose) of $\mathbf{X}$,

ω=Angular frequency in $\left[\frac{1}{s}\right]$ ($\omega = 2\pi f$),
$\mathbf{d}(\omega)$=Steering vector $\mathbf{d}(\omega) = [\mathrm{d}_1(\omega), \ldots, \mathrm{d}_\mathrm{M}(\omega)]^T$,
with

$$\underline{\varphi}_{NN}(\omega) = \begin{pmatrix} \varphi_{N_1N_1}(\omega) & \varphi_{N_1N_2}(\omega) & \cdots & \varphi_{N_1N_M}(\omega) \\ \varphi_{N_2N_1}(\omega) & \varphi_{N_2N_2}(\omega) & \cdots & \varphi_{N_2N_M}(\omega) \\ \vdots & \vdots & \ddots & \vdots \\ \varphi_{N_MN_1}(\omega) & \varphi_{N_MN_2}(\omega) & \cdots & \varphi_{N_MN_M}(\omega) \end{pmatrix} \tag{8.2}$$

where
M=Number of microphones,
and

$$d(\omega) = e^{-j\frac{\omega\left(\frac{M+1}{2}-n\right) d\cos(\Theta_0)}{c}} \tag{8.3}$$

where
$n \in [1, \ldots, M]$,
c=Speed of sound in $\left[\frac{m}{s}\right]$ $\left(c = 343\left[\frac{m}{s}\right] @ \vartheta = 20°\mathrm{C}\right)$,
Θ_0=Main receive direction, respectively direction, where the beam points in [rad].

In case, the sound source resides in the near field, the beam steering vector $\mathbf{d}(\omega)$ calculates to:

$$d(\omega) = \left[a_0 e^{-j(2\pi f \tau_0)}, a_1 e^{-j(2\pi f \tau_1)}, \ldots, a_{M-1} e^{-j(2\pi f \tau_{M-1})}\right] \tag{8.4}$$

where
a_0=Amplitude compensation value of the nth microphone signal,
τ_n=Time compensation value of the nth microphone signal,
with

$$a_n = \frac{\|q - p_{ref}\|}{\|q - p_n\|},$$

where
$\|q - p_{ref}\|$=Distance between the sound source q and the reference microphone p_{ref} in [m],
$\|q - p_n\|$=Distance between the sound source q and nth microphone p_n in [m].

Regarding a rule of thumb, one talk about far-field conditions if the sound source is located at a distance from the microphone array, exceeding twice its dimension, which is usually always the case, hence, (8.4) usually does not apply in practical applications.

8.3.2 Practical Modifications

According to Fig. 8.4 one usually excludes the beam steering vector $\mathbf{d}(\omega)$ from the design of the beamforming filter $\mathbf{A}(\omega)$. The beam steering is usually applied upstream the actual beamforming filter, i.e., first one calculates the delays and phase shifts, necessary for all microphones of the array, combined in the beam steering vector $\mathbf{d}(\omega)$, in order to let the resulting beam point to the desired direction, before the beamforming filter $\mathbf{A}(\omega)$ takes place. Thus the steering vector $\mathbf{d}(\omega)$ within (8.1) reduces to $\mathbf{d}(\omega) = \mathbf{1} = [1,1, \ldots,1]^T$.

In a further step, the cross-correlation matrix of the background noise $\underline{\varphi}_{NN}(\omega)$, which usually has to be measured continuously, or at least in situ, will be replaced by the complex coherence matrix of a diffuse noise field $\underline{\mathbf{\Gamma}}(\omega)$, for which a closed solution exists. This modification can be conducted, since measurements showed, that spatially homogeneous noise fields, as approximately apparent in automobiles, closely resemble a diffuse noise field.

Taking all these modifications into account, the design of the beamforming filter converts to:

$$\mathbf{A}(\omega) = \frac{\underline{\mathbf{\Gamma}}^{-1}(\omega)\mathbf{1}}{\underline{\mathbf{1}^{\mathrm{H}}\mathbf{\Gamma}}^{-1}(\omega)\mathbf{1}} \tag{8.5}$$

where

$\underline{\mathbf{\Gamma}}(\omega)$=Complex coherence of a diffuse noise field,

$\mathbf{1}$=Residual, respectively neutral steering vector $\mathbf{1} = M[1, 1, \ldots, 1]^T\ M$.

After the beam steering has been carried out, the complex coherence matrix of the diffuse noise field $\underline{\mathbf{\Gamma}}(\omega)$ calculates to:

$$\underline{\mathbf{\Gamma}}(\omega) = \begin{pmatrix} 1 & \Gamma_{X_1X_2}(\omega) & \cdots & \Gamma_{X_1X_M}(\omega) \\ \Gamma_{X_2X_1}(\omega) & 1 & \cdots & \Gamma_{X_2X_M}(\omega) \\ \vdots & \vdots & \ddots & \vdots \\ \Gamma_{X_MX_1}(\omega) & \Gamma_{X_MX_2}(\omega) & \cdots & 1 \end{pmatrix} \tag{8.6}$$

with

$$\Gamma_{X_iX_j}(\omega) = \mathrm{sinc}\left(\frac{\omega\,\mathrm{d_{ij}}}{\mathrm{c}}\right) e^{-j\frac{\omega\, d_{ij} \cos(\Theta_0)}{c}} \tag{8.7}$$

where

$i, j \in [1, \ldots,M]$,

sinc(x)=Sinc function $\left(\frac{\sin(x)}{x}\right)$,

d_{ij}=Element located at the ith row and jth column of the distance matrix $\underline{\mathbf{d}}$, and

$$\underline{\mathbf{d}} = \begin{pmatrix} 0 & d & \cdots & (M-1)d \\ -d & 0 & \cdots & (M-2)d \\ \vdots & \vdots & \ddots & \vdots \\ -(M-1)d & -(M-2)d & \cdots & 0 \end{pmatrix} \tag{8.8}$$

where
d=Inter-microphone distance in [m] of the equidistant microphone array.

8.3.3 Constrained Design

Both design rules, shown in (8.1) and (8.5) deliver the same, optimal beamforming filter $\mathbf{A}(\omega)$, in a diffuse noise field. Unfortunately neither the one nor the other can be applied without any further modifications, considering inevitable, practical limits, such as manufacturing tolerances, or variations in the placement of the microphones. These incertitudes have been considered in [8] by the addition of a small scalar μ to the elements at the main diagonal of the cross-correlation matrix $\underline{\varphi}_{NN}(\omega)$, or as proposed in [9] to the coherence matrix of a diffuse noise field $\underline{\mathbf{\Gamma}}(\omega)$. Another version, disclosed in [10], directly considers the inaccuracies in the design of the beamforming filter, leading to a constrained filter design as follows:

$$\mathbf{A}(\omega) = \frac{\left(\underline{\mathbf{\Gamma}}(\omega) + \mu(\omega)\underline{\mathbf{I}}\right)^{-1}\mathbf{d}(\omega)}{\mathbf{d}(\omega)^{\mathrm{H}}\left(\underline{\mathbf{\Gamma}}(\omega) + \mu(\omega)\underline{\mathbf{I}}\right)^{-1}\mathbf{d}(\omega)} \tag{8.9}$$

where
$\mathbf{d}(\omega)$=Steering vector (=**1**, if previously applied),
$\underline{\mathbf{I}}$=Identity matrix in the size of $\underline{\mathbf{\Gamma}}(\omega)$,
$\mu(\omega)$=Regularization parameter.

The value of the regularization parameter $\mu(\omega)$, which is now frequency dependent, and not a scalar as, e.g., in [8], depends on the MSE[1] of the imprecision of the placement of the microphones ($=\delta(\omega)^2$) within the array, but mainly on the MSE of the inter-microphone tolerances ($=\varepsilon(\omega,\Theta)^2$). The higher the quality of the microphone array, i.e., the lower the complete MSE ($=\Delta(\omega,\Theta)^2$), the smaller the value for the regularization parameter $\mu(\omega)$ can be. Practical values reside within a range of: $\mu(\omega) = [-40, \ldots, 40]$ [dB].

[1] MSE = Mean Squared Error.

The susceptibility $K(\omega)$ of a beamformer, given as:

$$\mathrm{K}(\omega) = \frac{\mathbf{A}(\omega)^H \mathbf{A}(\omega)}{\left|\mathbf{A}(\omega)^{\mathrm{H}} \mathbf{d}(\omega)\right|}, \tag{8.10}$$

describes the sensitivity of a beamformer regarding tolerances of the corresponding microphone array. Aim of the constraint algorithm is to design a robust beamformer by limiting the susceptibility to a maximal value $K_{\mathrm{Max}}(\omega)$. After [10], this upper limit $K_{\mathrm{Max}}(\omega)$ directly results from the total MSE of the microphone array $\Delta(\omega,\Theta)^2$ and the maximum tolerable deviations of the directional diagram $\Delta\Psi(\omega,\Theta)$, with the directional diagram $\Psi(\omega,\Theta)$ given as:

$$\Psi(\omega, \Theta) = \frac{\varphi_{y,y}(\omega, \Theta)}{\varphi_{x_{ref},x_{ref}}(\omega, \Theta)} = \left| \sum_{n=1}^{M} \mathbf{A}(\omega)\, e^{j \frac{\omega\, d \left(\frac{(M+1)}{2} - n\right) \left(\cos(\Theta_0) - \cos(\Theta)\right)}{c}} \right|^2 \tag{8.11}$$

where
$\varphi_{y,y}(\omega,\Theta)$=Auto power spectral density of the beamformer output signal y,
$\varphi_{x_{ref},x_{ref}}(\omega, \Theta)$=Auto power spectral density of the reference microphone signal x_{ref}.

For the total MSE of the microphone array holds:

$$\Delta(\omega, \Theta)^2 = \varepsilon(\omega, \Theta)^2 + \delta(\omega)^2 \tag{8.12}$$

with

$$\varepsilon(\omega, \Theta)^2 = E\left\{ \frac{\left|\Delta H_n^M(\omega, \Theta)\right|^2}{\left|H_0^M(\omega, \Theta)\right|^2} \right\} \tag{8.13}$$

where
$|H_0^M(\omega,\Theta)|^2$=Nominal, respectively mean transfer function of all microphones,
$|\Delta H_n^M(\omega,\Theta)|^2$=Deviation of the transfer function of the nth microphone from the nominal transfer function,
$E\{.\}$=Expectation operator,
and

$$\delta(\omega)^2 = \left(\frac{\omega}{c}\right)^2 \frac{\sigma^2}{3} \tag{8.14}$$

where
σ=Variance of the zero-mean, normally distributed positioning error of the microphone, equal for each dimension, hence the scaling by $\frac{1}{3}$.

Tolerances in the microphone array can be considered in the directional diagram $\Psi(\omega,\Theta)$ by addition of an error term, represented by $\Delta\Psi(\omega,\Theta)$ to it, resulting in:

$$E\left\{\widetilde{\Psi}(\omega,\Theta)\right\} = \Psi(\omega,\Theta) + \Delta\Psi(\omega,\Theta) \tag{8.15}$$

with

$$\Delta\Psi(\omega,\Theta) = \Delta(\omega,\Theta)^2 K(\omega) \leq \Delta\Psi_{\text{Max}}(\omega,\Theta), \tag{8.16}$$

which must not exceed a certain threshold, provided by $\Delta\Psi_{\text{Max}}(\omega,\Theta)$.

By inserting (8.12) and (8.14) in (8.16), it follows a maximally tolerable susceptibility $K_{\text{Max}}(\omega,\Theta)$ of:

$$K_{\text{Max}}(\omega,\Theta) = \frac{\Delta\Psi_{\text{Max}}(\omega,\Theta)}{\varepsilon(\omega,\Theta)^2 + \left(\frac{\omega}{c}\right)^2 \frac{\sigma^2}{3}} \tag{8.17}$$

The following practical simplifications can be applied to (8.17):

- Due to the fact that $\varepsilon(\omega,\Theta)^2$ hardly varies with Θ it suffices to determine $\varepsilon(\omega,\Theta)^2$ at a certain receive direction. Thereby the main receive direction Θ_0 is usually selected, which is $\Theta_0 = 90°$ in broadside and $\Theta_0 = 0°$ in endfire alignment of the beamformer.
- Inaccuracies in microphone placements, represented by $\delta(\omega)^2$, are much less probable then variations in the transfer functions of the array microphones, provided by $\varepsilon(\omega,\Theta)^2$. As such it suffices to consider mechanical deviations by a general value of, e.g., $\delta(\omega)^2 = 1\%$.
- A dependency on Θ of $\Delta\Psi_{\text{Max}}(\omega,\Theta)$ only makes sense, if one is interested in an exact reconstruction of the whole directional pattern, i.e., also of all side lobes, which is usually not the case. By taking a maximal, Θ-independent value $\Delta\Psi_{\text{Max}}(\omega)$, an almost perfect replication of the directional pattern in the main direction can still be obtained. Thereby, $\Delta\Psi_{\text{Max}}(\omega)$ can be determined by taking the maximum side lobe value of the ideal directional pattern. Furthermore, dependent on the use case, $\Delta\Psi_{\text{Max}}(\omega)$ could also be utilized as a frequency-independent threshold, e.g., $\Delta\Psi_{\text{Max}} = 15[\text{dB}]$.

Taking all above-mentioned items into consideration, (8.17) simplifies to:

$$K_{\text{Max}}(\omega) = \frac{\Delta\Psi_{\text{Max}}(\omega)}{\varepsilon(\omega)^2 + 1\%} \tag{8.18}$$

with

$$\varepsilon(\omega)^2 \geq 1\% \tag{8.19}$$

Based on the previous findings, the following, iterative, constrained algorithm for the design of the beamforming filter $\mathbf{A}(\omega)$, eventually leading to a robust, *superdirective beamformer* can be derived:

1. Preliminaries:
   (a) Determine $\Delta\Psi_{\mathrm{Max}}(\omega)$, based on the maximum values of the side lobes of the desired, ideal beamformer, over frequency (for $M = 3$: $\Delta\Psi_{\mathrm{Max}}(\omega) \approx -9.5$ [dB]).
   (b) Measure all transfer functions of the microphones $H_n^M(\omega)$ at the desired main direction Θ_0. Afterwards, use (8.13) to calculate $\varepsilon(\omega)^2$, thereby taking (8.19) into account.
2. Calculation of the maximum allowable susceptibility $K_{\mathrm{Max}}(\omega)$, utilizing (8.18).
3. As initialization for the iteration, use $\mu(\omega) = 1$.
4. Calculation of the beamforming filter $\mathbf{A}(\omega)$, utilizing (8.9).
5. Based on the beamforming filter $\mathbf{A}(\omega)$, calculated in step 4, calculate the current susceptibility $K(\omega)$, utilizing (8.10).
6. Increase the regularization parameter $\mu(\omega)$, if $K(\omega) > K_{\mathrm{Max}}(\omega)$, otherwise decrease $\mu(\omega)$, e.g., by $\Delta\mu = 10^{-5}$.
7. Repeat steps 4–6 until $K(\omega)$ approaches $K_{\mathrm{Max}}(\omega)$ as close as possible or if $\mu(\omega)$ drops below a certain lower threshold, given, e.g., by $\mu_{\mathrm{Min}} = 10^{-8}$, which is usually the case at higher frequencies $f \geq \frac{c}{2d}$.

8.4 Microphone Array

The susceptibility $K(\omega)$ of a beamformer mainly depends on the deviations of the inter-microphone transfer functions $\varepsilon(\omega)^2$, as discussed in Sect. 8.3.3. To enhance the quality of the beamformer, these differences have to be kept as small as possible. Therefore, so called *matched* or *paired microphones* have been used during the construction of the microphone array, which frame, by the way, is shown in Fig. 8.1. For this purpose the transfer functions of 100 microphone capsules (Panasonic WM-62a) have been measured at $\Theta_0 = 0°$ in an anechoic chamber, from which the 7, best matching capsules have been chosen, as shown in Fig. 8.5.

Since the sensitivity of a beamformer against tolerances decreases with increasing frequency, a frequency-dependent weighting function, provided by a nonlinear smoothing filter (e.g., $\frac{1}{3}$ octave filter) has been applied during the selection process, prior to the calculation of the difference matrices.

An analysis of the inter-microphone differences revealed, that due to the selection process, the deviation could be decreased from primarily $\pm$ 3 [dB], as provided by the data sheet of the manufacturer, to $\pm$ 0.5 [dB], as shown in Fig. 8.6, corresponding to an value of $\varepsilon(\omega)^2 < 0.7\%$, which is already below the lower limit as noted in (8.19).

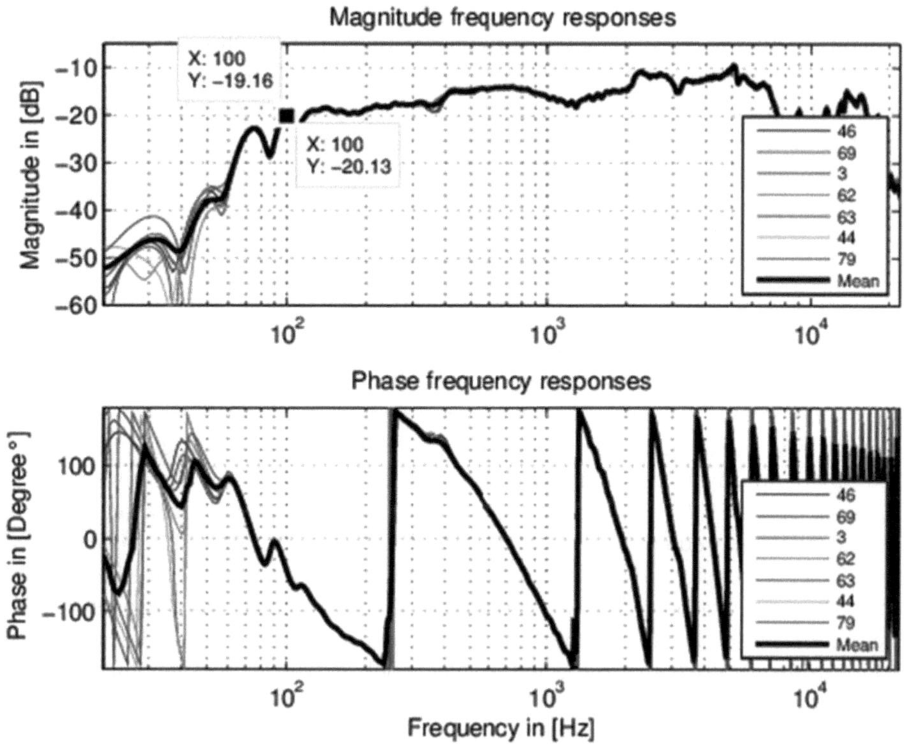

Fig. 8.5 Bode diagram of the 7 best matching microphone capsules

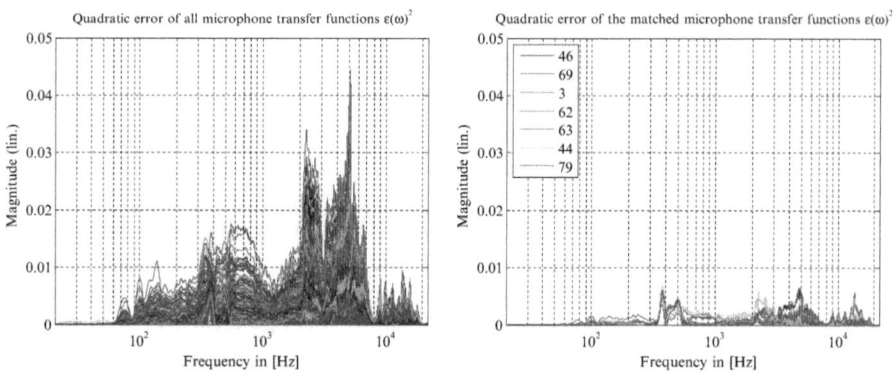

Fig. 8.6 Quadratic error $\varepsilon(\omega)^2$ of all microphones (*left* figure) and of the 7 best matching microphones (*right* figure)

Due to the fact, that the beamformer ought be used in an auralization method, it should ideally be able to work throughout the whole audio frequency range: $f \approx [20, \ldots, 20000]$ [Hz]. With only one compact microphone array, this task cannot be accomplished. The best probable compromise for this purpose, had been found by utilizing a superdirective beamformer, ideally showing a frequency-independent directivity pattern up to the *spatial aliasing frequency*, which calculates to:

$$f \leq \frac{c}{2d} \tag{8.20}$$

As can be seen in (8.20), the spatial aliasing frequency solely depends on the inter-microphone spacing d, thus the dimension of the microphone array should be kept small to enlarge the frequency range of operation. Since the utilized capsules already have a diameter of $\varnothing = 6$[mm] and the fact that the frame cannot be made too small, to ensure a minimum of mechanical robustness, a inter-microphone distance of $d = 1.25$[cm] has been applied for the microphone array, leading to a spatial aliasing frequency of $f = 13600$[Hz] for an array in endfire orientation, which can be considered as sufficient for our purpose. In order to let the beam point in any room direction, a 3D[2] arrangement of the microphones was mandatory. For that reason, three linear microphone arrays, each consisting of three microphones, were arranged along the X, Y, and Z axes, each sharing the center microphone, resulting in an array with 7 microphones. Depending on the direction where the beam shall point at, a beamformer for each of the three linear arrays will be calculated, either in endfire or broadside orientation, resulting in the desired beamformer by combination of the three individual beamformers. Hence, considering (8.3) and (8.5), the final superdirective beamformer calculates to:

$$\mathbf{A}_{\mathrm{Out}}(\omega) = \frac{1}{3}\left(\mathbf{A}_x^T(\omega)\mathrm{diag}\{\mathbf{d}_{\mathrm{x}}(\omega)\} + \mathbf{A}_y^T(\omega)\mathrm{diag}\{\mathbf{d}_{\mathrm{y}}(\omega)\} + \mathbf{A}_z^T(\omega)\mathrm{diag}\{\mathbf{d}_{\mathrm{z}}(\omega)\}\right) \tag{8.21}$$

where
diag{**X**}=Diagonal matrix of vector **X**.

Because all beams of the three linear arrays, point as close as possible, to the desired direction, the resulting beam will completely point to this location, as depicted in Fig. 8.7.

Each of the three beamformers exhibit a different aliasing pattern, which, as a matter of fact will also be combined, resulting in the positive effect, that the combined beamformer shows much less disturbing aliasing effects compared to each of the three individual beamformer, on which it is based on, as can be seen in Fig. 8.8.

[2] 3D = Three dimensional.

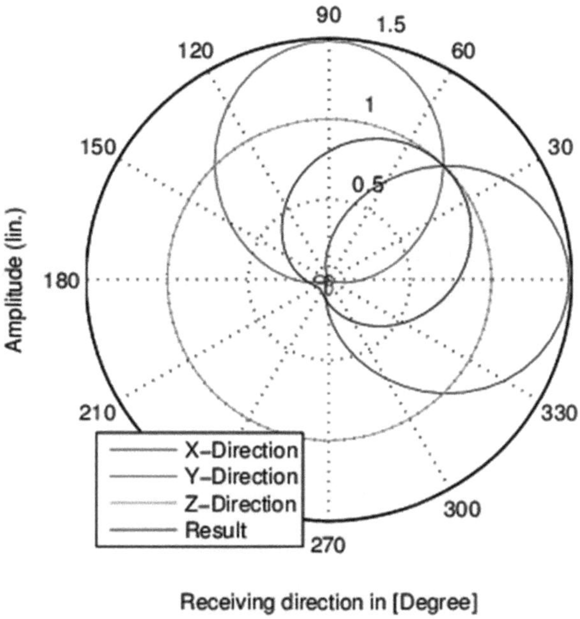

Fig. 8.7 Polar diagram along the *X*/*Y*-plane of the three linear microphone arrays as well as of the resulting beamformer at $f = 1$[kHz], steered to $\varphi = 0°$ and $\Theta = 45°$

Thereby the top plots of Fig. 8.8 affirm a spatial aliasing frequency of $f = 13600$ [Hz] on the one hand and reveal that only at those regions in the spatial–spectral domain where the aliasing products of the individual beamformers overlap, the resulting beamformer shows aliasing too, which as a matter of fact appears enhanced. All other regions in the spatial–spectral domain have been suppressed during the overlapping process of the individual beamformer, as indicated by (8.20), leading to a final beamformer, exhibiting a higher spatial aliasing frequency, smaller aliasing regions within the spatial–spectral domain as well as a narrower beam width, as any of the underlying beamformers. With a mean squared error of the microphone transfer functions of $\varepsilon(\omega)^2 = 1\%$, which could, as previously shown, be achieved by sorting of the microphone capsules, a fix mean squared error of the microphone placement of $\delta(\omega)^2 = 1\%$ and a maximum deviation of the directivity pattern of $\Delta\Psi_{\text{Max}}(\omega) \approx -9.5$[dB], for $M = 3$, a maximum susceptibility of $K_{\text{Max}}(\omega) \approx 16.75$ results, leading to a frequency from which on the beamformer can be considered as superdirective, of $f \approx 150$[Hz], which is regarded as acceptable, since small rooms, such as the interiors of automobiles, behave more like a pressure chamber, leading to a distinct modal acoustical behavior up to a certain frequency. This transition frequency, known as *Schröder frequency*, given as:

$$f_t = 2000\sqrt{\frac{T}{V}} \tag{8.22}$$

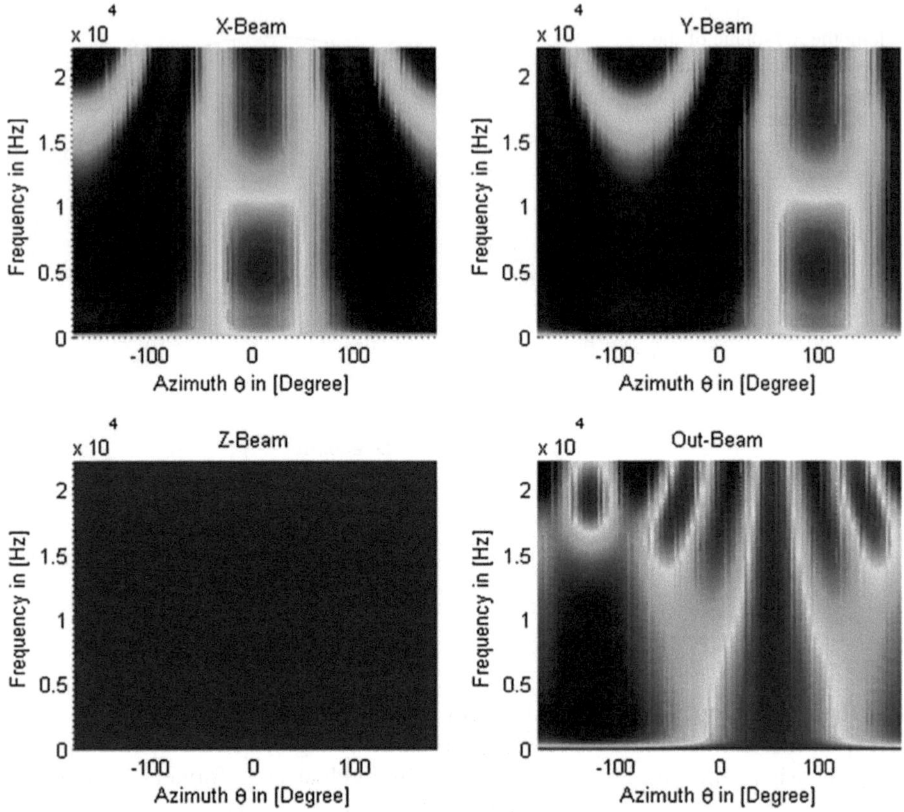

Fig. 8.8 Top view of the *X*-, *Y*-, and *Z*- as well as of the resulting, superdirective beamformer at $\varphi = 0°$ and $\Theta = 45°$

with
f_t=Schröder, or transition frequency in [Hz],
T=Reverberation time [s] (usually $T = T_{60}$),
V=Volume in the enclosure in [m^3],
calculates for a typical medium-class car environment, with $V \approx 3.5[\mathrm{m}^3]$ and $T_{60} \approx 0.08[\mathrm{s}]$ to:

$$f_t = 300[\mathrm{Hz}] \tag{8.23}$$

which is about twice the number of the previously determined, lower frequency bound of our final beamformer for superdirectivity. Hence it can be concluded that the novel, 3D microphone array, presents an adequate measuring device for broadband acoustical recordings.

8.5 Measurements

Measurements conducted in an anechoic chamber where used to verify the theory as described in the preceding chapters. Thereby, impulse responses from a broadband speaker with a membrane diameter of $\varnothing = 10$[cm], located 1[m] away from the center of the microphone array, to all 7 microphones had been gathered in the horizontal plane ($\varphi = 0°$) in 90° steps, i.e., for $\Theta = [0°, 90°, 180°, 270°]$, utilizing the exponential sine sweep technique as disclosed in [11].

In the first row of Fig. 8.9 one can see the behavior of the X beamformer, i.e., of a superdirective Beamformer in endfire direction, measured in four different orientations. At $\Theta = 0°$ its response should ideally follow the response of an omnidirectional microphone, represented by the reference microphone, located in the center of the microphone array, denoted as "RefMic" in Fig. 8.9, whereas at $\Theta = 180°$ the least amount of signal energy will be picked up. The second row shows the results of the Y beamformer, which forms a superdirective beamformer in broadside direction, measured at the same four orientations. Here one would expect equal responses with a maximum gain at $\Theta = [0°, 180°]$ and a minimum gain at $\Theta = [90°, 270°]$, with a gap, slowly increasing with frequency, which, in reality, is indeed the case.

The Z beamformer also shows this characteristics but along the vertical axis. Along the horizontal axis, ideally no deviation should occur, which again holds true, as the measurements show. In the last row the behavior of the novel beamformer, resulting out of the overlap of the X, Y, and Z beamformer is shown. In principle it exhibits a similar behavior to the X beamformer, but with much less directivity in the low- and mid-frequency region, relativizing the practicability of the new beamforming technique.

8.6 Conclusions

With the novel beamforming structure a beam, pointing at any desired position in a room, can easily be formed. This can solely be accomplished via software, i.e., by utilization of different beamforming filter. Following the preluding example, Fig. 8.10 shows the result of the novel beamforming technique.

Unfortunately the final beam shows a directivity factor, which is, compared to a superdirectional beamformer, directly pointing to a desired direction, inferior, which is true, especially at low and mid-frequencies. Doubtless, there will be applications for the novel beamforming structure, but for auralization purposes it appears logical to use robustly designed superdirectional beamformer for each speaker, as located in the target room, instead. In our example with 8 speakers,

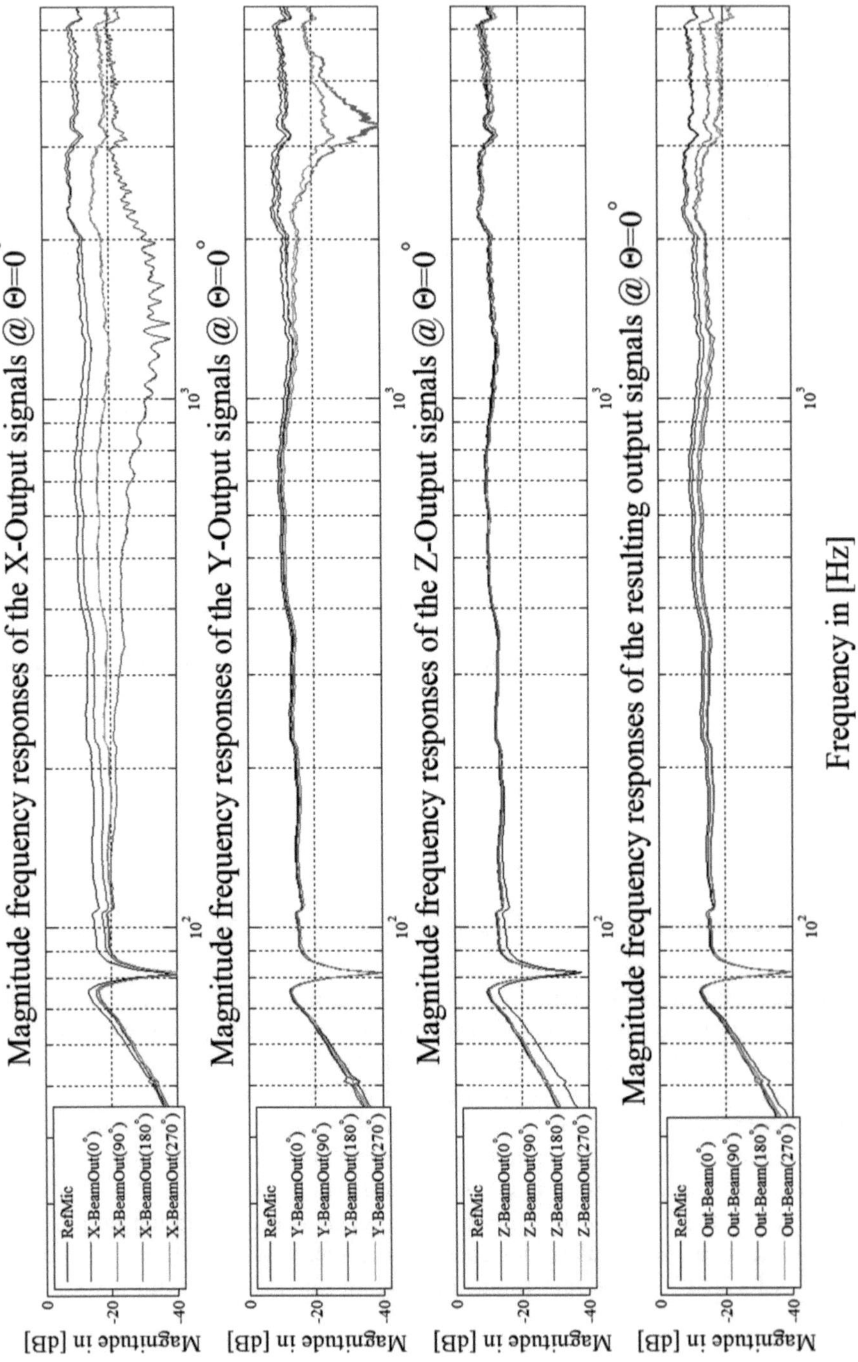

Fig. 8.9 Magnitude frequency responses of the *X*-, *Y*-, and *Z*- as well as of the resulting, beamformer at $\varphi = 0°$ and $\Theta = 0°$

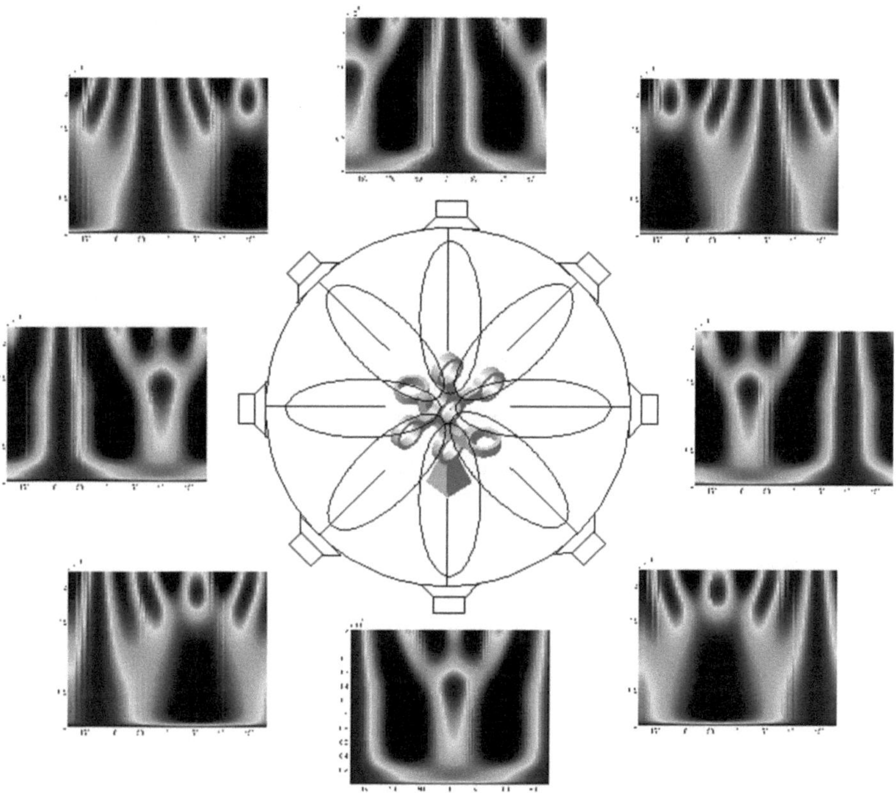

Fig. 8.10 Result of the novel beamforming technique, following the preluding example, i.e., beams pointing at $\varphi = 0°$ and $\Theta = [0°, 45°, 90°, 135°, 180°, 225°, 270°, 315°]$

regularly arranged in the target room, this would mean to measure the horizontal plane in the source room twice—one time with the X beam pointing at $\Theta = 0°$ and the other time oriented at $\Theta = 45°$. Applicability of this method, especially regarding the introduced "Room in a Room" concept, remains a task for the future.

References

1. D. Griesinger, Binaural techniques for music reproduction, *8th International Converence of the AES*, May 1990
2. P. Mackensen, U. Felderhoff, G. Theile, U. Horbach, R. Pellegrini, Binaural room scanning—a new tool for acoustic and psychoacoustic research. *DAGA*. (1999)
3. J. Garas, *Adaptive 3D sound systems* (Kluwer Academic Publishers, Dordrecht, The Netherlands, 2000)
4. B. Wiggins, An investigation into the real-time manipulation and control of three-dimensional sound fields, *Doctoral Thesis*, University of Derby, Derby, UK, 2004

5. J. Vilkamo, T. Lokki, V. Pulkki, Directional audio coding: virtual microphone-based synthesis and subjective evaluation. J. Audio Eng. Soc. **57**(9), 709–724 (2009)
6. S. Spors, H. Buchner, R. Rabenstein, W. Humboldt, Active listening room compensation for massive multichannel sound reproduction systems using wave-domain adaptive filtering. J. Acoust. Soc. Am. **122**(1), 354–369 (2007)
7. J. Daniel, R. Nicol, S. Moreau, Further investigations of high order Ambisonics and wave field synthesis for holophonic sound imaging, *114th AES Convention*, Springer, Amsterdam, The Netherlands, Mar 2003
8. E. Gilbert, S. Morgan, Optimum design of directive antenna arrays subject to random variations. *Bell Syst. Tech. J.* 637–663 (1955)
9. J. Bitzer, K. D. Kammeyer, K. U. Simmer. An alternative implementation of the superdirective beamformer, *IEEE Workshop on Applications of Signal Processing to Audio and Acoustics*, New Paltz, New York, Oct 1999
10. M. Dörbecker, Mehrkanalige Signalverarbeitung zur Verbesserung akustisch gestörter Sprachsignale am Beispiel elektronischer Hörhilfen, *Doctoral Thesis*, IND, RWTH Aachen, No. 10, ed. by P Vary, Aachener Beiträge zu Digitalen Nachrichtensystemen (ABDN), Mainz in Aachen, (1998)
11. A. Farina, Simultaneous measurement of impulse response and distortion with a swept-sine technique, *110th AES Convention*, Paris, France, Feb 2000

Chapter 9
Refinement and Temporal Interpolation of Short-Term Spectra: Theory and Applications

Mohamed Krini and Gerhard Schmidt

Abstract In this contribution, methods for spectral refinement (SR) and spectral interpolation (SI) are presented. These methods can be implemented as a post-processing stage after conventional frequency analyses such as overlap add-based decomposition schemes. The principle idea of SR is to individually refine each subband signal after frequency decomposition and to compute additional frequency-supporting points in between using a linear combination of a few neighboring (in terms of time and frequency) subband signals. For efficient implementation, a simplification of the SR method has been derived—it has been shown that the refinement can easily be implemented using short FIR filters in each subband, resulting in a very low computational complexity. The SI method exploits the redundancy of succeeding short-term spectra for computing interpolated temporal supporting points in between the originally generated frames. This is achieved by efficient approximations, and the whole method can be realized on weighted sums of subband signals. The new interpolation method can be applied in adaptive system identification schemes (e.g., echo cancellation or channel estimation), allowing for a significant increase of the frameshift (subsampling rate). This leads to a reduction of the computational complexity while keeping the convergence speed and the steady-state performance constant. Alternatively, the frameshift can be kept the same. In this case an improved steady-state convergence can be achieved. The proposed method for SR has been applied as a preprocessing stage for fundamental frequency (pitch frequency) estimation, and the SI method has been utilized for subband echo cancellation. Evaluations have shown that pitch frequency estimation

M. Krini (✉)
Acoustic Speech Enhancement—Research, Nuance Communications,
Soeflinger Str. 100, 89077 Ulm, Germany
e-mail: mokr@tf.uni-kiel.de

G. Schmidt
Digital Signal Processing and System Theory, Christian-Albrechts-Universität zu Kiel,
Kaiserstr. 2, 24143 Kiel, Germany
e-mail: gus@tf.uni-kiel.de

G. Schmidt et al. (eds.), *Smart Mobile In-Vehicle Systems: Next Generation Advancements*, DOI 10.1007/978-1-4614-9120-0_9,

can be improved significantly for all considered signal-to-noise ratios when employing the SR method. Real-time measurements performed on systems for acoustic echo cancellation have demonstrated that significant improvements in terms of echo reduction can be achieved while only marginally increasing the amount of required memory.

Keywords Echo cancellation • Filterbank • Pitch frequency • Spectral refinement • Speech enhancement

9.1 Introduction

In different applications such as automotive hands-free telephony or speech dialogue systems, the desired speech signal is disturbed by background noise (engine, wind noise, etc.) and by echoes (due to multipath propagation from a loudspeaker to a microphone). In order to reduce the disturbing components while keeping the speech signal as natural as possible speech enhancement algorithms are utilized. Most often enhancement algorithms like noise reduction or echo cancellation are applied in the subband domain to reduce computational complexity and to achieve a higher convergence for adaptive filtering [1, 2].

An echo cancellation unit within a speech processing system tries to estimate the impulse response of a loudspeaker-enclosure-microphone (LEM) system. For estimating the echo components in the subband domain, the microphone signal as well as the reference signal are usually first segmented into overlapping blocks of appropriate size (segments of 20–30 ms are often overlapped by 50–75 %) and subsequently weighted with a window function. Each block is transformed into the short-term frequency (subband) domain. The resulting reference subband signals are convolved with adaptively estimated LEM subband impulse response to obtain the subband echo signals [14]. These estimated signals are subtracted from the microphone subband signal to determine the error signals for the filter update.

Afterwards, a Wiener-type filter can be utilized to reduce remaining echo components (e.g., if the estimated LEM impulse response has estimation errors) as well as stationary background noise [1, 5]. After several subsequent signal-processing stages (e.g., feature extraction schemes such as pitch frequency for voice classification) the enhanced short-term spectrum (STS) is converted back to the time domain by an inverse DFT and appropriate windowing. The resulting overlapped signal blocks are added together to obtain the broadband output signal. Further details about subband signal processing can be found, e.g., in [9]. This type of overlap-add-based scheme is also well known as a DFT-modulated subsampled filterbank.

For windowing often Hann sequences are applied. They achieve perfect reconstruction for appropriate subsampling factors. In addition they show good aliasing properties which are important for adaptive subband filtering such as echo cancellation. However, the windowing of successive signal blocks has most often the negative effect that a significant frequency overlap of adjacent DFT subbands arises. Thus, adjacent fundamental frequency trajectories are sometimes hard to separate

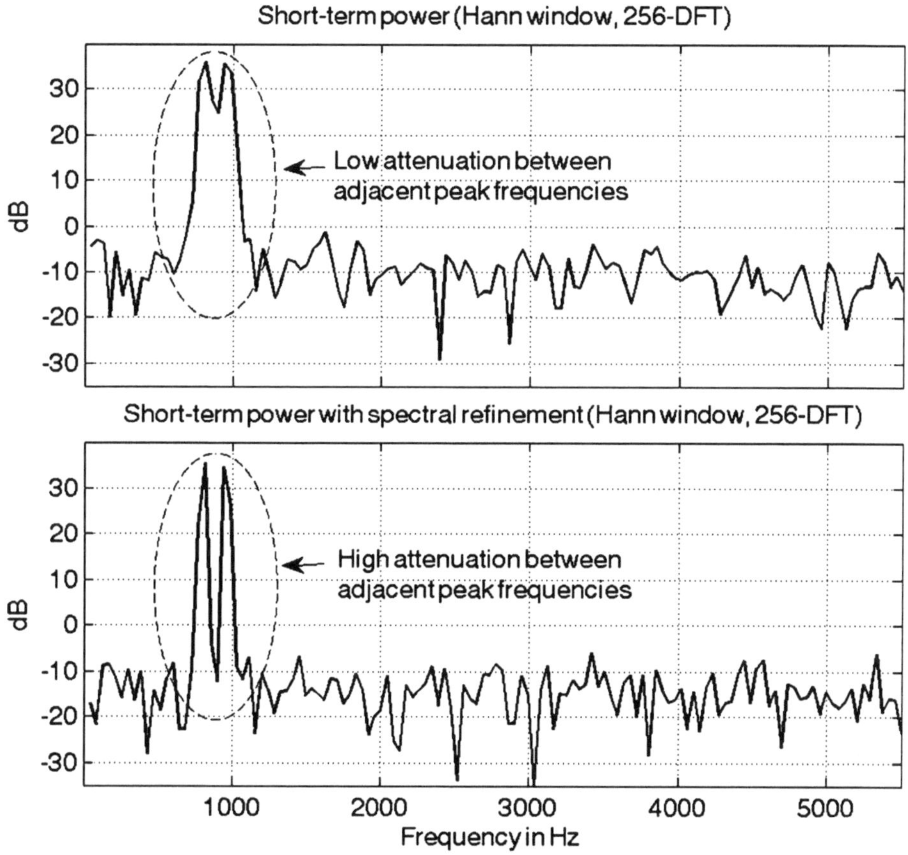

Fig. 9.1 Short-term power spectra of two sine signals added with white noise. The frequency distance between the two sine signals was set to 120 Hz

which is important for speech enhancement schemes that involve fundamental frequency estimation. In addition, due to the non-ideal filters large subsampling factors aliasing components within subbands can remarkably degrade the convergence behavior of the echo cancellation [7].

When increasing the DFT order to reduce the spectral overlap and aliasing effects one should consider that for hands-free telephone systems several restrictions have to be fulfilled. For example the front-end delay of a hands-free system connected to a public telephone network should not exceed 39 ms [6]. However, increasing, e.g., the DFT order from 1 $N = 256$ to $N = 512$ at a sampling frequency of $f_s = 11{,}025$ Hz results in a delay of approximately 46 ms in the signal path, which does not fulfill the ITU and the ETSI specifications any more. To overcome this, the herein proposed method for SR can be utilized. It is applied as a linear combination of a few weighted subband signal vectors at the output of a DFT.

Figure 9.1 shows, for illustration purposes, the estimated short-term power spectrum of two sine signals added with white noise, using a Hann window and a

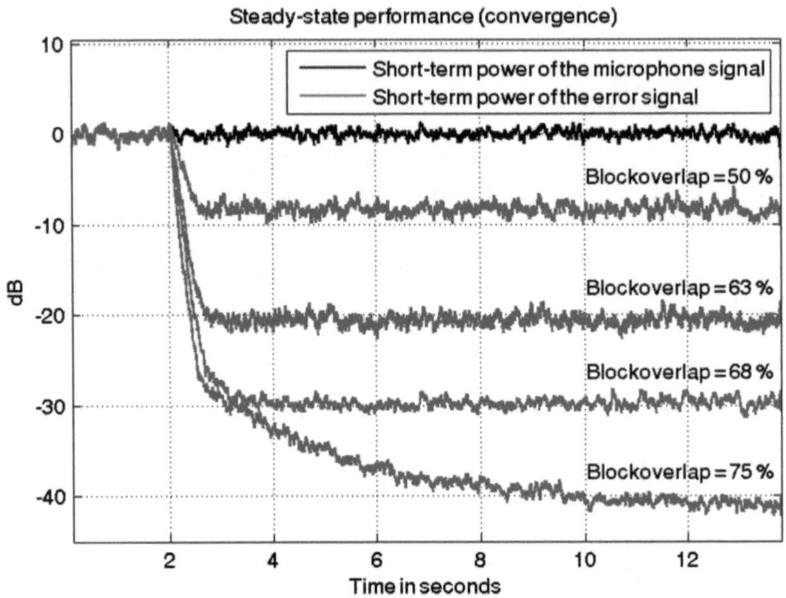

Fig. 9.2 Performance of echo attenuation for $N = 256$ and different subsampling rates (frameshifts) of 64, 82, 96, and 128

DFT of order 256. The distance between the two combined sine signals has been set to 120 Hz, corresponding to the average fundamental frequency of a male speaker. The upper graph illustrates the analysis without spectral refinement. However, the two sine signals cannot be separated in an effective manner; only a low attenuation between adjacent peak frequencies is achieved. The curve in the lower part of Fig. 9.1 depicts the short-term power with spectral refinement as a post-processing scheme after a conventional analysis filterbank/DFT. As a result a high attenuation between adjacent peak frequencies by about 45 dB is achieved. The derivation of the SR method is presented in Sect. 9.4.

For adaptive system identification in the subband domain—as it is used for echo cancellation—the analysis filterbanks are required to generate a very low amount of aliasing. Increasing the subsampling rate (i.e., frameshift) leads to reduced computational complexity. However in this case also the aliasing components within subband signals are increased—leading to low echo reduction. For practical purposes a compromise between performance and computational complexity has to be found. Figure 9.2 demonstrates the normalized power of the microphone signal and its corresponding error signal (the outcome after echo cancellation) for different block-overlap values within a range of 50–75 % (frameshift = 25–50 % of FFT order). As a test signal white noise was utilized and no local disturbances were considered for this experiment. However, for larger subsampling rates the required amount of echo attenuation (about 30 dB [12]) cannot be achieved anymore.

Therefore, a new temporal interpolation method is proposed in the following sections which are able to enhance significantly the convergence behavior for larger subsampling rates.

The rest of the chapter is organized as follows: First a brief overview about conventional methods will be given. After that the novel generalized method for SR and its simplifications will be presented. A first application example—showing how SR can be utilized to enhance fundamental frequency estimation schemes—will be described next. In the subsequent sections the spectral-temporal interpolation method and its simplified solution will be derived and applied to subband echo cancellation (a second application example). The chapter ends with some simulation results and a conclusion.

9.2 Different Types of Filterbanks

A DFT-modulated filterbank can be extended to a so-called non-critically subsampled polyphase filterbank to enhance the frequency selectivity of the analysis and to reduce aliasing effects [4]. In this case the length of the analysis and synthesis window function is allowed to be larger than the number of used subbands (determined by the DFT order N). A polyphase filterbank introduces much lower aliasing components, and the computational complexity is only increased marginally. However, depending on the used length of the prototype filters a frameshift close to the DFT order can be selected. In the literature (e.g., [22]) design procedures that achieve a frameshift of about $3/4\,N$ using filter orders of about $6\ldots8\,N$ are described. While a polyphase filterbank reduces the computational complexity for large frameshifts it has to be noted that a significant delay is introduced. Such a high delay is not tolerable in several applications such as hands-free telephony.

In [7] critically subsampled systems have been considered. It has been suggested to use adaptive cross filters in order to explicitly filter out the aliasing components. The use of such cross filters results in a significant increase of computational complexity and degrades the speed of convergence.

In [18] a delayless structure has been proposed where adaptive filter weights are computed in the subband domain and then transformed to an equivalent time-domain filter. With this structure the actual filtering is performed in the time domain, leading to an increase of computational complexity. A similar technique was developed in [3] and [21] for an acoustic echo canceller, whereas the adaptive processing part takes place in the frequency domain. However, as already mentioned before, all time-domain-based filtering approaches have the consequences of higher computation complexity.

The contribution [19] has addressed the issues of computational complexity and delay of subband adaptive filtering for applications of acoustic echo control. It has been suggested to use filterbanks based on all-pass polyphase IIR structures as an alternative to the FIR-based filterbanks. The use of all-pass polyphase IIR filterbanks achieves very high side-lobe attenuation, and it has been shown to be computational efficient while keeping the aliasing components low. However, it has to be noted that with this approach nonlinear-phase distortions and large aliasing terms appear at the filter boundaries.

In [8] an efficient prototype filter design method for an oversampled DFT filterbank has been proposed where the aliasing components are minimized while the total filterbank group delay is pre-specified. It has been shown that the estimation accuracy for non-critical decimated filterbanks is close to the fullband solution and significantly better than that of critically decimated cases. More recently, a method to improve the steady-state convergence has been reported in [10, 11]. The idea behind this method is to compute the FFT of the reference signal more often compared to all other FFTs/IFFTs used within a hands-free system.

In contrast to the state-of-the-art approaches, the herein proposed method for SR is employed as a postprocessor for analysis filterbanks. The enhanced frequency selectivity of the analysis is achieved either by reducing the spectral overlap of adjacent subbands or by computing additional subbands. The refinement procedure can easily be implemented using short FIR filters in each subband channel—resulting in a very low computational complexity and an insignificant additional delay in the signal path. In addition, a time-frequency interpolation method as a post-processing stage after a conventional frequency analysis is proposed. Using time-frequency interpolation a significant reduction of the aliasing terms can be achieved without inserting any additional delay in the signal path with only few operations by means of multiplications and additions.

9.3 Refinement and Interpolation of Short-Term Spectra

For the derivation of the spectral refinement and the temporal interpolation of STS, first the input signal is segmented into overlapping blocks of length N according to

$$\boldsymbol{y}(nR) = [y(nR), \ldots, y(nR - N + 1)]^{\mathrm{T}}, \tag{9.1}$$

where the parameter R corresponds to the used subsampling rate and the element n denotes the frame index. The subsampled input vector is windowed with a window function (e.g., Hann window), $h_k \in \mathbb{R}$ with $k \in \{0, \ldots, N-1\}$, and transformed into the frequency or subband domain using a DFT or a filterbank:

$$Y\left(e^{j\Omega}, n\right) = \sum_{k=0}^{N-1} y(nR - k) h_k e^{-j\Omega_\mu k}. \tag{9.2}$$

Note that (9.2) can also be interpreted as subsampled output signals of an analysis filterbank. The used frequency supporting points Ω_μ are equidistantly distributed over the normalized frequency range:

$$\Omega_\mu = \frac{2\pi}{N}\mu, \quad with \quad \mu \in \{0, \ldots, N-1\}. \tag{9.3}$$

For the sake of simplicity, the vector containing all subband signals are rewritten in a matrix-vector notation as

$$\boldsymbol{Y}\big(e^{j\Omega}, n\big) = \boldsymbol{D}\boldsymbol{H}\boldsymbol{y}(nR), \tag{9.4}$$

where the quantity $\boldsymbol{D}$ specifies a DFT matrix of order N and $\boldsymbol{H}$ characterizes a diagonal matrix consisting of the window sequence coefficients:

$$\boldsymbol{H} = \text{diag}\{\boldsymbol{h}\} = \begin{bmatrix} h_0 & 0 & 0 \\ 0 & \ddots & 0 \\ 0 & 0 & h_{N-1} \end{bmatrix}. \tag{9.5}$$

The principle idea of spectral refinement and temporal interpolation is to exploit the correlation of successive input signal blocks for refining originally generated signal frames and interpolating additional frames in between. The proposed method is performed in the frequency or the subband domain. Here the refined and interpolated subband signals are computed by weighted addition of the current and a number of previous input short-term spectra of lower order N according to

$$\begin{bmatrix} \widetilde{\boldsymbol{Y}}_0(e^{j\Omega}, n) \\ \vdots \\ \widetilde{\boldsymbol{Y}}_{M-1}(e^{j\Omega}, n) \end{bmatrix} = \boldsymbol{S} \begin{bmatrix} \boldsymbol{Y}_0(e^{j\Omega}, n) \\ \vdots \\ \boldsymbol{Y}_{M-1}(e^{j\Omega}, n) \end{bmatrix}, \tag{9.6}$$

with $\boldsymbol{S}$ describing a refinement and interpolation matrix and M being the amount of used short-term input spectra.

For the sake of simplicity the STS $\widetilde{\boldsymbol{Y}}_m(e^{j\Omega}, n)$ with $m \in \left\{0, \ldots, \widetilde{M} - 1\right\}$ are gathered in a vector as follows:

$$\widetilde{\boldsymbol{Y}}_{\text{Block}} = \left[\widetilde{\boldsymbol{Y}}_0^{\text{T}}(e^{j\Omega}, n), \widetilde{\boldsymbol{Y}}_1^{\text{T}}(e^{j\Omega}, n), \ldots, \widetilde{\boldsymbol{Y}}_{\widetilde{M}-1}^{\text{T}}(e^{j\Omega}, n)\right]^{\text{T}}, \tag{9.7}$$

where $\widetilde{Y}_0\left(e^{j\Omega_{\widetilde{\mu}}}, n\right)$ with $\Omega_{\widetilde{\mu}} = 2\pi\widetilde{\mu}/\widetilde{N}$ and $\widetilde{\mu} \in \left\{0, \ldots, \widetilde{N} - 1\right\}$ denotes a refined version of the originally computed STS $Y(e^{j\Omega_\mu}, n)$. The remaining elements $\widetilde{Y}_1\left(e^{j\Omega_{\widetilde{\mu}}}, n\right), \ldots, \widetilde{Y}_{\widetilde{M}-1}\left(e^{j\Omega_{\widetilde{\mu}}}, n\right)$ characterize interpolated spectra in between of the original input frames $\boldsymbol{y}((n - 1)R)$ and $\boldsymbol{y}(nr)$. Thus, the refined subband signals as shown in (9.5) are computed without the need for an additional DFT of higher order $\widetilde{N}$. Note that the interpolated subband signals correspond exactly to that signal blocks which would be computed with an analysis filterbank at a reduced rate $R/\widetilde{M}$. All quantities in this contribution which characterize a high order will be designated with a tilde symbol.

The refinement and interpolation matrix $\boldsymbol{S}$ introduced in (9.6) will be divided into two sub-matrices $\boldsymbol{S}_{\text{ref}}$ and $\boldsymbol{S}_{\text{int}}$ according to

$$\boldsymbol{S} = \begin{bmatrix} \boldsymbol{S}_{\text{ref}} \\ \boldsymbol{S}_{\text{int}} \end{bmatrix}. \tag{9.8}$$

$\boldsymbol{S}_{\text{ref}}$ of size $\widetilde{N} \times MN$ consists of coefficients for the refinement of the original STS, whereas $\boldsymbol{S}_{\text{int}}$ of size $\widetilde{N}\left(\widetilde{M} - 1\right) \times MN$ comprises weights for computing interpolated temporal supporting points in between of the original STS.

In the following sections we will try to find appropriate solutions for the unknown matrices $\boldsymbol{S}_{\text{ref}}$ and $\boldsymbol{S}_{\text{int}}$. Before doing this, we will first introduce the computation of the refined and interpolated STS based on higher order DFTs.

Alternatively to (9.6) the desired STS can also be computed using a higher order DFT of windowed and shifted input signal frames. By doing so, first a high-order input block is specified according to

$$\widetilde{\boldsymbol{y}}(nR) = \left[y(nR), \ldots, y\left(nR - \widetilde{N} - M' + 1\right)\right]^{\mathrm{T}}. \tag{9.9}$$

The extended input $\widetilde{\boldsymbol{y}}(nR)$ consists of the last $\widetilde{N} + M'$ samples of the input signal with $\widetilde{N} > N$ and $M' = (M - 1)R$. The higher order STS $\widetilde{Y}_0\left(e^{j\Omega_{\widetilde{\mu}}}, n\right)$ can be formulated in a matrix vector notation as follows:

$$\widetilde{\boldsymbol{Y}}_0\left(e^{j\Omega_{\widetilde{\mu}}}, n\right) = \widetilde{\boldsymbol{D}}\,\widetilde{\boldsymbol{H}}_0\widetilde{\boldsymbol{y}}(nR), \tag{9.10}$$

where $\widetilde{\boldsymbol{D}}$ denotes a DFT matrix of higher order $\widetilde{N}$. Furthermore, $\widetilde{\boldsymbol{H}}_0$ specifies an extended matrix of the filter coefficients:

$$\widetilde{\boldsymbol{H}}_0 = \left[\widetilde{\boldsymbol{H}}\ \boldsymbol{0}^{\left(\widetilde{N}\times M'\right)}\right] = \begin{bmatrix} \widetilde{h}_0 & \cdots & 0 & 0 & \cdots & 0 \\ \vdots & \ddots & \vdots & \vdots & \ddots & \vdots \\ 0 & \cdots & \widetilde{h}_{\widetilde{N}-1} & 0 & \cdots & 0 \end{bmatrix}. \tag{9.11}$$

The aim of $\widetilde{\boldsymbol{H}}_0$ is to add $\widetilde{N} \times M'$ zeros after the extended diagonal window matrix $\widetilde{\boldsymbol{H}} = \text{diag}\left\{\widetilde{\boldsymbol{h}}\right\}$, with $h_k \in \mathbb{R}$ and $\in\left\{0, \ldots, \widetilde{N} - 1\right]$.

The interpolated STS can be computed in a matrix-vector notation as follows:

$$\begin{bmatrix} \widetilde{\boldsymbol{Y}}_1(e^{j\Omega}, n) \\ \vdots \\ \widetilde{\boldsymbol{Y}}_{\widetilde{M}-1}(e^{j\Omega}, n) \end{bmatrix} = \widetilde{\boldsymbol{D}}_{\text{block}}\widetilde{\boldsymbol{H}}_{\text{block}}\widetilde{\boldsymbol{y}}(nR). \tag{9.12}$$

$\widetilde{\boldsymbol{D}}_{\text{block}}$ describes an extended transformation matrix (block-diagonal DFT matrix of size $\widetilde{N}\left(\widetilde{M}-1\right) \times \widetilde{N}\left(\widetilde{M}-1\right)$) defined by

$$\widetilde{\boldsymbol{D}}_{\text{block}} = \begin{bmatrix} \widetilde{\boldsymbol{D}} & \cdots & \boldsymbol{0} \\ \vdots & \ddots & \vdots \\ \boldsymbol{0} & \cdots & \widetilde{\boldsymbol{D}} \end{bmatrix}, \tag{9.13}$$

where $\widetilde{\boldsymbol{D}}$ denotes a DFT matrix of higher order $\widetilde{N}$. The extended matrix $\widetilde{\boldsymbol{H}}_{\text{block}}$ consisting of filter coefficients with a dimension of $\widetilde{N}\left(\widetilde{M}-1\right) \times \widetilde{N}+M'$ is defined according to

$$\widetilde{\boldsymbol{H}}_{\text{block}} = \left[\widetilde{\boldsymbol{H}}_1^{\text{T}}, \widetilde{\boldsymbol{H}}_2^{\text{T}}, \ldots, \widetilde{\boldsymbol{H}}_{\widetilde{M}-1}^{\text{T}}\right]^{\text{T}}. \tag{9.14}$$

The first element matrix $\widetilde{\boldsymbol{H}}_1$ adds $\widetilde{N} \times \left(R/\widetilde{M}\right)$ zeros before the extended diagonal window matrix and $\widetilde{N} \times \left(M' - R/\widetilde{M}\right)$ after, whereas the remaining matrices $\widetilde{\boldsymbol{H}}_2, \widetilde{\boldsymbol{H}}_3$, etc. represent cyclic shifts of $\widetilde{\boldsymbol{H}}_1$. This means that equal row indices of adjacent sub-matrices are rotated by $R/\widetilde{M}$ elements. Thus, the first and the last element matrices are defined according to

$$\widetilde{\boldsymbol{H}}_1 = \left[\boldsymbol{0}^{\left(\widetilde{N}\times R/\widetilde{M}\right)} \; \widetilde{\boldsymbol{H}} \; \boldsymbol{0}^{\left(\widetilde{N}\times M'-R/\widetilde{M}\right)}\right] \tag{9.15}$$

and

$$\widetilde{\boldsymbol{H}}_{\widetilde{M}-1} = \left[\boldsymbol{0}^{\left(\widetilde{N}\times R\left(\widetilde{M}-1\right)/\widetilde{M}\right)} \; \widetilde{\boldsymbol{H}} \; \boldsymbol{0}^{\left(\widetilde{N}\times M'-R\left(\widetilde{M}-1\right)/\widetilde{M}\right)}\right]. \tag{9.16}$$

Once the principle idea of the spectral refinement and temporal interpolation is introduced we will continue in the next chapter in finding appropriate solutions for the refinement and interpolation matrices $\boldsymbol{S}_{\text{ref}}$ and $\boldsymbol{S}_{\text{int}}$.

9.4 Spectral Refinement Method and Its Application

In this section we concentrate on finding a solution for the unknown matrix $\boldsymbol{S}_{\text{ref}}$ from (9.6) and (9.8). Since we will only focus on the refinement of the original STS in this section, the interpolation of STS is disregarded for simplicity. The number of processed STS is therefore set to $\widetilde{M}-1$. For spectral refinement it is assumed that the lower order STS are already available. They might be used, e.g., to estimate the noise power for speech enhancement within a hands-free system. However, in some situations it is desired to determine an improved STS in order to enhance feature extraction schemes such as pitch frequency for increasing the performance of a speech recognizer. For that purpose we suggest to apply a linear combination of the lower order STS as stated in (9.6). The derivation of the SR matrix $\boldsymbol{S}_{\text{ref}}$ as well as its simplified version will be explained in the following.

For the derivation of the spectral refinement method (first published in [17]), we assume that the increased DFT of order $\widetilde{N}$ is a function of the basic DFT of order N according to

$$\widetilde{N} = N + k_0 R \quad \textit{with} \quad k_0 \in \{1, 2, 3, \ldots\}, \tag{9.17}$$

where R corresponds to the used subsampling factor. Before calculating the SR matrix $\boldsymbol{S}_{\text{ref}}$ a constraint for the higher order window function $\widetilde{\boldsymbol{h}}$ is introduced as

$$\boldsymbol{A}[\boldsymbol{h}, \boldsymbol{h}, \ldots, \boldsymbol{h}]^{\mathrm{T}} = \widetilde{\boldsymbol{h}}. \tag{9.18}$$

The matrix $\boldsymbol{A}$ of size $\widetilde{N} \times MN$ consists of appropriate weights $a_k^{(m)}$ for each m-th window $\boldsymbol{h}$ with $k \in \{0, \ldots, N-1\}$ and $m \in \{0, \ldots, M-1\}$. The design of such a matrix is defined by

$$\boldsymbol{A} = \left[\widetilde{\boldsymbol{A}}_0, \widetilde{\boldsymbol{A}}_1, \ldots, \widetilde{\boldsymbol{A}}_{M-1}\right]. \tag{9.19}$$

The first element matrix $\widetilde{\boldsymbol{A}}_0$ consists of diagonal weight coefficients followed by $(M-1)R \times N$ zero values. To be precise, the matrix $\widetilde{\boldsymbol{A}}_0$ is composed as follows:

$$\widetilde{\boldsymbol{A}}_0 = \begin{bmatrix} \boldsymbol{A}_0 \\ \boldsymbol{0}^{((M-1)R \times N)} \end{bmatrix} = \begin{bmatrix} a_0^{(0)} & \cdots & 0 \\ \vdots & \ddots & \vdots \\ 0 & \cdots & a_{N-1}^{(0)} \\ 0 & \cdots & 0 \\ \vdots & \ddots & \vdots \\ 0 & \cdots & 0 \end{bmatrix}. \tag{9.20}$$

The remaining submatrices $\widetilde{\boldsymbol{A}}_1, \widetilde{\boldsymbol{A}}_2$, etc. with individual weight coefficients have the same structure as $\widetilde{\boldsymbol{A}}_0$ except that the equal column indices of adjacent submatrices are rotated by R, etc. elements. Thus, the element matrices $\widetilde{\boldsymbol{A}}_1, \widetilde{\boldsymbol{A}}_2$, etc. are defined according to

$$\widetilde{\boldsymbol{A}}_1 = \begin{bmatrix} \boldsymbol{0}^{(R \times N)} \\ \boldsymbol{A}_1 \\ \vdots \\ \boldsymbol{0}^{(R \times N)} \end{bmatrix}, \ldots, \widetilde{\boldsymbol{A}}_{M-1} = \begin{bmatrix} \boldsymbol{0}^{(R \times N)} \\ \boldsymbol{0}^{(R \times N)} \\ \vdots \\ \boldsymbol{A}_{M-1} \end{bmatrix}. \tag{9.21}$$

The main task of $\boldsymbol{A}$ is to weight M window sequences of lower order N and shift subsequently adjacent window sequences by the chosen subsampling factor R. The modified window sequences are summed up to obtain a desired higher order window sequence. Consequently, the window sequence $\widetilde{\boldsymbol{h}}$ consists of a weighted sum of shifted vectors $\boldsymbol{h}$. The coefficients $a_k^{(m)}$ can be designed in such a way that an arbitrary window sequence of lower order is transformed into a desired window sequence of higher order. The resulting order of the window sequence $\widetilde{\boldsymbol{h}}$ from (9.18) is given by

$$\widetilde{N} = N + (M-1)R. \tag{9.22}$$

In the upper part of Fig. 9.2 an example of weighted and shifted Hann windows each of lower order (dashed lines with $N = 256$, $M = 5$, $R = 64$) as well as the resulting window function of higher order (solid line with $\widetilde{N} = 512$) is shown. The coefficients used for weighting the window functions have been chosen as follows:

$$\begin{aligned} a_k^{(0)} &= a_k^{(M-1)} = 0.3K_0, \quad \textit{for all } k, \\ a_k^{(1)} &= a_k^{(M-2)} = 0.7K_0, \quad \textit{for all } k, \\ \text{and} \quad & a_k^{((M-1)/2)} = K_0, \quad \textit{for all } k. \end{aligned} \tag{9.23}$$

As normalization constant $K_0 = 3$ has been applied. In the lower part of Fig. 9.3 the corresponding analyses of the STS are depicted. Comparing the results one can see that the main-lobe width as well as the side-lobe amplitudes are reduced when using the weighted sum of shifted window sequences $\boldsymbol{h}$.

Once the constraint for the window sequences is defined the next step is to solve for the SR matrix $\boldsymbol{S}_{\text{ref}}$. By doing so, first (9.6) for $\widetilde{M} = 1$ is rewritten as follows:

$$\widetilde{Y}_0\left(e^{j\Omega}, n\right) = \boldsymbol{S}_{\text{ref}} \begin{bmatrix} \boldsymbol{D} & \cdots & \boldsymbol{0} \\ \vdots & \ddots & \vdots \\ \boldsymbol{0} & \cdots & \boldsymbol{D} \end{bmatrix} \begin{bmatrix} \boldsymbol{H} & \cdots & \boldsymbol{0} \\ \vdots & \ddots & \vdots \\ \boldsymbol{0} & \cdots & \boldsymbol{H} \end{bmatrix} \begin{bmatrix} \boldsymbol{y}(nR) \\ \vdots \\ \boldsymbol{y}((n-M+1)R) \end{bmatrix}. \tag{9.24}$$

Using the abovementioned constraint (9.18) the improved STS from (9.10) can be expressed as

$$\widetilde{Y}_0\left(e^{j\Omega}, n\right) = \widetilde{\boldsymbol{D}}\ \boldsymbol{A} \begin{bmatrix} \boldsymbol{H} & \cdots & \boldsymbol{0} \\ \vdots & \ddots & \vdots \\ \boldsymbol{0} & \cdots & \boldsymbol{H} \end{bmatrix} \begin{bmatrix} \boldsymbol{y}(nR) \\ \vdots \\ \boldsymbol{y}((n-M+1)R) \end{bmatrix}. \tag{9.25}$$

Inserting the expression from (9.25) in (9.24) results in several solutions for the matrix $\boldsymbol{S}_{\text{ref}}$ that in general depends on the input signal vectors $\boldsymbol{y}((n - m)R)$. A solution that is independent of the input signal can be obtained by

$$\boldsymbol{S}_{\text{ref}} = \widetilde{\boldsymbol{D}}\ \boldsymbol{A} \begin{bmatrix} \boldsymbol{D}^{-1} & \cdots & \boldsymbol{0} \\ \vdots & \ddots & \vdots \\ \boldsymbol{0} & \cdots & \boldsymbol{D}^{-1} \end{bmatrix}. \tag{9.26}$$

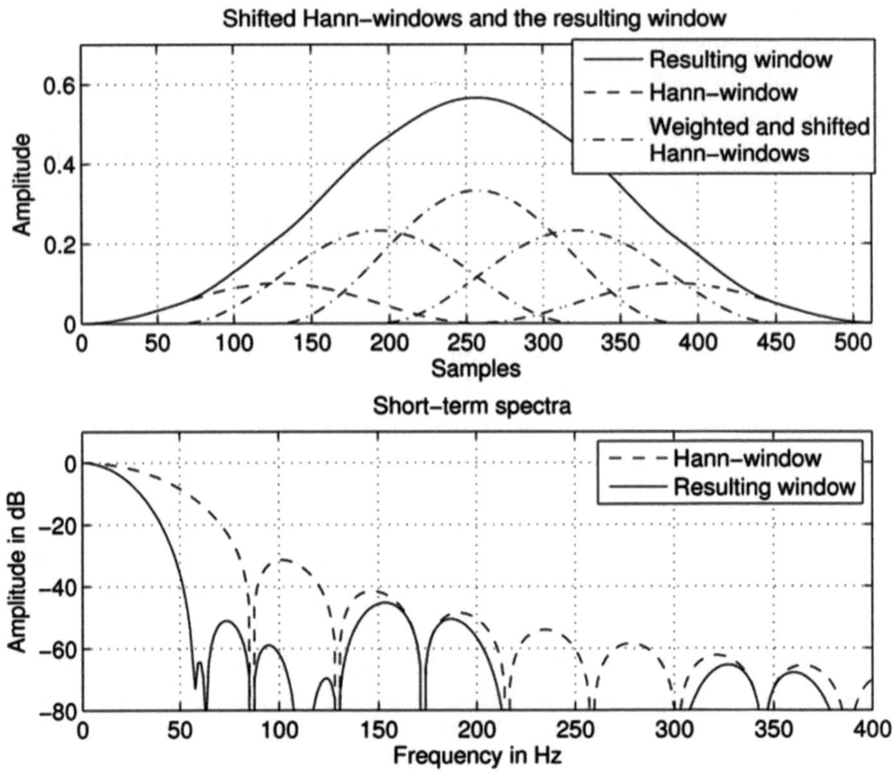

Fig. 9.3 *Upper plot* shows the weighted and shifted Hann windows and the resulting window sequence, *lower plot* depicts the corresponding short-term spectra

After inserting the definitions of the matrices in (9.24) the SR matrix S_{ref} can finally be rewritten in the following way:

$$S_{\text{ref},i,mN+l} = \frac{1}{N} e^{-j\frac{2\pi}{N}imR} \sum_{k=0}^{N-1} a_k^{(m)} e^{-j2\pi\left(\frac{i}{N}-\frac{l}{N}\right)k}. \tag{9.27}$$

The parameter i in (9.27) specifies the row index and the quantity $nN + l$ the column index of the SR matrix.

9.4.1 Simplified Spectral Refinement

Once the general solution for the spectral refinement matrix S_{ref} is formulated we can try to simplify and approximate the matrix. This can be done mainly due to sparseness of the refinement matrix that appears if certain conditions—as described in the following—are introduced. First it is assumed that the weighting coefficients for each m-th window function are identical, meaning that $a_k^{(m)} = a^{(m)}$. In order to analyze the SR matrix quantitatively, (9.27) is rewritten as follows:

$$S_{\text{ref},i,mN+l} = \frac{a^{(m)}}{N} e^{-j\frac{2\pi}{N}imR} \sum_{k=0}^{N-1} e^{-j2\pi\left(\frac{i}{N}-\frac{l}{N}\right)k}. \tag{9.28}$$

Further, the geometric series on the right-hand side of (9.28) can be simplified in the following way:

$$S_{\text{ref},i,mN+l} = \frac{a^{(m)}}{N} \frac{\sin\left(\pi\frac{iN-lN}{N}\right)e^{-j\pi\frac{iN-lN}{N}}}{\sin\left(\pi\frac{iN-lN}{NN}\right)e^{-j\pi\frac{iN-lN}{NN}}} e^{-j\frac{2\pi}{N}imR}. \tag{9.29}$$

If the condition holds, that the higher filter order is a multiple of the lower filter order

$$\widetilde{N} = k_0 N \quad with \quad k_0 \in \{2, 3, \ldots.\}, \tag{9.30}$$

then specific rows and columns of the SR matrix can be further simplified to

$$S_{\text{ref},i,mN+l} = \begin{cases} 0, & \text{if} \left(\frac{i}{k_0} \in \mathbb{Z}\right) \wedge \left(l \neq \frac{i}{k_0}\right), \\ a^{(m)} e^{-j\frac{2\pi}{N}imR}, & \text{if} \left(\frac{i}{k_0} \in \mathbb{Z}\right) \wedge \left(l = \frac{i}{k_0}\right), \\ S_{i,mN+l}, & \text{else.} \end{cases} \tag{9.31}$$

The symbol $\mathbb{Z}$ denotes the set of integers. Thus, each k_0-th row of $\boldsymbol{S}_{\text{ref}}$ is sparsely populated, i.e., the elements of each k_0-th row are zero except for the column indices that are multiples of N. Furthermore, if the filter order N is chosen to be a multiple of the used frameshift, e.g., $2R$ or $4R$, then those elements of the sparsely populated rows of the SR matrix are either real or imaginary.

For illustration purposes a simple example of the SR matrix is shown in Fig. 9.4 with $M = 3$, $R = 2$ and $N = 4$. As a result, each second row (even indices of the SR matrix) is sparsely populated. The elements in white color indicate values equal zero, whereas the ones in black values unequal zero. However, these rows are related to that frequency-supporting points, which would be computed with a basic DFT/FFT of order N as well as with a higher DFT/FFT of order $\widetilde{N}$.

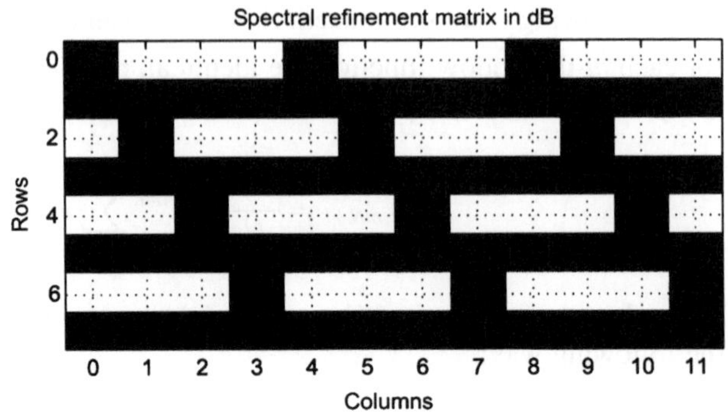

Fig. 9.4 SR matrix: *White color* indicates values equal zero and *black* element values unequal zero

9.4.2 Realization Aspects of the Spectral Refinement

The proposed method for spectral refinement can either be applied to refine only the original frequency resolution of the input signal or to compute additional frequency-supporting points in between.

9.4.2.1 Refinement While Keeping the Original Frequency Resolution

If it is desired to calculate a spectral refinement only for the original frequency resolution—i.e., each k_0-th frequency-supporting point of the vector $\widetilde{Y}\left(e^{j\widetilde{\Omega_\mu}}, n\right)$ is refined—the realization of the spectral refinement can be performed in an efficient and robust manner. Due to the sparse population of the matrix $\boldsymbol{S}_{\text{ref}}$ the refinement can be realized by short FIR filters applied in each subband after the frequency decomposition of the input signal $y(n)$. The FIR filter coefficients

$$\boldsymbol{g}_{i,i\cdot k_0} = \left[g_{i,i\cdot k_0,0}, g_{i,i\cdot k_0,1}, \ldots, g_{i,i\cdot k_0,M-1}\right]^{\mathrm{T}} \tag{9.32}$$

are extracted from the sparsely populated refinement matrix by

$$g_{i,i\cdot k_0,m} = S_{\text{ref},i\cdot k_0,i+mN}. \tag{9.33}$$

The refined spectrum of the i-th subband is determined by

$$\begin{aligned} \widetilde{Y}_0\left(e^{j\Omega_{i\cdot k_0}},\right)n &= g_{i,i\cdot k_0,0} Y_0(e^{j\Omega_i}, n) + \ldots \\ &\quad + g_{i,i\cdot k_0,M-1} Y_0(e^{j\Omega_i}, n-(M-1)). \end{aligned} \tag{9.34}$$

Often analysis-synthesis schemes use a frameshift which is a fraction (e.g., half or quarter) of the DFT order. For such cases the filter coefficients $g_{i,i\cdot k_0,m}$ are either

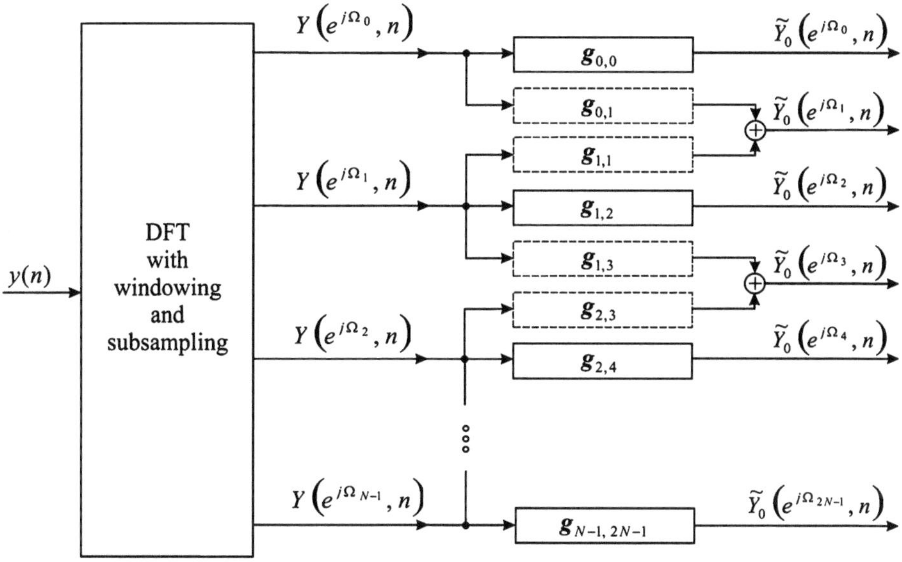

Fig. 9.5 Analysis filterbank with spectral refinement as a postprocessor by means of FIR filters for $k_0 = 2$

real or imaginary which in turn results in a further reduction of the computational complexity.

Figure 9.5 shows a realization of an analysis filterbank with spectral refinement as a postprocessor by means of FIR filters for $k_0 = 2$. The so-called *auto* FIR filters $g_{i,i \cdot k_0}$ are applied to refine the original frequency-supporting points (solid black frames).

9.4.2.2 Computing Additional Frequency-Supporting Points

Besides refinement of the original frequency resolution it is also possible to calculate frequency points in between of the original spectrum. At a first glance, however, it is computationally intensive due to the non-sparseness of the remaining rows of the $\boldsymbol{S}_{\text{ref}}$ matrix. In order to reduce the computational complexity, one can approximate the non-sparse rows of the $\boldsymbol{S}_{\text{ref}}$ matrix by the M largest coefficient pairs. The largest coefficient pairs correspond exactly to those weighting values around the desired frequency-supporting points of the STS. Analyses have confirmed that the resulting spectrum shows low errors even if only $M = 3 \ldots 5$ filter coefficients are used. The complete system of spectral refinement for $k_0 = 2$ is depicted in Fig. 9.5.

The refinement of the original frequency resolution is accomplished using *auto* FIR filters (solid black frames), and the computation of the additional

Table 9.1 Computational complexity of a higher order FFT and of a basic FFT with additional spectral refinement

Complex multiplications and additions ($N = 512$, $N = 256$, $M = 3 \ldots 5$)		
$\widetilde{N}$-order FFT	$\widetilde{N}\ ld\left(\widetilde{N}\right) = 4{,}608$	100 %
N-order FFT with spectral refinement	$N\ ld(N) + MN/2 = 2,\ 432 \ldots 2,\ 688$	53–58 %
N-order FFT with spectral refinement and additional frequencies	$N\ ld(N) + MN/2 + MN = 3,\ 200 \ldots 3,\ 968$	69–86 %

frequency-supporting points is performed using *cross* FIR filters (dashed gray frames). The *cross* as well as the *auto* filters can be calculated as

$$g_{i,l,m} = S_{\text{ref},l,i+mN}, \tag{9.35}$$

and the refined STS $\widetilde{Y}_0(e^{j\Omega_l}, n)$ is finally determined by

$$\widetilde{Y}_0\left(e^{j\Omega_l}, n\right) = \begin{cases} \displaystyle\sum_{m=0}^{M-1} g_{l/k_0,l,m} Y\left(e^{j\Omega_{l/k_0}}, n - m\right), & \text{if } \dfrac{l}{k_0} \in \mathbb{Z}, \\ \displaystyle\sum_{m=0}^{M-1} g_{\lfloor l/k_0 \rfloor,l,m} Y\left(e^{j\Omega_{l/k_0}}, n - m\right) & \\ \displaystyle+\sum_{m=0}^{M-1} g_{\lceil l/k_0 \rceil,l,m} Y\left(e^{j\Omega_{l/k_0}}, n - m\right), & \text{else,} \end{cases} \tag{9.36}$$

where $\lfloor \ldots \rfloor$ and $\lceil \ldots \rceil$ denote rounding to the next smaller and larger integer, respectively.

9.4.3 Computational Complexity of the Spectral Refinement

After the simplified version of SR and its efficient realization were described, as a next stage we analyze its overall complexity. Hence, the computational complexity of a 256-FFT order with additional refinement is compared with a 512-FFT order by means of complex multiplications and additions as shown in Table 9.1.

It should be noted that only few operations for refining the original frequency-supporting points are required. Using SR as a post-processing stage of a basic 256-FFT only about 2,688 complex multiplications and additions are required while doubling the basic 256-FFT order to 512-FFT about 4,608 operations are needed. It has to be mentioned that in many applications a basic FFT is already available, needed for the estimation of several parameters, like pitch frequency which can be used for speech recognition. In such situations, however, only minor additional operations are required for performing SR. If it is desired to calculate also additional frequency-supporting points as presented in Sect. 9.4.2.2 only few complex multiplications and additions have to be added to the refinement system as seen in Table 9.1.

9.4.4 Spectral Refinement for Pitch Frequency Estimation

The refinement method can be utilized in a variety of speech, audio, and other signal processing applications. In this section the principle idea of using spectral refinement as a pre-processing stage for enhanced fundamental frequency estimation is presented.

A broad variety of different algorithms for estimating the fundamental frequency of speech signals exist: e.g., methods based on the harmonic product-spectrum [20] or on short-term autocorrelation [2]. For the following evaluations a method based on the latter approach has been employed. In a first stage the corrupted speech signal $y(n)$ is divided into overlapping blocks and subsequently windowed. Once the FFT as well as the spectral refinement method are applied to the input signal block according to Fig. 9.5, the short-term power spectral density (PSD) is estimated. Applying the IFFT to a normalized version of the PSD results in the autocorrelation function (ACF). By performing a maximum search of the ACF in a selected range of indices, the normalized pitch period is estimated by using the argument of the maximum. Finally, the pitch frequency is obtained using the inverse of the pitch period. Further details can be found in [15].

To show the performance and the accuracy of the proposed method, the estimated fundamental frequencies without and with spectral refinement at different signal-to-noise ratio (SNR) conditions have been compared with a clean speech laryngograph database. The reference database consists of a multitude of pitch frequencies. Figure 9.6 shows the histogram of the used reference pitch frequencies out of the interval $f_{\mathrm{p}}(n) \in [60\ \mathrm{Hz}, 350\ \mathrm{Hz}]$. About 7×10^5 reference pitch frequencies have been used for the experiments extracted from 38 different speakers (18 female and 20 male).

In addition to the reference pitch frequencies extracted from the laryngograph signals the corresponding Lombard [13] speech utterances measured at different noise conditions in an anechoic room have been stored within the reference database. To generate a noisy speech signal, the Lombard speech signals were convolved with different impulse responses (mouth-to-microphone) and added with stationary background noise. The impulse responses as well as the background noises were measured in advance in a multitude of cars. Furthermore the Lombard speech signals have been adjusted in power to achieve the desired SNR.

For evaluation the correctness and false detections have been considered. To analyze the correctness of the estimated pitch frequencies three ranges of values

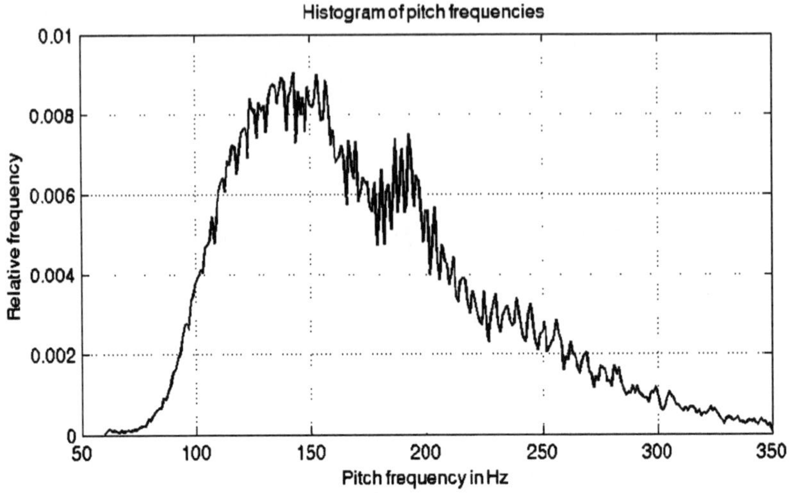

Fig. 9.6 Histogram of reference pitch frequencies out of the interval 60–350 Hz (visualized in 1 Hz steps)

Table 9.2 Correctness of estimated fundamental frequencies without and with spectral refinement for different SNR and tolerance ranges

		Correctness (%)		
	Accepted tolerance	High SNR	Medium SNR	Low SNR
Standard method	<3 %	62.1	70.1	47.2
	<10 %	65.1	70.9	48.4
	<20 %	65.5	71.4	49.3
Method with SR up to 1 kHz	<3 %	82.1	80.3	55.9
	<10 %	88.1	85.3	58.2
	<20 %	88.8	86.2	58.7
Method with SR up to 3 kHz	<3 %	83.2	80.1	53.4
	<10 %	89.8	85.3	56.9
	<20 %	90.6	86.3	57.5

have been defined: the estimated error lies within a tolerance range of ± 3, ± 10, and $\pm 20\,\%$. False detection means that the algorithm under test detects a pitch frequency where no reference pitch is available.

Table 9.2 summarizes the correctness of the pitch estimation method without and with spectral refinement for high SNR (20–25 dB), medium SNR (9–15 dB), and low SNR (0–6 dB).

Pitch frequency was only detected if the normalized ACF at maximum lag exceeds a predefined threshold of $p_0 = 0.25$. Note that the refinement was only performed at lower frequencies up to 1 and 3 kHz, respectively.

Table 9.3 False detection of pitch frequency without and with spectral refinement up to 1 and 3 kHz for different SNR ranges

	False detection (%)		
	High SNR	Medium SNR	Low SNR
Standard method	18.1	11.4	8.6
Method with SR up to 1 kHz	18.2	11.3	8.2
Method with SR up to 3 kHz	17.2	11.4	8.1

The results show that by applying spectral refinement an increase of correctness by about 20–25 % (abs.) at high SNR is achieved, approx. 10–15 % (abs.) at medium SNR, and about 8–10 % (abs.) at low SNR. Moreover, it can be observed that nearly the same performance is achieved when using SR up to 1 and 3 kHz. Hence, for pitch estimation it is sufficient to refine the input spectrum only at lower frequencies up to 1 kHz which in turn results in a significant reduction of the computational complexity. The measured results for false detections are listed in Table 9.3. From the evaluations one can see that the false detection rates can be kept nearly constant for all SNR levels considered while the correctness rates are increased at the same time.

9.5 Temporal Interpolation of Short-Term Spectra and Its Application

The principle idea of the interpolation method which was first published in [16] is to exploit the correlation of successive input signal blocks for computing interpolated temporal supporting points in between of the originally generated frames. For the derivation it is assumed that the chosen subsampling rate R is even-valued. For simplicity reasons, the following condition is defined: $\widetilde{N} = N$. Meaning that the interpolated STS have the same order as the original input; to be more precise, no spectral refinement is performed for the interpolated STS.

Reformulating (9.6) by using the interpolation matrix $\boldsymbol{S}_{\text{int}}$ and the extended input vector $\widetilde{\boldsymbol{y}}(nR)$, the following expression is obtained:

$$\begin{bmatrix} \widetilde{Y}_1(e^{j\Omega}, n) \\ \vdots \\ \widetilde{Y}_{\widetilde{M}-1}(e^{j\Omega}, n) \end{bmatrix} = \boldsymbol{S}_{\text{int}} \boldsymbol{D}_{\text{block}} \boldsymbol{H}_{\text{block}} \widetilde{\boldsymbol{y}}(nR). \tag{9.37}$$

$\boldsymbol{D}_{\text{block}}$ describes a block-diagonal matrix of size $MN \times MN$ consisting of DFT matrices of order N (analogue to (9.10)) and $\boldsymbol{H}_{\text{block}}$ denotes an extended window matrix with a dimension of $MN \times (N + M')$:

$$\boldsymbol{H}_{\text{block}} = \left[\boldsymbol{H}_0^{\text{T}}, \boldsymbol{H}_1^{\text{T}}, \ldots, \boldsymbol{H}_{M-1}^{\text{T}}\right]^{\text{T}}. \tag{9.38}$$

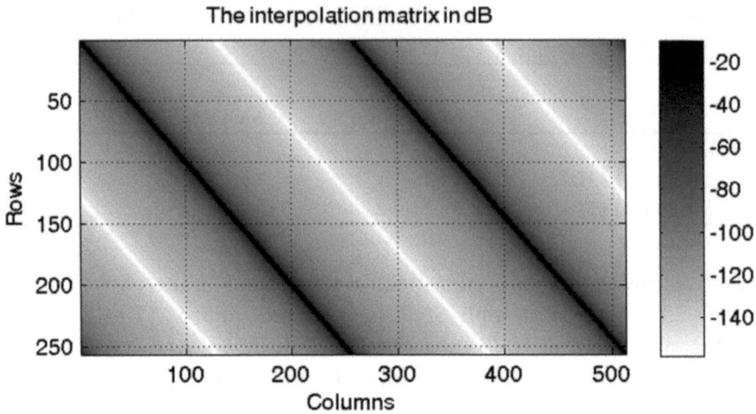

Fig. 9.7 Magnitude of the elements of the interpolation matrix $\boldsymbol{S}_{\text{int}}$ in dB with $\widetilde{N} = N = 256$ and $\widetilde{M} = M = 2$

The first element matrix $\boldsymbol{H}_0$ adds $N \times M'$ zero values after the diagonal window matrix $\boldsymbol{H}$, whereas the remaining matrices $\boldsymbol{H}_1$, $\boldsymbol{H}_2$, etc. represent cyclic shifts of $\boldsymbol{H}_0$. This means that equal row indices of adjacent submatrices are rotated by R elements. Thus, the first and the last element matrices are defined according to

$$\boldsymbol{H}_0 = \left[\boldsymbol{H} \ \boldsymbol{0}^{(N\times M')}\right] \quad \text{and} \quad \boldsymbol{H}_{M-1} = \left[\boldsymbol{0}^{(N\times M')} \ \boldsymbol{H}\right]. \tag{9.39}$$

The definitions (9.12) and (9.37) result in several solutions for the interpolation matrix $\boldsymbol{S}_{\text{int}}$. The solutions depend in general on the input signal vector. A solution that is independent of the input signal can be obtained as follows:

$$\boldsymbol{S}_{\text{int}} = \widetilde{\boldsymbol{D}}_{\text{block}} \widetilde{\boldsymbol{H}}_{\text{block}} \boldsymbol{H}_{\text{block}}^{\Xi} \boldsymbol{D}_{\text{block}}^{-1}, \tag{9.40}$$

where $\boldsymbol{D}_{\text{block}}^{-1}$ describes the inverse of the block-diagonal DFT matrix and $\boldsymbol{H}_{\text{block}}^{\Xi}$ characterize the Moore–Penrose inverse which is defined for real matrices as

$$\boldsymbol{H}_{\text{block}}^{\Xi} = \left[\boldsymbol{H}_{\text{block}}^{\text{T}} \ \boldsymbol{B}_{\text{block}}\right]^{-1} \boldsymbol{H}_{\text{block}}^{\text{T}}. \tag{9.41}$$

9.5.1 Approximated Interpolation

Once the general solution for the interpolation matrix $\boldsymbol{S}_{\text{int}}$ is formulated, we can try to simplify and approximate the matrix. In Fig. 9.7 the log-magnitudes of the elements of the interpolation matrix are shown for $M = 2$ and $N = 256$. From this result, one can observe that the matrix $\boldsymbol{S}_{\text{int}}$ contains only few coefficients being significantly different from zero. This results from the diagonal structure of the matrix $\boldsymbol{H}$, the sparseness of the extended window matrix $\boldsymbol{H}_{\text{block}}$, and the orthogonal eigen functions included in the transformation matrices. Thus, the computation of the temporally interpolated spectra can be approximated very efficiently as described below.

Since $\boldsymbol{S}_{\text{int}}$ is a sparse matrix, the interpolation can be realized as a post-processing stage after an analysis filterbank using a weighted sum of subband signals. The weighting coefficients for the i-th subband can be easily extracted from the interpolation matrix according to

$$g_p^{(i,l,m)} = \boldsymbol{S}_{\text{int},\, i+(l-1)\widetilde{N},\, L_i - mN + p}. \tag{9.42}$$

The parameter $i + (l-1)\widetilde{N}$ in (9.42) specifies the row index and the quantity $L_i - nM + p$ the column index of the interpolation matrix entries, with $m \in \{0, \ldots, M-1\}$, $p \in \{0, \ldots, K_i - L_i\}$, and $l \in \left\{0, \ldots, \widetilde{M}-1\right\}$. The interpolated STS for the i-th subband are then determined by

$$\widetilde{Y}_l\left(e^{j\Omega_i}, n\right) = \sum_{m=0}^{M-1} \sum_{k=L_i}^{K_i} g_{k-L_i}^{(i,l,m)} Y\left(e^{j\Omega_k}, n-m\right). \tag{9.43}$$

Experiments have shown that it is sufficient to use only five to ten complex multiplications and additions for computing one interpolated subband signal. The filter order of $\boldsymbol{g}^{(i,k,m)}$ for the i-th subband is defined by the difference $K_i - L_i$ with

$$L_i = \max\left\{0, i - \left\lfloor \frac{P}{2} \right\rfloor\right\} \tag{9.44}$$

and

$$K_i = \min\left\{i + \left\lceil \frac{P}{2} \right\rceil, N-1\right\}, \tag{9.45}$$

with P being the maximal filter order used for the interpolation. Figure 9.8 shows the principle realization of an analysis filterbank with time-frequency interpolation as a postprocessor by means of weighted sum of subband signals for $\widetilde{M} = M = 2$.

9.5.2 Application to Echo Cancellation

The use of echo cancellation by means of adaptive filters offers the possibility of a full-duplex communication in hands-free telephony. Due to computational complexity often adaptive filters in the subband domain are used to estimate a digital replica of a LEM system [9]. However, when increasing the subsampling

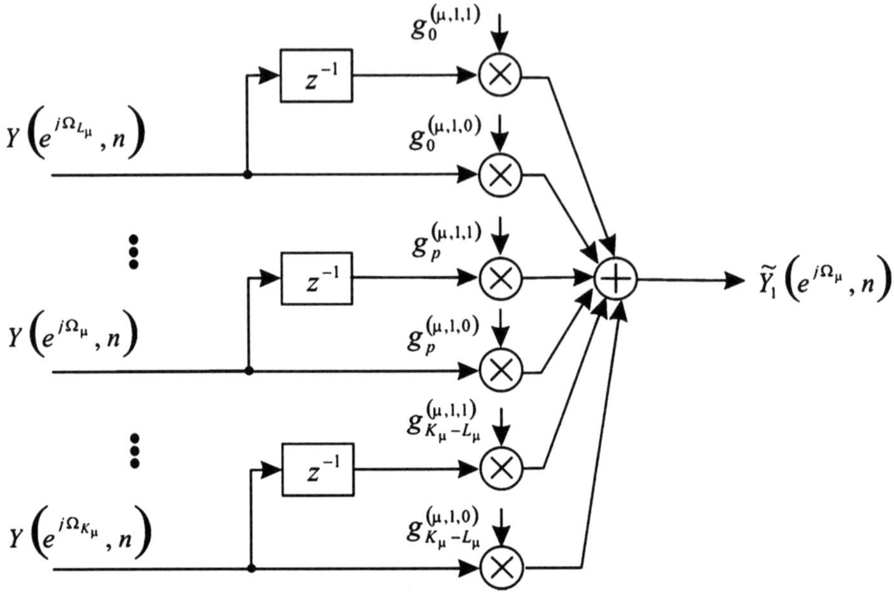

Fig. 9.8 Time-frequency interpolation by means of weighted sum of subband signals applied as a postprocessor after a conventional analysis filterbank

rate R, the computational load is decreased while at the same time aliasing components within subband signals are increased. It is well known that subband echo cancellation requires nearly aliasing free subband signals. Therefore, a compromise between performance and computational cost has to be found.

To overcome a low steady-state convergence of echo cancellation at large subsampling rates, the new interpolation method can be implemented. Figure 9.9 depicts the proposed structure for subband echo cancellation with additional temporal interpolation applied only in the reference channel for $\widetilde{M} = M = 2$.

First we suggest applying the same subsampling rate R for the reference and the microphone path. The resulting reference subband signals $Y(e^{j\Omega_\mu}, n)$—after decomposition of $y(n)$ using an analysis filterbank (anti-aliasing filtering and downsampling)—are subsequently fed to a time-frequency interpolation unit that includes temporally interpolating the time series of the STS. The original reference subband signals as well as the output of the time-frequency interpolation $\widetilde{Y}_1(e^{j\Omega_\mu}, n)$ are fed to the echo cancellation for estimating the subband echo signals. The usage of both the reference subband signals as well as its interpolated version reduces the unwanted effects of aliasing. The subband echo signals are estimated by a convolution of the input subband signals with the estimated LEM subband impulse response according to

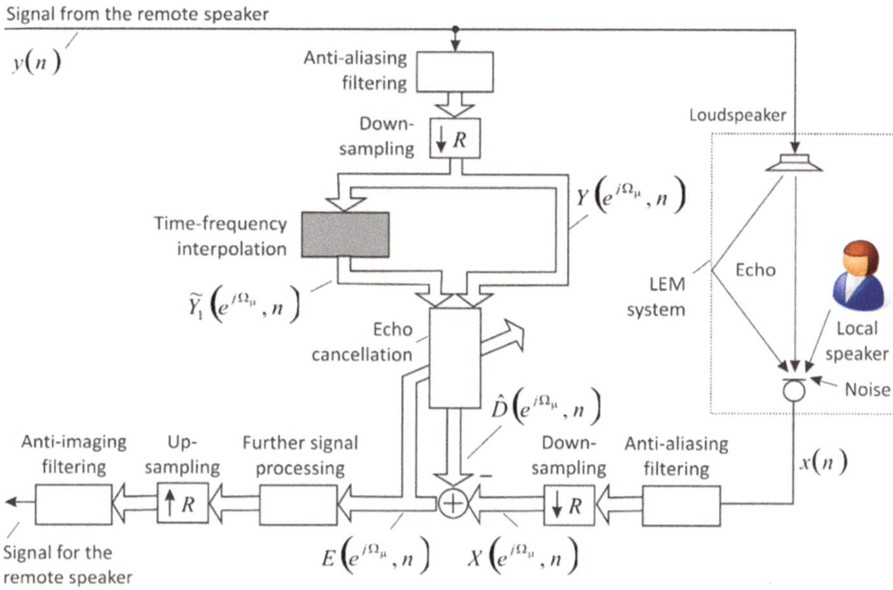

Fig. 9.9 Proposed system for subband echo cancellation with additional time-frequency interpolation in the reference path

$$\hat{D}\left(e^{j\Omega_\mu}, n\right) = \sum_{i=0}^{V-1} W_{i,\mu}(n) Y\left(e^{j\Omega_\mu}, n-i\right) + \sum_{i=0}^{V-1} \widetilde{W}_{i,\mu}(n) \widetilde{Y}_1\left(e^{j\Omega_\mu}, n-i\right). \tag{9.46}$$

The quantities $W_{i,\mu}(n), \widetilde{W}_{i,\mu}(n)$, $i \in \{0, \ldots, V-1\}$ are the subband filter coefficients for the interpolated and non-interpolated part, respectively. The parameter V denotes the number of filter coefficients. It has to be noticed that the convolution is still only operating at the original subsampling rate R. The estimated echoes are subtracted from the microphone subband signals $X(e^{j\Omega_\mu}, n)$ to determine the error $E(e^{j\Omega_\mu}, n)$ for the filter update. For adaptation of the filter coefficients a typical gradient-based optimization procedure (e.g., the NLMS algorithm) can be utilized:

$$W_{i,\mu}(n+1) = W_{i,\mu}(n) + \beta \frac{Y(e^{j\Omega_\mu}, n-i) E^*(e^{j\Omega_\mu}, n-i)}{\sum_{i=0}^{V-1} \left|Y(e^{j\Omega_\mu}, n-i)\right|^2 + \sum_{i=0}^{V-1} \left|\widetilde{Y}_1(e^{j\Omega_\mu}, n-i)\right|^2} \tag{9.47}$$

and

$$\widetilde{W}_{i,\mu}(n+1) = \widetilde{W}_{i,\mu}(n) + \beta \frac{\widetilde{Y}_1(e^{j\Omega_\mu}, n-i) E^*(e^{j\Omega_\mu}, n-i)}{\sum_{i=0}^{V-1} \left|Y(e^{j\Omega_\mu}, n-i)\right|^2 + \sum_{i=0}^{V-1} \left|\widetilde{Y}_1(e^{j\Omega_\mu}, n-i)\right|^2} \tag{9.48}$$

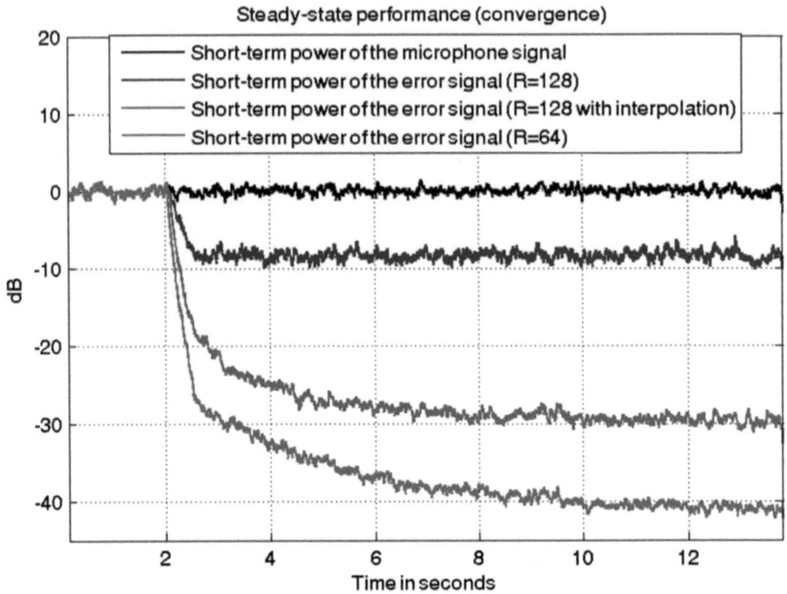

Fig. 9.10 Performance of echo cancellation without and with additional interpolation for $\widetilde{M} = M = 2, \widetilde{N} = N = 256$ and $P = 7$

where β characterizes the step-size and the symbol (*) denotes the conjugate complex. The amount of samples and filter coefficients involved in the convolution and the adaptation, however, is now larger than in a basic scheme: twice as many as before. All other components used within a complete hands-free system like the adaptation control of the echo canceller, residual echo suppression, and beamforming are still operating at the original subsampled rate R. Real-time experiments have shown that with the new interpolation method a much higher subsampling rate can be used, and thus a significant reduction of the computational complexity can be achieved (even if twice as many filter coefficients are required for echo cancellation).

9.5.3 Experimental Results

To show the performance and the accuracy of the proposed time-frequency interpolation method two simulation examples in terms of steady-state convergence are shown in Figs. 9.10 and 9.11. For these measurements, white noise as reference excitation has been used. For the analysis decomposition an FFT of size 256 and a Hann window have been utilized. Local speech and background noise are not considered in these simulations. For the filter update the Normalized Least Mean Square (NLMS) algorithm with a step-size of $\beta = 1.0$ has been employed.

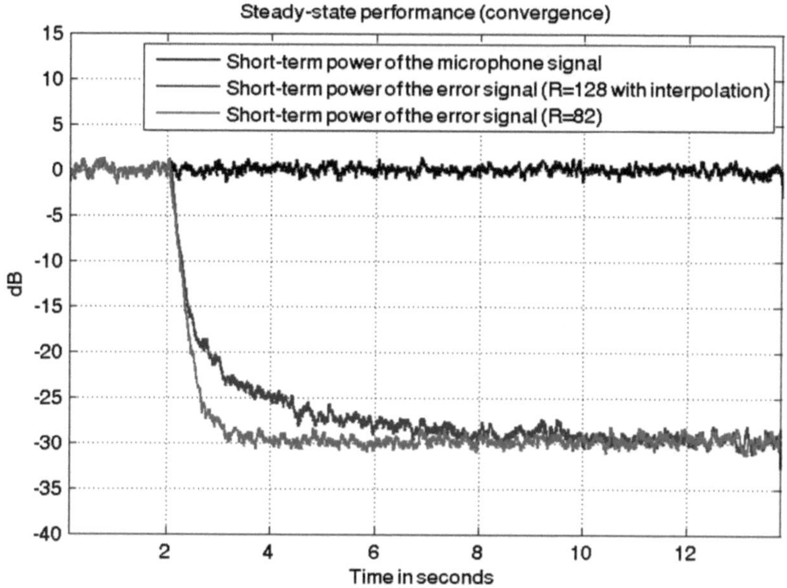

Fig. 9.11 Performance of echo cancellation without and with additional interpolation $\left(\widetilde{M} = M = 2, \widetilde{N} = N = 256, P = 7\right)$. Different subsampling rates (frameshifts) are used

To model the entire tail of the LEM impulse response in this experiment, sufficient echo cancellation filter orders have been applied.

First graph from the top of Fig. 9.10 shows the normalized power of the microphone signal, whereas the second and the third curves depict the power of the error signal (steady-state performance) without and with additional temporal interpolation at a subsampling rate of 128 (frameshift = 50 % of FFT order). As a result, the performance of echo cancellation without additional interpolation (second graph) is strongly degraded due to increased aliasing terms—only about 9 dB echo attenuation can be obtained. Using the proposed interpolation method with $M = 2$ and $P = 7$ at the same subsampling rate of $R = 128$ a significant improvement of about 22 dB in terms of echo reduction is achieved (third graph). However, the performance of a 75 % blockoverlap (frameshift = 25 % of FFT order) as it is shown in the fourth curve cannot be achieved, but it has to be noticed that in a real application the performance is in the majority of cases limited to about 30 dB (e.g., due to the background noise).

Figure 9.11 shows the performance of echo cancellation in terms of echo attenuation for different subsampling rates. Second curve (from the top) demonstrates that at maximum a frameshift of $R = 82$ (blockoverlap of about 68 %) can be chosen in order to achieve a sufficient echo attenuation of about 30 dB.

As already mentioned, when the frameshift is increased, the echo attenuation decreases. Increasing the frameshift to $R = 128$ would cause severe problems for the echo reduction performance. The achieved echo attenuation performance with activated time-frequency interpolation at an increased frameshift of ($R = 128$) is visualized in the second graph. The visualization of the error signal without interpolation and at a reduced subsampling rate ($R = 82$) shows nearly the same overall performance as with interpolation and at an increased subsampling rate ($R = 128$). In both cases an echo attenuation of about 30 dB is achieved (after the echo cancellation filter has converged). Compared to a standard method without interpolation only small performance degradation in terms of speed of convergence is obtained. It has to be noticed that using the temporal interpolation for echo cancellation the overall system complexity can be significantly reduced (with the setup mentioned above by about 35 %) while the overall performance almost remains unchanged.

9.6 Conclusions

In this contribution a spectral refinement method and a temporal interpolation method were presented. Both methods are applied as a post-processing stage of a conventional analysis filterbank for speech signals.

In a first stage, a general solution on how to individually refine subband signals was derived. For its realization, a computationally efficient method was proposed based on a linear combination of weighted subband signal vectors—the refinement procedure can easily be implemented using short FIR filters in each subband. The SR method is particularly suitable for speech processing systems with integrated analysis filterbanks or DFTs—thus, by applying SR as a post-processing stage, specific feature estimation schemes such as pitch frequency or noise power estimation can be further improved. The calculation of SR introduces an additional delay in the signal path, which can be kept low using short FIR filters for the refinement. In this contribution, the SR method has been applied for fundamental frequency estimation. Evaluations demonstrated that pitch frequency estimation was improved considerably for all considered SNR levels. For pitch estimation, only a refinement of the input signal at lower frequencies (up to 1 kHz) is needed. This results in a very low computational complexity.

In a second step, a post-processing scheme for analysis filterbanks applied in the reference path of an adaptive system identification scheme has been presented. It is well known that by increasing the frameshift, the computational complexity is reduced while the overall performance achieved is decreased. Speech enhancement algorithms such as noise suppression or residual echo suppression still operate well with higher frameshifts. The most critical component within a hands-free system is the subband echo cancellation due to increased aliasing distortions. The proposed time-frequency interpolation method is able to significantly reduce aliasing terms caused by a subsampling unit within an analysis filterbank. It has been shown that the post-processing scheme can be realized in an effective and efficient way based on a weighted sum of subband signals and without inserting any additional delay in the signal path. The new time-frequency interpolation method has been applied for subband echo cancellation in the reference path. Experimental results have shown that the post-processing stage allows for an improved steady-state convergence if the subsampling is kept unchanged. Alternatively, the frameshift can be increased significantly with the proposed interpolation method, leading to a reduction of the computational complexity while keeping the overall performance in terms of convergence speed and steady-state performance constant.

References

1. C. Beaugeant, V. Turbin, P. Scalart, A. Gilloire, New optimal filtering approaches for hands-free telecommunication terminals. Signal Process. **64**(1), 33–47 (1998)
2. A. de Cheveigne, H. Kawahara, Yin, a fundamental frequency estimator for speech and music. J. Acoust. Soc. Am. **111**(4), 1917–1930 (2002)
3. J.M. Cioffi, J.A.C. Bingham, Data-driven multitone echo canceller. In *Proc. GLOBECOM 1991*, vol 1, pp. 57–61, Phoenix, Arizona, 1991
4. R.E. Crochiere, L.R. Rabiner, *Multirate Digital Signal Processing* (Prentice-Hall, Upper Saddle River, NJ, 1983)
5. Y. Ephraim, D. Malah, Speech enhancement using a MMSE short-time spectral amplitude estimator. IEEE Trans. Acoust. Speech Signal Process. **32**(6), 1109–1121 (1984)
6. ETS 300 903 (GSM 03.50), *Transmission Planning Aspects of the Speech Service in the GSM Public Land Mobile Network (PLMS) System* (ETSI, France, 1999)
7. A. Gilloire, M. Vetterli, Adaptive filtering in subbands with critical sampling. IEEE Trans. Signal Process. **40**(8), 1862–1875 (1992)
8. N. Grbic, J.M. de Haan, I. Claesson, S. Nordholm, Design of oversampled uniform DFT filter banks with reduced inband aliasing and reduced inband aliasing and delay constraints, in *Proc. ISSPA 2001*, vol 1 (2001), pp. 104–107
9. E. Hänsler, G. Schmidt, *Acoustic Echo and Noise Control: A Practical Approach* (Wiley, Hoboken, NJ, 2004)
10. P. Hannon, M. Krini, G. Schmidt, A. Wolf, Reducing the complexity or the delay of adaptive subband filtering, in *Proc. ESSV 2010*, Berlin, Germany, 2010
11. Harman/Becker Automotive Systems, *Low Complexity Echo Compensation*, EPO patent application, EP 1936939 A1, 2006
12. ITU-T Recommendation P.340, in *Transmission characteristics and speech quality parameters of hands-free terminals* (Geneva, Switzerland, 2001)
13. J.C. Junqua, The influence of acoustics on speech production, a noise-induced stress phenomenon known as the Lombard reflex. Speech Commun. **20**(1), 13–22 (1996)
14. W. Kellermann, Analysis and design of multirate systems for cancellation of acoustic echoes, in *Proc. ICASSP 1988*, vol 32, no 2 (1988) pp. 2570–2573
15. M. Krini, G. Schmidt, Model-based speech enhancement, in *Speech and Audio Processing in Adverse Environments*, ed. by E. Hänsler, G. Schmidt (Springer, Berlin, Germany, 2008), pp. 89–134
16. M. Krini, G. Schmidt, Method for temporal interpolation of short-term spectra and its application to adaptive system identification, in *Proc. ICASSP 2012*, Kyoto, Japan, 2012
17. M. Krini, G. Schmidt, Spectral refinement and its application to fundamental frequency estimation, in *Proc. WASPAA 2007*, New York, USA, 2007
18. D.R. Morgan, J.C. Thi, A delayless subband adaptive filter architecture. IEEE Trans. Signal Process. **43**(8), 1819–1830 (1995)
19. P.A. Naylor, O. Tanrikulu, A.G. Constantinides, Subband adaptive filtering for acoustic echo control using allpass polyphase IIR filterbanks. IEEE Trans. Speech Audio Process. **6**(2), 143–155 (1998)

20. M.R. Schroeder, Period histogram and product spectrum: new methods for fundamental frequency measurements. J. Acoust. Soc. Am. **43**(4), 829–834 (1968)
21. P.J. VanGerwen, F.A. Van de Laar, J. Kotmans, *Digital Echo Canceller*, U.S. Patent 4,903,247, 1990
22. G. Wackersreuther, On the design of filters for ideal QMF and polyphase filter banks. Arch Elektronik Übertragungstech **39**(2), 123–130 (1985)

Part III
Driver Distraction

Chapter 10
Effects of Multitasking on Drivability Through CAN-Bus Analysis

Amardeep Sathyanarayana, Pinar Boyraz, and John H.L. Hansen

Abstract Humans try their best to maximize their abilities to handle various kinds of tasks, be it physical, auditory, visual, or cognitive. The same is true when a person is driving a vehicle—while driving is the primary task of a driver, he/she will attempt to accomplish secondary tasks such as speaking over a cell phone, checking and creating text messages, and selecting music or viewing/accessing news. Though the driver's primary intention is a safe drive, as previous studies have shown (Wilde, Target risk: dealing with the danger of death, disease and damage in everyday decisions, 1994), drivers elevate their risk-taking ability to an optimal level. While performing various tasks this balance between drivability and risk taking can vary, leading to driver distraction and possible accidents. The automotive industry has taken special care to reduce the complexity of operating in-vehicle infotainment systems. Better ergonomics and haptic (tactile) systems have helped achieve comfortable usability. Advances in driver assistance systems have also resulted in increased use of audio-based feedback (Forlines et al. Comparison between spoken queries and menu-based interfaces for in-car digital music selection, 2005) from navigation and other systems. It is very important to understand how these secondary tasks and feedback systems affect the driver and his/her drivability. This chapter focuses on understanding how drivers react to various secondary tasks. An analysis on driving performance using vehicle dynamics and sensor information via CAN-bus shows interesting results on how performing secondary tasks affect some drivers. Previous studies

A. Sathyanarayana (✉) • J.H.L. Hansen
Center for Robust Speech Systems (CRSS), Erik Jonsson School of Engineering and Computer Science, University of Texas at Dallas, Dallas, TX, USA
e-mail: amardeep@utdallas.edu; john.hansen@utdallas.edu

P. Boyraz
Center for Robust Speech Systems (CRSS), Erik Jonsson School of Engineering and Computer Science, University of Texas at Dallas, Dallas, TX, USA

Istanbul Technical University & MERC, Istanbul, Turkey
e-mail: boyraz.pinar@googlemail.com

G. Schmidt et al. (eds.), *Smart Mobile In-Vehicle Systems: Next Generation Advancements*, DOI 10.1007/978-1-4614-9120-0_10,

(Sathyanarayana et al. Driver behavior analysis and route recognition by hidden Markov models, 2008 I.E.E.E International Conference on Vehicular Electronics and Safety, 2008) have shown how maneuvers can be segmented into preparatory, maneuver, and recovery phases. Initial results presented in this chapter show a similar trend in how drivers handle secondary tasks. Even if secondary tasks do not distract the driver, results show driver variations in anticipation or preparation for the task, performing the task itself, and post completion (recovery) of the task.

Keywords Active safety • CAN-bus • Distraction detection • Drivability

10.1 Introduction

Today's fast-paced world places increased demands on sophisticated modes of transport. The sophistication is not just in safety but also in infotainment systems. People spend more time in their cars and therefore are trying to "do more" within that time. An average American spends more than 300 h in a vehicle each year [1]. Business, shopping, ordering food, searching for places, exchanging emails, eating, speaking over a cell phone, texting, and many more tasks happen on the move while people are driving. Many engineering fields have come together to make this possible in the car. The automotive industry also recognizes and accepts the needs of today's society. However, the auto industry stops short of placing restrictions on secondary tasks performed in the car. Human error has been the cause of 57 % of accidents, and in 95 % of the accidents, it was a contributing factor [2]. Though the automotive industry is currently focusing on fuel efficiency and green vehicles to protect the environment, safety of the occupants has always been an integral part of evolution of the automobile. Though technologies such as voice interactive systems, navigation systems, and hands-free mobile communication have proven to achieve better and safer driving than their manual interfaces [3], it is important to understand the impact of adding new infotainment features in the car on the cognitive driver load. Handling more than one system at a time could increase the physical and cognitive load-handling capacity of the driver, causing distraction while driving. In the context of driving, distraction could be defined as anything which diverts the attention of the driver causing any deviation from a normal driving pattern [4]. Causes of such distractions could be broadly classified into visual, cognitive, biomechanical/physical, and auditory. These distractions have a varying impact on normal driving patterns and could result in slight, severe, or fatal accidents [5].

Even with new laws prohibiting the use of infotainment systems in various regions, the number of accidents has not shown a decline. An alternate and more feasible option could be in the development of intelligent vehicles which help manage the cognitive workload of the driver. Understanding and modeling driver behavior is a major component of such systems. Though this modeling approach is not new [6–8], researchers worldwide have begun to recognize its advantages.

Infotainment systems are employing learning algorithms to understand the user's (driver's) needs, and suggestions are prompted to the driver for easy access, hence reducing the effect of secondary tasks while driving.

Even if these secondary tasks do not divert the driver's attention from the road, and might not pose as a major threat, they can still expose drivers to increased cognitive workload. In this chapter we try to identify such risky situations when the driver is most vulnerable and understand how drivers handle the vehicle and perform secondary tasks.

10.2 Signal Acquisition

For the safety of both driver and vehicle, it is important to understand how drivers perform while driving. Environmental conditions, vehicle condition, and driver state are important factors which affect driving performance. Various sensors and systems such as sun load sensors (to identify illumination), gyro sensors (for vehicle and road banking), head distance sensors, and cameras are used to assess and provide assistance for environmental variations. There are also sensors which provide monitoring of the vehicle condition and provide control feedback to compensate for any unnecessary vehicle variations.

A driver could be cautious in general, but due to secondary tasks, fatigue, or urgency in reaching a destination, he/she might drive erratically. So accessing information on how the driver controls and maneuvers the vehicle could provide an improved understanding of driver status and long-term driving characteristics. In an auto transmission vehicle, the driver's primary contacts are the steering wheel, gas, and brake pedals. He/she uses these to maneuver the vehicle in response to the route and regulate the vehicle speed.

Over the past few decades, automobiles have transitioned from pure mechanical systems to electromechanical systems with extensive sensors, actuators, and embedded systems controlling the core vehicle functionality. Communication between these systems mostly happens via a network called the Controller Area Network (CAN) [9]. The CAN-bus carries vital information such as engine temperature, air pressure, and fuel monitoring, which reflects current vehicle conditions. Along with these, there are signals such as gas pedal pressure, brake pedal pressure, and steering wheel angle which are the driver's direct controls to maneuver the vehicle and vehicle speed, which is the driver's main feedback. Some CAN signals are made available to the outside world through the On-Board Diagnostic (OBD) port.

Since the main focus of this study is to understand how drivers handle the vehicle while performing secondary tasks, CAN-bus data provides a reliable and sufficient source of information. Rather than adding extra sensors, CAN-bus data was tapped from the OBD port and deciphered to obtain valuable vehicle dynamic information. Figure 10.1 shows the engine rpm, brake pedal pressure, steering wheel angle, and vehicle speed information obtained from the CAN-bus.

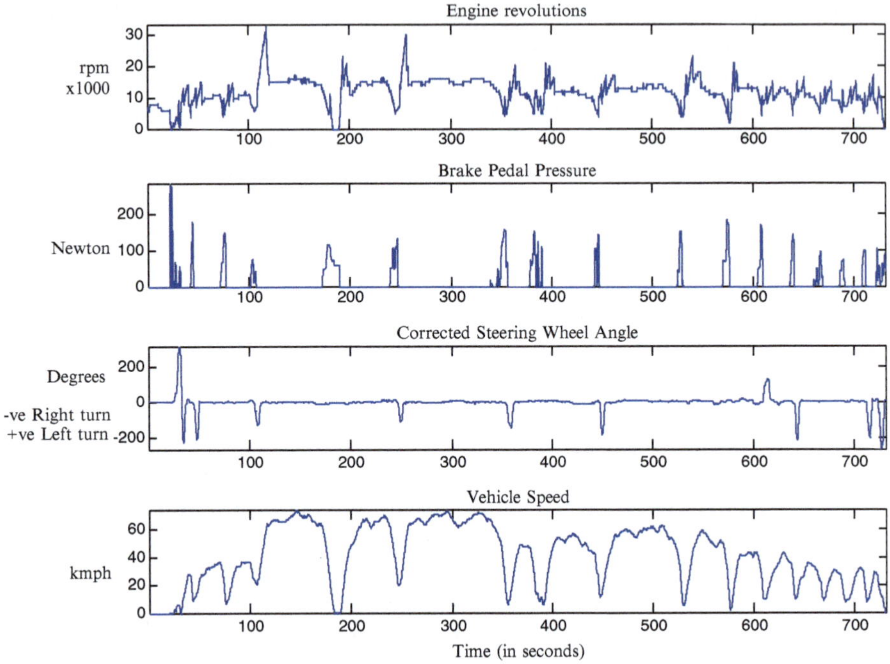

Fig. 10.1 Vehicle CAN-bus signals extracted from OBD port

10.2.1 A Brief Note on UTDrive Project

To build effective driver dependent systems, developing mathematical models capable of explaining and predicting driver behavior is important. In order to obtain modeling parameters to build driver models, a multimodal data acquisition platform is used to collect data using a vehicle in real traffic conditions.

Efforts have been made to build a rich multimodal database of real-world driving data which is demographically well balanced with a wide range of drivers from different nationalities, age, gender, and varied levels of driving experience. This UTDrive Project [10] is part of a 3-year NEDO-supported international collaboration between universities in Japan, Italy, Singapore, Turkey, and the USA. The UTDrive Project has been collecting and researching on multimodal data for developing a framework for driver behavior modeling and driver–vehicle interactions for safe driving. The data was collected using a Toyota RAV4, instrumented with various sensors (i.e., audio, video, gas/brake pedal pressures, forward distance, GPS information, and CAN-bus signals) as shown in Fig. 10.2.

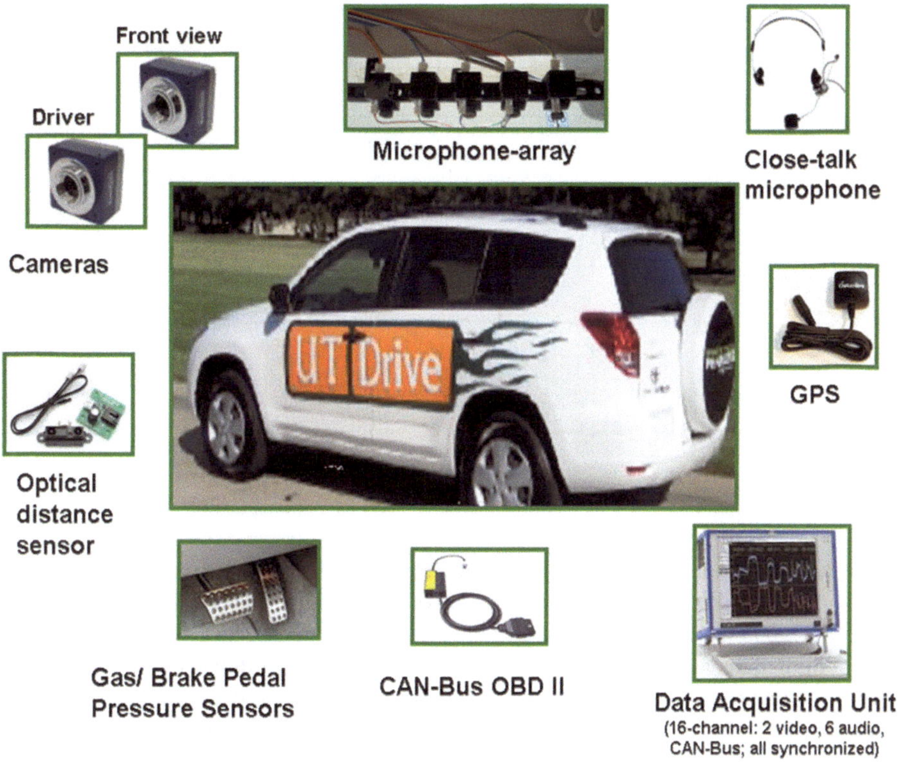

Fig. 10.2 Instrumented UTDrive data collection vehicle along with the sensors [10]

10.2.2 Data/Route Description

A subset of the UTDrive Project corpus is used in this study to analyze the effect of secondary tasks on drivers. Signals from CAN-bus, such as engine rpm, brake pedal pressure, steering wheel angle, and vehicle speed (as seen in Fig. 10.1), are used to model driver behavior. Other sensors including camera and microphone data are used to transcribe the CAN-bus signals. Data transcription plays a crucial part in developing mathematical models as it labels the driver's actions. This not only provides the basis for further signal processing, but also serves in evaluating results as a ground truth. Since real-world traffic scenarios and driving are highly dynamic, transcribers process the entire route to label events which occur during driving. To remove any ambiguity due to the subjective nature of manual transcription, transcription is performed more than once by different transcribers. Feedback is also collected from drivers after every driving session regarding their experience in performing maneuvers and secondary tasks. Hence driver activity is labeled using multiple sensor information; two cameras—one facing the road and another facing the driver, microphone array—listening to in-vehicle conversations; and CAN-bus signals—looking at the vehicle dynamics.

Fig. 10.3 Secondary tasks performed in different legs of the route [4]

A small subset of eight drivers' data from the UTDrive corpus is used in this study. Each driver is required to drive through the route twice which is shown in Fig. 10.3. The route takes approximately 10 min to complete and passes through residential areas and school zone. The drivers are made familiar with the vehicle, its controls, and the route itself. In the first run, drivers do not perform any secondary tasks and drive in a neutral/normal driving scenario. In the second run, they are expected to perform secondary tasks while driving. The selected secondary tasks are commonly performed/attempted by drivers on a regular basis (labeled in Fig. 10.3, against each leg of the route). The four different tasks performed are conversation with a co-passenger, random lane changing, speaking over a cell phone to an automated dialog system, and performing some common tasks such as tuning the radio, selecting a particular song in a music player, and adjusting the AC/heater levels. Though lane changing is not a secondary task and a part of normal driving maneuver, it has been included in this study to benchmark cognitive loads for secondary tasks versus a typical driving task.

In this study the drivers are numbered from 1 to 8 (in no particular order), and the secondary tasks performed are labeled as LC (Lane Changing), CO (Conversation with co-passenger), MP (speaking on mobile phone), and CT (Common Tasks) [4].

10.3 Previous Work

The main focus of the UTDrive Project during the last few years has been to collect and perform research on multimodal in-vehicle data to understand and model driver behavior in developing intelligent vehicles. Among several areas, one key focus has been to best utilize the available CAN-bus signals. Previous work [10–13] has shown how CAN-bus signals can be used to identify maneuvers and distraction in those maneuvers. Two different approaches—driver independent and driver dependent—were adopted for the analysis.

10.3.1 Driver-Independent Approach [11]

In this approach, the signal patterns are used to identify maneuvers and routes. Driving signals and maneuvers are considered to be analogous to speech signals in their structure. As in speech recognition, where phonemes form words which form phrases which complete a sentence, in route recognition, sub-maneuvers (drivemes) form maneuvers which form multi-maneuvers and finally form a route. Hidden Markov Models (HMM) were used to model maneuvers for both neutral and distracted versions of the maneuvers. Using this HMM framework, right turns, left turns, and lane changes were detected with 100, 93, and 81 % accuracy [11]. However, the accuracies were not high in classifying these detected maneuvers into neutral and distracted driving.

10.3.2 Driver-Dependent Approach [10, 13]

From the driver independent approach, it was observed that a fast and robust system would need to remove inter-driver variations that depend on driver's individual traits. This is especially true in detecting driver distractions, as one person's distracted driving pattern could be a normal driving pattern for another. For example, a cautious driver's average speed could be 40 mph against a more aggressive driver whose average speed is 50 mph. A classic three-stage strategy was adopted to identify the driver, prune the search space to only driver-specific maneuver models, and once maneuvers were recognized, further prune the search to driver's comfort levels, to identify if he/she was distracted or not. With a 100 % accuracy in identifying the driver using audio data, and a high accuracy in maneuver recognition, this GMM–UBM framework was used for distraction detection. Though distraction detection performance was 71.2 %, it was noted that the false alarm rate (distracted driving detected as normal driving) was 28.2 %, which is not acceptable for any active safety application [10].

Further analysis was performed based on high-frequency content, entropy, and standard deviation of signals, and distraction detection was improved to 95 % [12].

10.4 Effect of Secondary Tasks on Drivers

From previous work, it is clear that by using GMMs or HMMs, it is possible to identify maneuvers and also identify if these maneuvers are executed normally or if the driver was distracted. It is also noted that some drivers are more distracted than others. If some drivers are not distracted by performing secondary tasks, in this study we try to understand how their driving differs from those who are distracted while performing the same task.

As noted in Sect. 10.2.2, drivers are expected to perform secondary tasks such as LC (Lane Changing), CO (Conversation with co-passenger), MP (speaking over mobile phone), and CT (Common Tasks) on a particular route which would have been driven once normally without any secondary tasks. The data is carefully transcribed to identify instances when the drivers are performing secondary tasks and are found distracted. Since the effect of driver variability and driving context should be minimized to assess driver intent and identify distraction detection, a driver-dependent GMM–UBM framework is adopted as shown in Fig. 10.4.

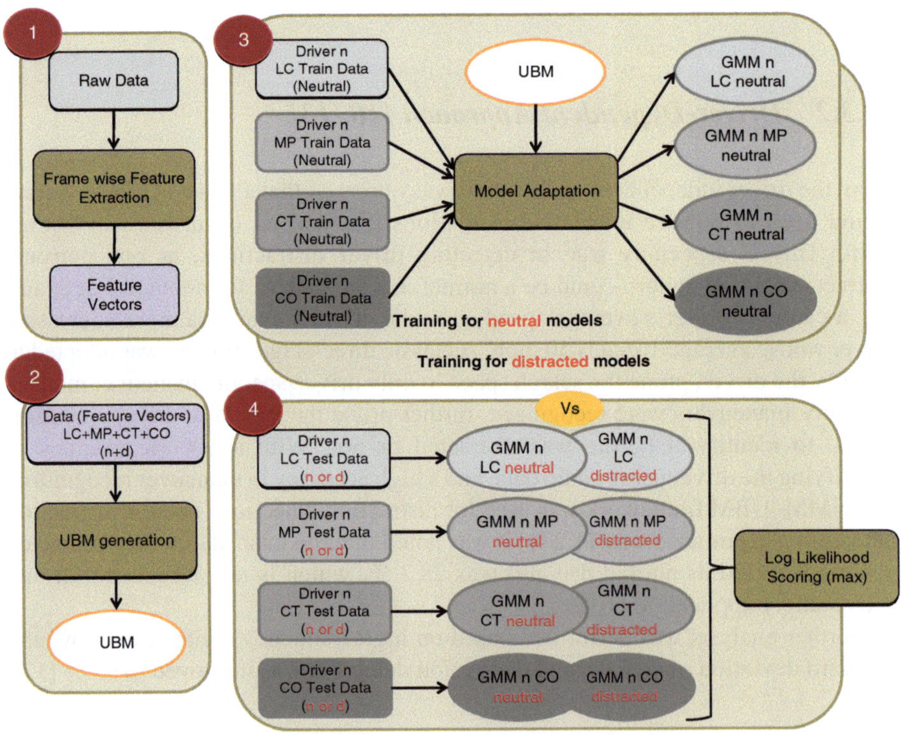

Fig. 10.4 Distraction detection system based on GMM–UBM framework (*1*) feature extraction (*2*) UBM generation (*3*) MAP adaptation (*4*) maximum log-likelihood scoring

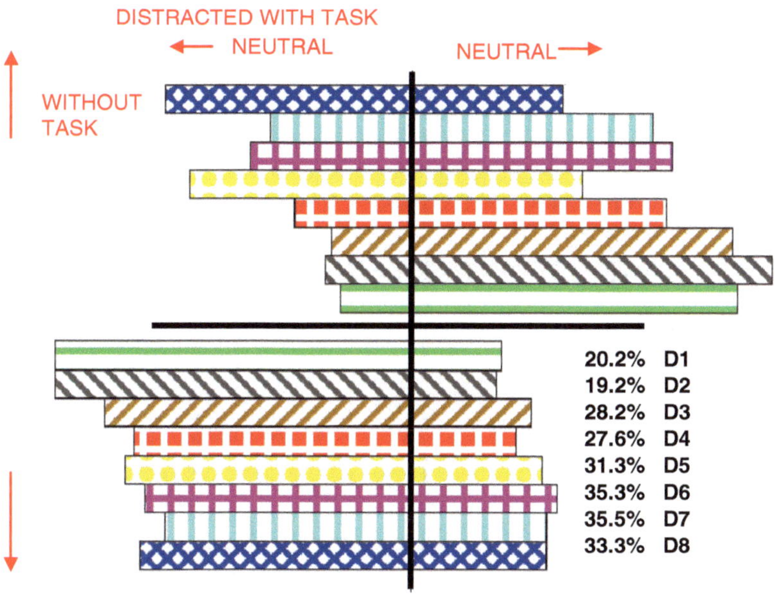

Fig. 10.5 Neutral versus distracted driving of all drivers while performing and while not performing secondary tasks

A detailed description of Gaussian Mixture Models (GMM) and Speaker Recognition can be found in [14]. Incorporating the same approach, eight drivers' CAN-bus data was segmented on the basis of secondary tasks. There are four stages in the GMM/UBM framework: (1) feature extraction, (2) universal background model (UBM) generation (development), (3) driver-dependent model adaptation (training), and (4) testing as shown in Fig. 10.4. Some salient features such as raw signals of vehicle acceleration, brake pedal pressure, steering wheel angle, vehicle speed, their derivates, and standard deviation are extracted from the CAN-bus signals to form feature vectors. A UBM is developed using a large number of drivers' CAN-bus data other than the eight used here for training and test. Two sets of driver-dependent GMMs (neutral and distracted) are obtained by MAP (Maximum A Posteriori) adapting the UBM using neutral and distracted feature vectors. Using log-likelihood scoring, each test data is scored against both GMMs representing neutral and distracted models for a particular driver. The results are plotted in Figs. 10.5 and 10.6.

Similar to the study in [4], the Kullback–Leibler (KL) distance is computed between neutral and distracted GMMs for every driver. The results are tabulated in Table 10.1. KL distance measures the difference between a reference and another arbitrary probability distribution [15]. In this case, if the distance is small, the neutral and distracted GMMs do not differ in CAN-bus signal structure. However, if the distances are large, this implies significant changes in the underlying GMM structure for distraction.

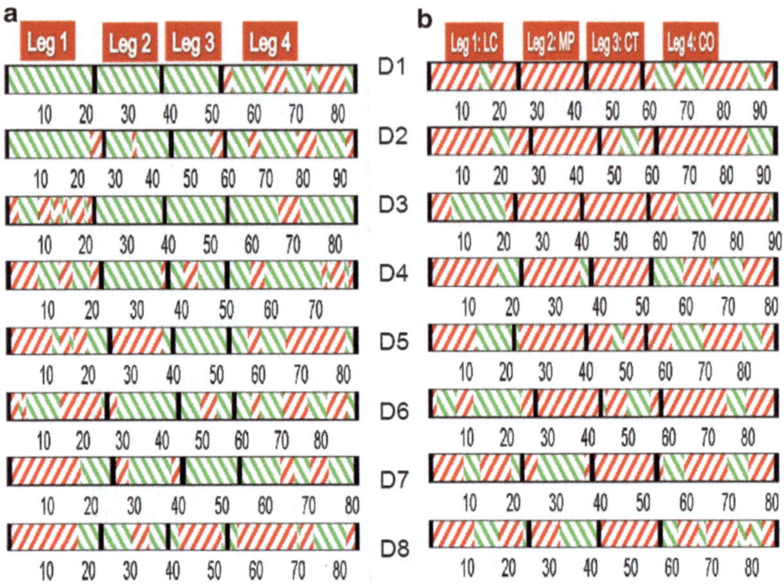

Fig. 10.6 Neutral (*green* "\" *hashed*) and distracted (*red* "/" *hashed*) decision over every 5 s frame of driving data. *Black bars* indicate leg boundaries. (**a**) Decision when no secondary tasks were performed (**b**) Decision when secondary tasks were performed

Table 10.1 KL distance between neutral and distraction driving GMM

KL	LC	MP	CT	CO
D1	8.45	26.62	21.86	22.8
D2	12.01	27.39	17.04	13.2
D3	18.7	30.21	20.2	16.44
D4	8.93	23.02	19.61	14.03
D5	11.65	12.96	20.3	12.1
D6	16.33	24.82	19.29	20.87
D7	9.48	14.47	24.07	14.7
D8	9.2	15.1	10.8	13.94
Avg	11.84	21.82	19.14	16.01
Classify	No	High	Medium	Low

10.5 Results and Conclusions

The results obtained show the effect of performing secondary tasks on each driver's driving pattern. Adding extra features has proven useful in identifying distractions, and KL distance still shows similar trends compared to the results obtained in [4]. When test data (at 5 s per frame) is classified into either neutral or distracted, it is found to have some unique characteristics. The per-frame decision for all eight drivers for each leg of the data is shown in Fig. 10.6. If the data frame was classified as neutral, it is represented by a green "\" hashed region, and a distracted frame was represented by a red "/" hashed region. Figure 10.6a was the result obtained for data when the route was driven normally without performing any secondary tasks, and Fig. 10.6b was obtained for data when secondary tasks were performed in each leg of the route. The black bars in Fig. 10.6 indicate different leg boundaries.

Figure 10.5 shows a graphical representation of the percentage of time each driver was distracted while performing a secondary task and the percentage of time the same driver was distracted while not performing any secondary task. It can be observed that the first three drivers (D1, D2, D3) are quite comfortable driving the car without performing any secondary task, but get highly distracted while performing some secondary tasks. It can also be observed that drivers D5 and D8 are generally distracted while driving. The video playback shows that these drivers were cautious all the time while driving and did not drive in their normal driving habit. Hence the system marked a majority of their driving as distracted.

A deeper analysis of Fig. 10.6 shows that when the KL distance between neutral and distracted models is smaller than the average, it is difficult to distinguish between neutral and distracted states. In such cases, the test result per frame can be seen toggling between the two states.

Also, when the KL distance between neutral and distracted models is greater than the average, it is easier to classify the test data as neutral or distracted. There are fewer or no transitions between the states. It is noted that some drivers are comfortable handling some in-vehicle controls while struggling with others. So, even though they are performing some in-vehicle task while driving, their comfort level in handling the tasks determines whether they are distracted or not. Conversation with co-passengers is very erratic, and the authors believe that a separate study is required to analyze whether there exists a pattern in drivers being distracted while taking or listening or both.

A graphical representation of how neutral and distracted GMMs could be separated is shown in Figs. 10.7 and 10.8. As can be seen from Fig. 10.8, there could be a large overlapping region between the models, or they could be very close making transitions between models possible. Drivers generally do not stay in one state and often toggle between models/states. A deeper analysis on how often they toggle would give new insight into understanding the driver and their driving traits. This tendency is seen by observing the results from Fig. 10.6. Drivers have a tendency to adjust to small changes over time. These corrective actions should not be wrongly flagged as distraction. In fact, this is the driver's effort to recover towards normal driving. An example of such a scenario is when the driver is trying to tune the radio to any channel and starts drifting away but stays within the lane boundaries. As soon as he switches his attention towards the road, he pulls the steering wheel to recover the vehicle back to the center of the lane. This momentary lapse in attention could be a stray occurrence and should not be flagged as distracted. If the driver drifts out of the lane or for a longer time or continues to drift a few times within the lane, then it could be that the driver is really distracted and could be helped with some assistance.

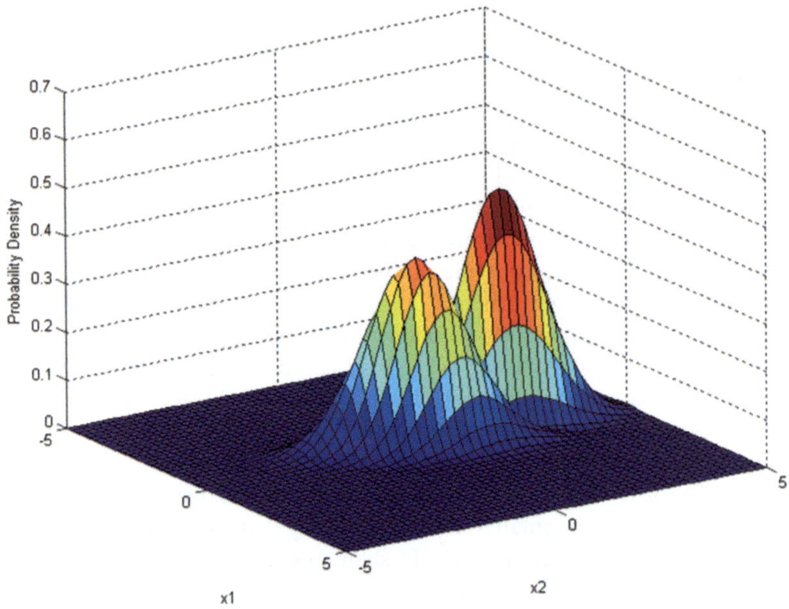

Fig. 10.7 Graphical representation of neutral and distracted GMMs

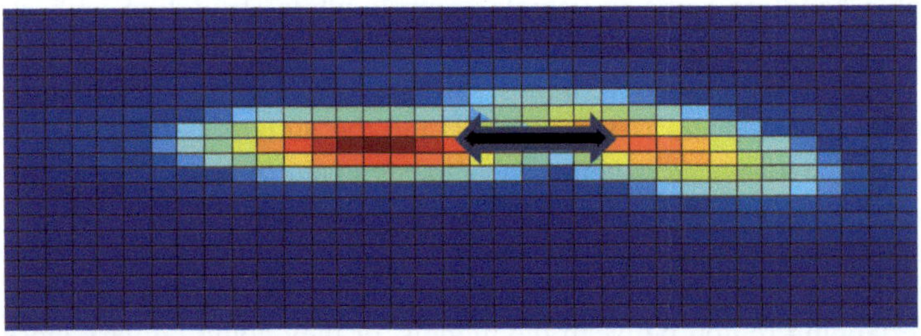

Fig. 10.8 Top view of the GMM graphical representation showing the area covered and overlapping regions

On careful observation of results in Fig. 10.6b, it can be seen that more than 70 % of the drivers in this set follow a particular pattern while performing different secondary tasks. Most of the observation frames flagged as distractions (not their normal driving pattern) can be grouped into three sequential events. Initially during the start of a task, most of the drivers are distracted. This is justified as they give more *attention* towards the task, gauge/assess the surroundings, and get ready to perform the task. This phase can be termed as Anticipation or Preparatory Phase. Once the drivers are prepared with their secondary task, 50 % of the drivers fall into a comfort zone where they feel confident and comfortable in executing the task. Hence they focus more on driving and return back to their normal driving pattern. This phase is the actual execution of the task. Similar to the preparatory phase, several drivers are slightly distracted at the end of the task or immediately after completion of the secondary task. This phase is called the Recovery or Post Completion Phase. On observing the videos it is noted that these drivers reassess their surroundings during this phase and so deviate momentarily. The duration of each of these phases is based on individual driver's comfort and confidence level. It can be seen from Fig. 10.6b that most drivers are distracted even while performing the secondary tasks in Leg 2 and Leg 3 (speaking over mobile phone and performing in-vehicle common tasks). This shows that these tasks are inherently distractive, and drivers are not able to return to normal driving while performing these secondary tasks and hence are suggested to be avoided.

The results from this study suggest a viable means of modeling and assessing driver performance with and without the presence of distraction-based tasks. Further research building on this study would clearly be justified to formulate real-time in-vehicle detection/assessment methods that tune to the specific driver traits.

References

1. National Highway Traffic Safety Administration official website (Online source, 2011, March) http://www.nhtsa.dot.gov
2. J.R. Treat, N.S. Tumbas, S.T. McDonald, D. Shinar, R.D. Hume, R.E. Mayer, R.L. Stanisfer, N.J. Castellan, Tri-level study of the causes of traffic accidents, *Report No. DOT-HS-034-3-535-77 (TAC)* (1977)
3. C. Carter, R. Graham, Experimental comparison of manual and voice controls for the operation of in-vehicle systems, in *Proceedings of the IEA2000/HFES2000 Congress*, Santa Monica, CA, USA
4. A. Sathyanarayana, P. Angkititrakul, J.H.L. Hansen, Detecting and classifying driver distraction, in *Biennial on DSP for In-Vehicle and Mobile Systems*, Istanbul, Turkey, 17–19 June 2007
5. M. Pettitt, G. Burnett, A. Stevens, Defining driver distraction, in *World Congress on Intelligent Transport Systems*, San Francisco, November 2005
6. D. McRuer, D. Weir, Theory of manual vehicular control. Ergonom **12**, 599–633 (1969)
7. C. MacAdam, Application of an optimal preview control for simulation of closed-loop automobile driving. IEEE Trans. Syst. Man. Cybern. **SMC-11**, 393–399 (1981)
8. J.A. Michon, A critical view of driver behavior models: what do we know, what should we do? in *Human Behavior and Traffic Safety*, ed. by L. Evans, R.C. Schwing (Plenum Press, New York, 1985), pp. 485–520
9. CAN-Bus technical specifications (Online source, Bosch, 2011, March), http://www.semiconductors.bosch.de/pdf/can2spec.pdf
10. A. Sathyanarayana, P. Boyraz, J.H.L. Hansen, Information fusion for context and driver aware active vehicle safety systems. Inf. Fusion (Elsevier) **12**, 293–303 (2011)
11. A. Sathyanarayana, P. Boyraz, J. H. L. Hansen, in *Driver Behaviour Analysis and Route Recognition by Hidden Markov Models*, IEEE International Conference on Vehicular Electronics and Safety, Ohio, USA, 22–24 September, 2008
12. H. Boril, P. Boyraz, J. H. L. Hansen, Towards multi-modal driver's stress detection, in *4th Biennial Workshop on DSP for In-Vehicle Systems and Safety*, TX, USA, 25–27 June 2009

13. A. Sathyanarayana, P. Boyraz, Z. Purohit, R. Lubag, J.H.L. Hansen, driver adaptive and context aware active safety systems using CAN-bus signals, in *IEEE Intelligent Vehicle Symposium*, San Diego, CA, USA, 21–24 June 2010
14. D.A. Reynolds, Speaker identification and verification using Gaussian Mixture Models. Speech Commun. **17**, 91–108 (1995)
15. S. Kullback, R.A. Leibler, On information and sufficiency. Ann. Math. Stat. **22**, 79–86 (1951)

Chapter 11
Using Perceptual Evaluation to Quantify Cognitive and Visual Driver Distractions

Nanxiang Li and Carlos Busso

Abstract Developing feedback systems that can detect the attention level of the driver can play a key role in preventing accidents by alerting the driver about possible hazardous situations. Monitoring drivers' distraction is an important research problem, especially with new forms of technology that are made available to drivers. An important question is how to define reference labels that can be used as ground truth to train machine-learning algorithms to detect distracted drivers. The answer to this question is not simple since drivers are affected by visual, cognitive, auditory, psychological, and physical distractions. This chapter proposes to define reference labels with perceptual evaluations from external evaluators. We describe the consistency and effectiveness of using a visual-cognitive space for subjective evaluations. The analysis shows that this approach captures the multidimensional nature of distractions. The representation also defines natural modes to characterize driving behaviors.

Keywords Driver distraction • Active safety • Driver perception • Subjective evaluation • Driving performance

11.1 Introduction

The development of new in-vehicle technology for communication, navigation, and infotainment has significantly changed the drivers' experience. However, these new systems can negatively affect the drivers' attention, exposing them to hazardous situations leading to motor-vehicle accidents [17]. According to the study reported by *The National Highway Traffic Safety Administration* (NHTSA), over 25 % of police-reported crashes involved inattentive drivers [28]. This finding is not

N. Li • C. Busso (✉)
The University of Texas at Dallas, Richardson, TX 75080, USA
e-mail: nxl056000@utdallas.edu; busso@utdallas.edu

G. Schmidt et al. (eds.), *Smart Mobile In-Vehicle Systems: Next Generation Advancements*, DOI 10.1007/978-1-4614-9120-0_11,

surprising since it is estimated that about 30 % of the time that drivers are in a moving vehicle, they are engaged in secondary tasks [27]. Therefore, it is important to develop active safety systems able to detect distracted drivers. A key step in this research direction is the definition of reference metrics or criteria to assess the attention level of the drivers. These reference labels can be used as ground truth to train machine-learning algorithms to detect distracted drivers.

A challenge in defining driver distraction measure methods is the multidimensional nature of the distractions caused by different tasks. Performing secondary tasks while driving affects the primary driving task by inducing visual, cognitive, auditory, psychological, and physical distractions. Each of these distractions has distinct effects on the primary driving performance [9]. During visual distractions, the drivers have their eyes off the road, compromising their situation awareness. During cognitive distractions, the drivers have their mind off the road, impairing their decision making process and their peripheral vision [27] (*looking but not seeing* [32]). A driver distraction measure should capture these facets to reflect the potential risks induced by new in-vehicle systems.

Some studies have used direct measurements derived from the driving activity. These measures include lateral control measures (e.g., lane-related measures), longitudinal control measures (e.g., accelerator-related measures, brake, and deceleration-related measures), obstacle and event detection (e.g., probability of detection measures), driver response measures (e.g., stimulus-response measures), vision-related measures (e.g., visual allocation to roadway), and manual-related measures (e.g., hands-on-wheel frequency) [4, 19, 36–38]. Other studies have used measurements from the drivers including *electroencephalography* (EEG), size of eye pupils and eye movement [4, 22, 23, 26]. Unfortunately, not all these metrics can be directly used to define labels to train machine-learning algorithms to predict distracted drivers. Some of these metrics can only be estimated in simulated conditions (e.g., event detection tasks) while others require intrusive sensors to reliably estimate their values (e.g., bio-signals).

The study addresses the problem of describing driver distraction through perceptual assessments. While common subjective evaluations such as the *NASA task load index* (NASA-TLX), *driving activity load index* (DALI), *subjective workload assessment technique* (SWAT), and *modified Cooper Harper* (MCH) scale rely on self-evaluations [39], we propose the use of external observers to separately evaluate the perceived visual and cognitive distractions—a two-dimensional space to characterize distractions. Subjects, who were not involved in the driving experiments, are invited to observe randomly selected video segments showing both the driver and the road. After watching the videos, they rate the distraction level based on their judgment. Notice that the external observers are required to have driving experience such that they can properly relate to the drivers' actions. The study uses a database recorded in real driving conditions collected with the UTDrive platform—a car equipped with multiple nonintrusive sensors [2]. The recordings include drivers conducting common secondary tasks such as interacting with another passenger, operating a phone, GPS, or radio [6, 15, 16].

Building upon our previous work [16], the chapter analyzes the consistency and effectiveness of using the proposed visual-cognitive space in subjective evaluations to characterize driver distraction. First, the scores are analyzed in terms of the secondary tasks considered in the recordings. The analysis shows high consistency with previous findings describing the detrimental effect of certain secondary tasks. The visual-cognitive space captures the multidimensional nature of driver distractions. Then, the scores provided by different external observers are compared. The inter-evaluator agreement shows very strong correlation for both visual and cognitive distraction scores. The evaluations from external observers are also compared with self-evaluations provided by the drivers. The comparison reveals that both subjective assessments provide consistent descriptions of the distractions induced by secondary tasks. Likewise, the scores from the subjective evaluation are compared with eye glance metrics. The recordings in which the drivers have their eye off the road are consistently perceived with higher visual and cognitive distraction levels. Finally, we highlight the benefits of using the visual-cognitive space for subjective evaluations. This approach defines natural distraction modes to characterize driving behaviors.

The chapter is organized as follows. Section 11.2 summarizes previous work describing metrics to characterize distracted drivers. Section 11.3 describes the experiment framework used to record the audiovisual database and the protocol to obtain the subjective evaluations. Section 11.4 analyzes the subjective evaluation in terms of secondary tasks, and the consistency in the evaluations between external raters. The section also compares the subjective evaluations of external observers with the ones collected from the drivers (e.g., self-evaluations). Section 11.5 studies the deviations observed in eye glance metrics when the driver is engaged in secondary tasks. The section discusses the consistency between perceptual evaluations and eye glance features. Section 11.6 highlights the benefits of using the proposed visual-cognitive space for subjective evaluations to characterize distraction modes. Section 11.7 concludes the chapter with discussion, future directions, and final remarks.

11.2 Related Work

Several studies have proposed and evaluated measurements to characterize driver distractions. This section summarizes some of the proposed metrics.

11.2.1 Secondary Task Performance

A common distraction metric is to measure secondary task performance [4]. In some studies, the recordings in which the driver was performing secondary tasks are directly labeled as distracted while the controlled recordings are labeled as normal [3, 20, 40]. In other studies, the drivers are asked to complete artificial detection tasks not related to the primary driving task, such as identifying objects or events, and solving mathematical problems. The performance is measured as the effectiveness (accuracy) and efficiency (required time) to complete the task. There are various approaches that fall under this category. Examples include *peripheral detecting task* (PDT), *visual detection task* (VDT), *tactile detection task* (TDT), and *signal detection task* (SDT) [9, 23, 26, 36]. Most of the studies are conducted using car simulators, in which the stimulus can be controlled.

11.2.2 Surrogate Distraction Measurements

Studies have proposed surrogate schemes to evaluate the distraction level when the driver operates an in-vehicle technology. These methods are particularly suitable for early stages in the product design cycle of a device that is intended to be used inside the car. The *lane change test* (LCT) is one example [21]. Using a car simulator, the driver is asked to change lanes according to signals on the road while operating a particular device. The distraction level is measured by analyzing the driving performance. Another example is the *visual occlusion* approach, which has been used by automotive human factor experts as a measure of the visual demand of a particular task [11]. In this approach, the field of view is temporally occluded mimicking the eye off the road patterns for visual or visual-manual tasks. During the occlusion interval (usually set equal to 1.5 s), the subject can manipulate the controls of the device, but cannot see the interface or the control values. The time to complete the task provides an estimation of the required visual demand. However, these metrics are not suitable for our goal of defining ground truth labels to describe the distraction level of recordings collected in real traffic conditions.

11.2.3 Direct Driving Performance

Another type of attention measurement corresponds to primary task performance metrics [10, 14, 20, 22, 23, 33, 35]. They determine the attention level of the driver by directly measuring the car response [4]. These measures include *lateral control* such as lane excursions, and steering wheel pattern, *longitudinal control*, such as speed maintenance and brake pedal pattern, and *car following performance*, such as the distance to the leading car. Notice that these measurements may only capture distractions produced by visual intense tasks, since studies have shown that metrics such as lane keeping performance are not affected by cognitive load [9]. Lee et al. [19] suggested that it is important to study the entire brake response process. In this direction, they considered the *accelerator release time* (i.e., the time between the leading car brakes and the accelerator is released), the *accelerator to brake* (i.e., the movement time from accelerator release to initial brake depress), and the *brake to maximum brake* (i.e., the time from the initial brake depress to maximum deceleration). From these measurements, they found that the accelerator release time was the most sensitive metric of braking performance.

11.2.4 Eye Glance Behavior

Movement of the eyes usually indicates where the attention is allocated [36]. Therefore, studies have proposed eye glance behavior to characterize inattentive drivers [4, 22]. This is an important aspect since tasks with visual demand require foveal vision, which forces the driver to take the eyes off the road [36]. The proposed metrics range from detailed eye-control metrics, such as within-fixation metrics, saccade profiles, pupil control, and eye closure pattern, to coarse visual behavior metrics, such as head movement [36]. The total eye off the road to complete a task is accepted as a measure of visual demand associated to secondary tasks. It is correlated with the number of lane excursions committed during the task [38]. The farther away from the road that a driver fixes his/her eyes, the higher the detrimental effect on his/her driver performance [36]. Also, longer glances have higher repercussions than few short glances [38]. In fact, when the eye off the road duration is greater than 2 s, the chances of accidents increases [4, 17]. Another interesting metric is the *percent road center* (PRC), which is defined as the percentage of time within 1 min that the gaze falls in the 8° radius circle centered at the center of the road. While visual distraction is the prominent factor that forces drivers to take their eye off the road, cognitive distractions can also have an impact on eye glance behavior. As the cognitive load increases, drivers tend to fix their eye on the road center, decreasing their peripheral visual awareness [27, 29, 30]. Therefore, lack of eye glances may also signal driver distractions.

One important aspect that needs to be defined in many of the aforementioned driver distraction measurements is the corresponding values or thresholds that are considered acceptable for safe driving [39]. In some cases, organizations have defined those values. For example, the *Alliance of Automobile Manufacturers* (AAM) stated that the total duration required to complete a visual-manual task should be less than 20 s. Metrics such as total glance duration, glance frequency, and mean single glance duration have been standardized by the *International Organization for Standardization* (ISO). In other cases, a secondary task such as manual radio tuning is used as a reference task. When a new in-vehicle task is evaluated, the driving behaviors are compared with the ones observed when the driver is performing the reference task. To be considered as an acceptable, safe task, the deviation in driving performance should be lower than the one induced by the reference task.

11.2.5 Physiological Measurements

Physiological measurements provide useful information about the internal response of the drivers' body when they are conducting secondary tasks. Although the information is collected with intrusive sensors, they provide objective, consistent, and continuous measurements describing drivers' attention (e.g., increased mental workload) [9, 23, 26]. Engstrom et al. [9] used cardiac activity and skin conductance as the physiological measurements for their study on visual and cognitive load. They showed that secondary tasks have an impact on physiological signals. Mehler et al. [23] used physiological measurements including heart rate, skin conductance, and respiration rate to study young adult drivers in a simulator. They found physiological measurements are sensitive to mental workload. Putze et al. [26] considered labeling the workload using subjective evaluation, secondary task performance and multiple physiological measurements (skin conductance, pulse, respiration, and EEG). The results suggested a strong correlation between the three measurements. If these physiological metrics are used to label whether a driver is distracted, appropriate thresholds need to be established to determine acceptable driving behaviors. The challenge is that these thresholds may vary across drivers.

11.2.6 Subjective Assessments

Subjective assessments have been proposed to measure driver distraction. The most common techniques are self-evaluation scales for subjective mental workload. Examples include the *NASA task load index* (NASA-TLX), *driving activity load index* (DALI), *subjective workload assessment technique* (SWAT), *Modified Cooper Harper scale* (MCH), and *rating scale mental effort* (RSME) [39]. For assessment of fatigue, studies have used the *Karolinska sleepiness scale* (KSS) [7]. The NASA-TLX is commonly used to rate self-perceived workload [1, 14, 18, 26]. It includes rating on six different subscales: *mental demand*, *physical demand*, *temporal demand*, *performance*, *effort*, and *frustration*. In addition to the six NASA-TLX scales, Lee et al. [18] included a modified version of the questionnaires to assess situation awareness and perceived distraction. These self-reported evaluations were used to evaluate the workload introduced by a speech-based system to read email. Some studies use a subset of these subscales. For example, Aguilo [1] included only the *mental demand*, *temporal demand*, and *frustration scales* as part of the guidelines in designing *in-vehicle information system*s (IVISs). Harbluk et al. [14] combined eye glance behavior, braking performance, and subjective evaluations (NASA-TLX scales) to study cognitive distraction. They concluded that the drivers' ratings were closely related to the task demands.

Along with self-evaluations, subjective evaluations by external observers have also been used to characterize driver distractions [25, 31]. Sathyanarayana et al. [31] relied on perceptual evaluations to label the videos of the drivers as either distracted or not distracted. Four raters were asked to observe video recordings, and the consensus labels were used as labels for pattern recognition experiments. Piechulla et al. [25] used objective and subjective methods to assess the drivers' distraction. Their study proposed an adaptive interface to reduce the drivers' workload.

This study analyzes the consistency and effectiveness of perceptual assessments of visual and cognitive distractions provided by external evaluators. We demonstrate that the use of subjective evaluations is a valid approach that can overcome the limitations of other measurements to characterize driving behaviors.

11.3 Methodology

11.3.1 UTDrive Platform

To collect a corpus in real driving conditions, this study relies on the UTDrive car (Fig. 11.1a). This is a research platform developed at *The Center for Robust Speech Systems* (CRSS) at *The University of Texas at Dallas* (UT Dallas) [2]. Its goal is to serve as a research platform to develop driver behavior models that can be deployed into human-centric active safety systems. The UTDrive car has been custom fit with data acquisition systems comprising various modalities. It has a frontal facing video camera (PBC-700H), which is mounted on the dashboard facing the driver (see Fig. 11.1b). The placement and small size of the camera are suitable for recording frontal views of the driver without obstructing his/her field of vision. The resolution of the camera is set to 320 × 240 pixels and records at 30 fps. Another camera is

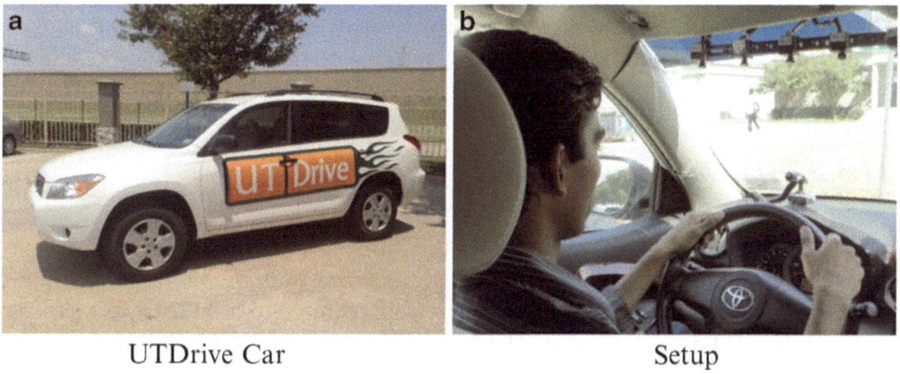

Fig. 11.1 Car platform used for the recording. (**a**) Picture of the UTDrive car (**b**) Placement of the frontal camera and the microphone array

Fig. 11.2 Dewesoft software used for recording and exporting the data. The figure shows the frontal and road videos. It also shows the instantaneous values of various CAN-bus signals

placed facing the road, which records at 15 fps at 320 $\times$ 240 resolution. The video from this camera can be used for lane tracking. Likewise, the UTDrive car has a microphone array placed on top of the windshield next to the sunlight visors (see Fig. 11.1b). The array has five omnidirectional microphones to capture the audio inside the car. We can also extract and record various CAN-bus signals, including vehicle speed, steering wheel angle, brake value, and acceleration. A sensor is separately placed on the gas pedal to record the gas pedal pressure.

The modalities are simultaneously recorded into a Dewetron computer, which is placed behind the driver's seat. A Dewesoft software is used to retrieve synchronized information across modalities. Figure 11.2 shows the interface of the Dewesoft software, which displays the frontal and road videos and various CAN-bus signals. For further details about the UTDrive car, readers are referred to [2].

11.3.2 Database and Protocol

A multimodal database was recorded for this study, using the UTDrive car. Twenty students or employees of the university were asked to drive while performing a number of common secondary tasks. They were required to be at least 18 years old and have a valid driving license. The average and standard deviation of the age of the participants are 25.4 and 7.03, respectively. The recordings were conducted during dry days with good light condition to reduce the impact of the environment variables. Although wet weather can lead to different challenges for the driver,

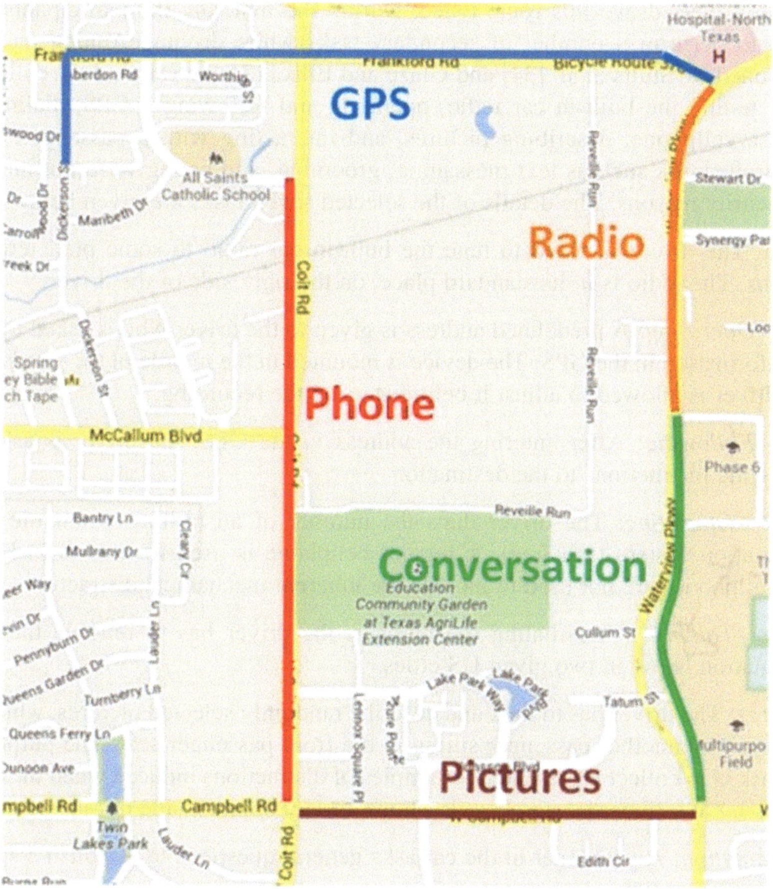

Fig. 11.3 Route used for the collection of the data. The subjects drove this 5.6 miles-long route twice. First, they were asked to perform a series of tasks starting with operating the radio and ending with a conversation with a passenger. Then, they drove the route without any in-vehicle distractions

studies have shown that crashes related to distractions are more likely to occur during dry days with less traffic congestion [13]. By collecting the data during dry days, we have relevant information for the study. The subjects were advised to take their time while performing the tasks to prevent potential accidents.

We defined a 5.6 mile route in the vicinities of the university (see Fig. 11.3). The route includes traffic signals, heavy and low traffic zones, residential areas, and a school zone. We decided to exclude streets with high speed limit (e.g., highways or freeways) from the analysis to minimize the risks in the recording. The participants took between 13 and 17 min to complete the route.

The drivers drove this route twice. During the first run, the participants were asked to perform a number of secondary tasks while driving. Among the tasks mentioned by Stutts et al. [34] and Glaze and Ellis [12], we selected the following tasks: tuning the built-in car radio, operating and following a GPS, dialing and using a cellphone, describing pictures, and interacting with a passenger. Some dangerous tasks such as text messaging, grooming, and eating were not included for security reasons. The details of the selected seven tasks are given below.

Radio: The driver is asked to tune the built-in car radio to some predetermined stations. The radio is in its standard place, on the right side of the driver.

GPS—Operating: A predefined address is given to the driver who is asked to enter the information in the GPS. The device is mounted in the middle of the windshield. The driver is allowed to adjust it before starting the recording.

GPS—Following: After entering the address in the GPS, the driver is asked to follow the instructions to the destination.

Phone—Operating: The driver dials the number of an airlines automatic flight information system (toll-free). A regular cellphone is used for this task. Hands-free cellphones are not used to include the inherent mechanical distraction.

Phone—Talking: After dialing the number, the driver has to retrieve the flight information between two given US cities.

Pictures: The driver has to look and describe randomly selected pictures, which are displayed by another passenger sitting in the front passenger seat. The purpose of this task is to collect representative samples of distractions induced when the driver is looking at billboards, sign boards, shops, or any object inside or outside the car.

Conversation: A passenger in the car asks general questions to establish a spontaneous conversation.

According to the driver resources-based taxonomy defined by Wierwille et al. [37], the selected secondary tasks include visual-manual tasks (e.g., *GPS—Operating* and *Phone—Operating*), visual-only tasks (e.g., *GPS—Following* and *Pictures*), and manual primarily task (e.g., *Radio*). The set also includes tasks characterized by cognitive demand (e.g., *Phone—Talking*) and auditory/verbal demands (e.g., *Conversation*). Therefore, they span a wide spectrum of distractions, meeting the requirements imposed by this study.

During the second run, the drivers were asked to drive the same route without performing any of the aforementioned tasks. This data is collected as a normal reference to compare the deviation observed in the driver behaviors when he/she is engaged in secondary tasks. Since the same route is used to compare normal and task conditions, the analysis is less dependent on the selected road. Overall, the database for this study consists of over 12 h of real driving recordings. More details about this corpus are provided in [6, 15].

11.3.3 Perceived Driver Distraction Using Subjective Evaluations

This study evaluates the use of subjective evaluations to quantify the level of distraction perceived from the driver. The underlying assumption is that the previous driving experience of the external evaluators will allow them to accurately identify and rank the distracting scenarios or actions, as observed in the video recordings showing the driver and the road. One advantage of this approach is that a number quantifying the perceived distraction level is assigned to localized segments in the recording. Therefore, it is possible to identify various multimodal features that correlate with this distraction metric. Using these features, regression models can be designed to directly identify inattentive drivers [16]. Another advantage is that many raters can assess the videos so the aggregated values are more accurate (see Sect. 11.4.2).

As described in Sect. 11.3.2, the database contains over 12 h of data. However, only a portion of the corpus was considered for the study to limit the evaluation time. The corpus was split into 10 s, nonoverlapped recordings. Each set contains synchronized audio and videos showing the driver and the road. For each driver, three videos were randomly selected for each of the seven secondary tasks (Sect. 11.3.2). Three videos from normal condition were also randomly selected. Therefore, 24 videos per driver are considered, which give altogether 480 unique videos (3 videos $\times$ 8 conditions $\times$ 20 drivers $=$ 480). Eighteen students at UT Dallas with valid driver's license were invited to participate in the subjective assessment. None of the evaluators participated as drivers in the recording of the corpus. A *graphical user interface* (GUI) was built for the subjective evaluation with a sliding bar that takes continuous values between 0 and 1 (see Fig. 11.4). The extreme values are defined as *less distracted* and *more distracted*. In our previous work, we used a single, general metric to describe distraction using a similar GUI [16]. The study concluded that using a single metric for distraction was not enough to properly characterize tasks that increase the driver's cognitive load (e.g., *Phone—Talking*). To overcome this limitation, this study proposes a two-dimensional space to explicitly describe visual and cognitive distractions, separately. First, the evaluators assessed the perceived visual distraction of 80 video segments. In average, the evaluation lasted for 15 min. After a break, they assessed the perceived cognitive distraction of a different set of 80 video segments (nonoverlapped set of videos from the visual distraction evaluation). The average duration of the evaluation was 25 min.

The evaluators were instructed to relate themselves to the scenarios observed in the videos before assigning the perceived metric. We carefully instructed the evaluators with the definition of cognitive and visual distractions to unify their understanding. We follow the description given by Ranney et al. [27]. Visual distraction is defined as eye off the road—drivers looking away from the roadway. The evaluators were asked to rate the visual distraction level based on the glance behavior of the drivers. The road camera was included to help the evaluators to assess whether the observed head motions or eye glances were related to the

Fig. 11.4 GUI for the subjective evaluation of cognitive and visual distractions (0—*less distracted*, to 1—*more distracted*)

primary driving task. Cognitive distraction is defined as mind off the road—drivers being lost/busy in thought. For cognitive distraction, the evaluators were asked to give ratings based on his/her own judgment. However, we highlighted that facial expressions (stress level, eye pupil size, eye movements), secondary task performance (talking speed, phone dialing speed), and driving performance (vehicle in-lane position, driving speed, distance to front vehicle) can all be used to assess the cognitive distraction level. In total, each video was assessed by six independent evaluators, three for visual distractions and three for cognitive distractions.

11.4 Reliability and Consistency of Subjective Evaluations

This section validates the use of perceptual evaluations to characterize driver behaviors. We argue that employing the perceived visual and cognitive distraction assessments is a valid approach to characterize distractions. This scheme is particularly useful for cognitive distractions. While internal physiological measures can provide consistent indication of the driver's cognitive workload [23], observable driver's behaviors can only provide indirect cues [40]. We expect that evaluators can infer the expected cognitive load of the driver after observing and judging these external behaviors. First, we analyze the results of the perceptual evaluation in terms of secondary tasks (Sect. 11.4.1). Then, we study the consistency of the subjective evaluations by estimating the inter-evaluator agreement (Sect. 11.4.2). The results of the subjective evaluation are compared with self-reports from the drivers that participated in the recording (Sect. 11.4.3).

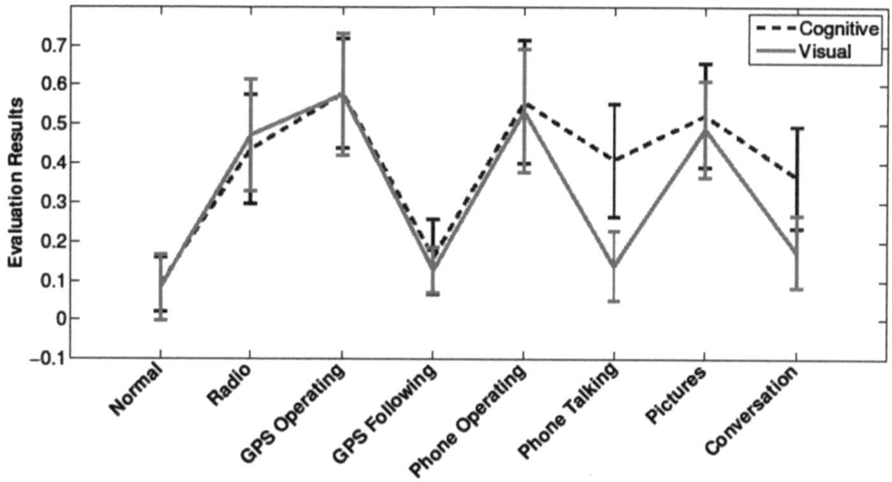

Fig. 11.5 Means and standard deviations of the perceived visual and cognitive distraction scores across secondary tasks collected with subjective evaluations

11.4.1 Analysis of Subjective Evaluations

Figure 11.5 shows the means and standard deviations of the perceived visual (solid line) and cognitive (dashed line) distractions across secondary tasks and normal conditions. The result suggests that secondary tasks identified as visually intensive activities such as *GPS—Operating*, *Phone—Operating* and *Pictures* received the highest scores for visual distractions. The cognitive distraction scores for secondary tasks that are known to increase the cognitive workload of the driver (e.g., *Phone—Talking* and *Conversation*) are higher than the corresponding visual distraction scores. These results are consistent with previous studies reporting that conversation is intrinsically a cognitive task [24]. The perceptual evaluations also agree with Bach et al. [4] who suggested that the cognitive distraction induced by using a cellphone is more detrimental than the mechanical distraction associated with operating the device.

Although Fig. 11.5 suggests that the recordings received similar cognitive and visual distraction scores for most of the secondary tasks, a closer look at the evaluation reveals that the proposed two-dimensional space captures their distinction. Figure 11.6 shows a scattering plot of the subjective evaluation across tasks and normal conditions in the visual-cognitive space. The figure shows samples covering much of the two-dimensional space. The only empty area corresponds to recording with low cognitive distractions but with high visual distractions. Notice that visual demanding tasks also induce cognitive demands. Therefore, this finding is expected. These results suggest that the subjective evaluation is effective in capturing both visual and cognitive distractions. A further discussion about the scattering plot defined by the visual-cognitive space is given in Sect. 11.6.

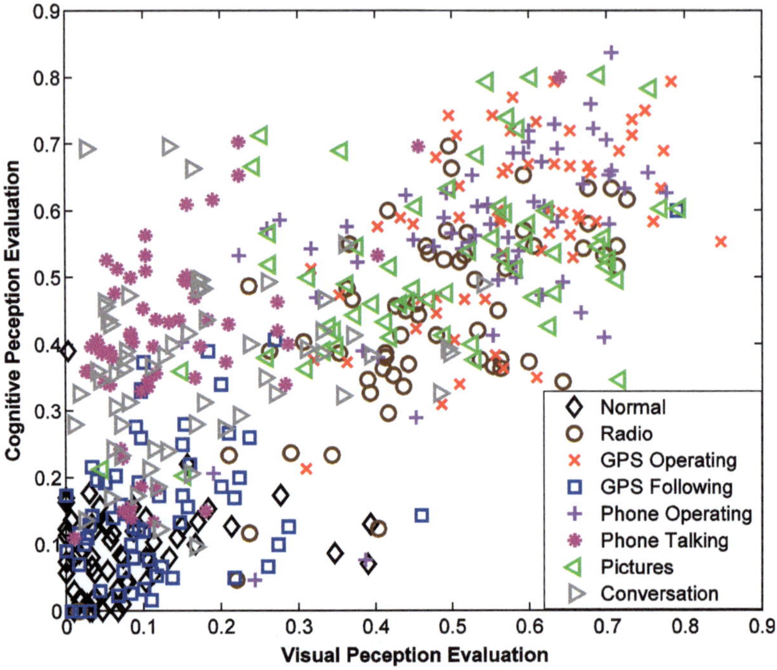

Fig. 11.6 Scattering plot of the subjective evaluation across secondary tasks in the visual-cognitive space

11.4.2 Inter-Evaluator Agreement

Since each video segment is separately assessed by three different evaluators for cognitive and visual distractions, the agreement between raters is a useful indicator of the reliability of these metrics. Stronger agreement suggests higher consistency among the evaluators, which validates the proposed approach. The analysis consists in measuring the correlation between the provided scores. For each evaluator, we calculated the average scores provided by the remaining two raters. Then, we estimate the Pearson correlation between his/her scores and the average scores. We repeat this approach for each of the three evaluators. The average correlation across evaluators is $\rho^v = 0.75$ for visual distractions, and $\rho^c = 0.70$ for cognitive distractions. These correlation values represent very strong positive relationship between the scores provided by raters. Figure 11.7 gives the correlation values for cognitive and visual distractions for each of the 18 evaluators. The correlation values are always above $\rho = 0.5$. These findings reveal high consistency for both visual and cognitive distraction evaluations. In general, visual distraction scores have higher values than cognitive distraction scores. This result and the fact that the duration of the cognitive distraction evaluation was in average 10 min longer than visual distraction evaluations (Sect. 11.3.3) suggest that assessing cognitive distractions is harder than assessing visual distractions.

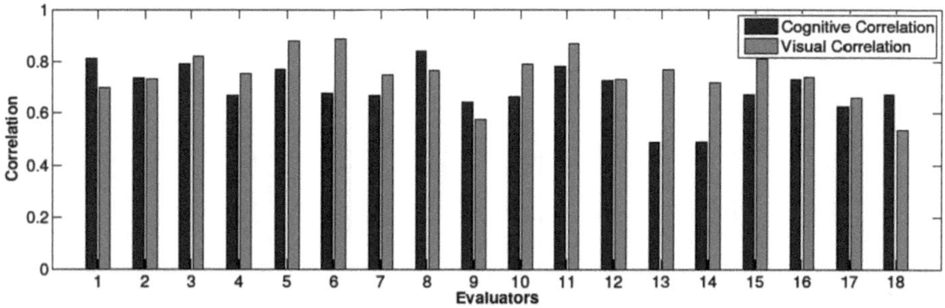

Fig. 11.7 Correlation values for cognitive and visual distractions. The results are given for each of the 18 evaluators

11.4.3 Self-Evaluations Versus External Evaluations

The most common questionnaires used to assess mental workload are based on self-evaluations conducted after the experiments [1, 14, 18, 26]. The underlying assumption in self-evaluations is that drivers are aware of the distraction level felt when they were performing secondary tasks. Therefore, they can rank the tasks that were more distracting to them. This section compares self-evaluations with the assessments provided by external observers.

A self-evaluation was collected from the drivers after recording the data to rate how distracted they felt while performing each of the secondary tasks. Unfortunately, the subjects participating in the driving recordings were not available to provide detailed assessments over small video segments. Therefore, we use a simplified methodology for this self-evaluation. First, the drivers self-evaluated their perceived distraction, without distinguishing between cognitive and visual distractions. Second, instead of evaluating several localized segments in the recording, the drivers provided a single coarse value for each secondary task without watching videos of the recordings. They used a Likert scale with extreme values corresponding to 1—*less distracted*, and 5—*more distracted.* Figure 11.8 presents the average and standard deviation values of the perceived distraction scores. The result suggests that, on average, *GPS—Operating* is regarded as the most distracted task, while *Conversation* is considered as the least distracted task. The fact that *Phone—Talking* is perceived as more distracting than *Conversation* is consistent with the conclusions by Drews et al. [8]. They claimed that the situational awareness of being in the same vehicle makes conversation with a passenger a less distracted task than a conversation with someone who is unaware of the surrounding traffic (e.g., avoiding increasing the driver's cognitive demands during decision making times).

Although the setting for the drivers' self-evaluation differs from the one used to collect evaluations from external observers (Sect. 11.3.3), the global patterns can be compared. Figures 11.5 and 11.8 show consistent patterns across secondary tasks.

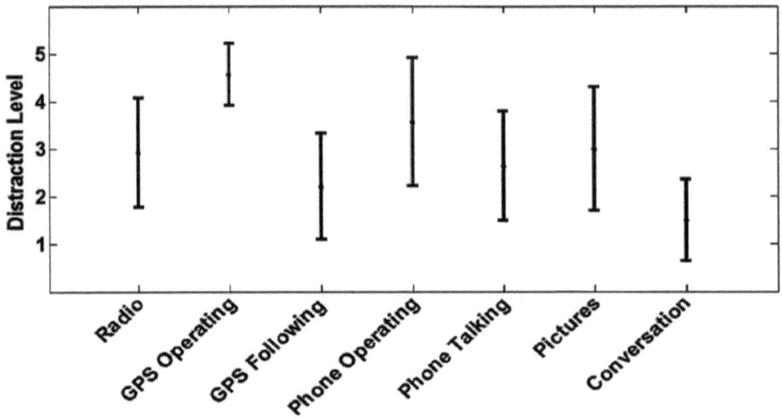

Fig. 11.8 Average distraction levels based on self-evaluations across the drivers. The figure shows the mean and standard deviation of the values assigned to each task

The ranked order of the four tasks that are perceived as the most distracting for self-evaluations are exactly the same as the corresponding ones in the cognitive and visual evaluations from external observers: *GPS—Operating*, *Phone—Operating*, *Picture*, and *Radio*. The main differences are observed in the cognitive evaluations for the tasks *Conversation* and *Phone—Talking*, which received higher values by external observers than by the drivers. Since we requested the external evaluators to specifically assess the perceived cognitive load of the driver, higher values for these tasks are expected.

The average values of self-evaluations provide coarse indicators to represent the distraction level induced by the corresponding task. Depending on the scenario, certain actions associated with secondary tasks can be more distracting than others (e.g., having a conversation in a busy traffic intersection). Self-evaluations fail to capture this inherent within-task variability. Also, drivers may fail to notice the adjustments made to complete secondary tasks (e.g., jittery steering wheel behavior, reduced speed). We believe that perceptual evaluations collected by multiple external evaluators over small segments of driving recordings can overcome these limitations.

11.5 Subjective Evaluations and Eye Glance Behavior

As discussed in Sect. 11.2.4, eye glance behaviors provide useful metrics to characterize distractions [4, 22]. This approach gives unbiased metrics to describe driver behaviors. This section compares perceptual evaluation scores provided by external observers with eye glance behavior measurements. The analysis shows that both approaches provide consistent patterns. First, we describe the eye glance metrics used in the analysis, which are automatically extracted from the videos (Sect. 11.5.1). Then, we compare the cognitive and visual distraction scores from recordings with extreme eye glance behaviors (Sect. 11.5.2).

11.5.1 Metrics Describing Eye Glance Behavior

The drivers' glance is a reliable indicator of attention. This study relies on two glance metrics that have been previously used to characterize driver distraction: the *total eye off the road* duration (EOR), and the *longest eye off the road duration* (LEOR). These features are automatically estimated over the videos evaluated by the external observers. Given that evaluators assessed 10 s videos, we set the window analysis accordingly. EOR measures the total time within 10 s in which the drivers' glance is not on the road. As mentioned in Sect. 11.2.4, this is an important metric that is considered to assess the visual demand of IVIS. LEOR captures tasks that require longer glances, which are known to increase the chances of accidents [38].

The glance metrics are automatically extracted from the frontal camera facing the drivers using the *computer expression recognition toolbox* (CERT) [5]. CERT is a robust system that extracts facial expression features and head pose. Given the challenges in recognizing the driver's gaze in real recordings, we approximate glance behavior with the drivers' head pose, parameterized with three rotation angles (yaw, pitch, and roll). Certain videos present adverse illumination conditions or occluded faces due to the driver's hands. In these cases, CERT fails to recognize the face producing empty values. If this problem was observed over half of the duration of a video (5 s), the recording was discarded from the analysis. Otherwise, we approximate the head pose by interpolating missing values.

Head yaw (horizontal rotation) and pitch (vertical rotation) are used for eye off the road detection. We define thresholds on these angles to decide whether the driver is looking at the road. Due to the differences in the drivers' height and in their sitting preference, the thresholds are separately calculated for each individual from his/her normal driving recordings. The thresholds for head yaw and head pitch are set at their mean plus/minus two times their standard deviation, defining in average a $16° \times 16°$ rectangular region. This approach aims to replicate the 8° radius circle defined in the *percent road center* (PRC) calculation [36]. The frames detected as eye of the road are accumulated over the video sequence to estimate EOR. LEOR is calculated by counting the longest consecutive eye off the road frames. Both measurements are divided by the video frame rate to convert the metrics into seconds. Notice that this approach may detect as eye off the road action glances associated to the primarily driving task (e.g., checking mirrors).

Figure 11.9 shows the average values for EOR and LEOR for normal and task conditions across 20 drivers. The task condition includes the recordings from the

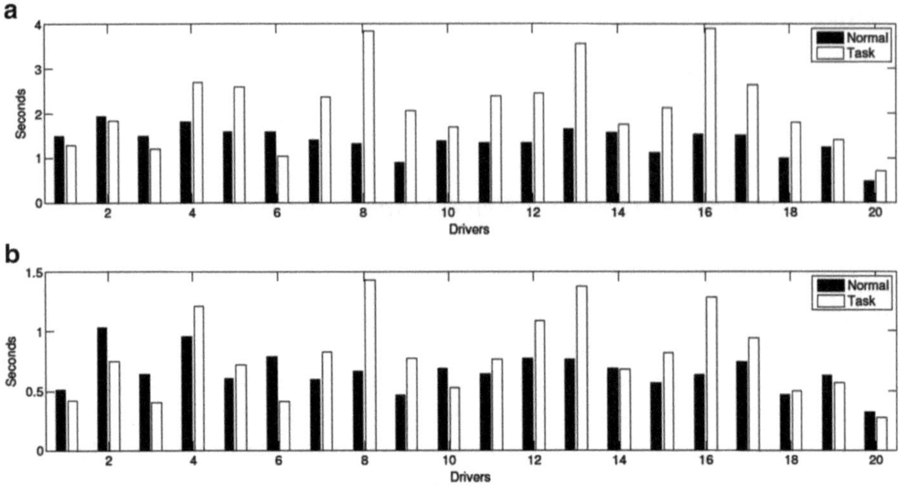

Fig. 11.9 Eye glance features extracted from 20 drivers for normal and task driving conditions. (**a**) Total eye off-road (EOR) duration in 10 s. (**b**) Longest eye off-road (LEOR) duration in 10 s

seven secondary tasks considered in this study (Sect. 11.3.2). The figure reveals that most of the drivers glance longer and more frequently when they are involved in secondary tasks. Therefore, these glance metrics are appropriate to evaluate the effectiveness of subjective evaluations.

11.5.2 Eye Glance Metrics and Subjective Evaluations

Given that EOR and LEOR have been used to characterize drivers' distractions, we expect to observe agreement between extreme values of these metrics and the subjective evaluations. We follow the approach presented by Liang et al. [20], which defined distracted recordings when the considered metrics have higher values (e.g., the upper quartile of steering error values). For each glance metric, we select a subset of the video recordings to form two extreme groups: driving recordings with low EOR or LEOR values (e.g., "normal" class), and driving recordings with high EOR or LEOR values (e.g., "distracted" class). Figure 11.10 shows the distributions for EOR and LEOR values estimated from the 10 s videos used for the subjective evaluation. The vertical lines are the thresholds defined to create the two groups, which are set so that each group has at least 72 samples to estimate reliable distributions (Figs. 11.11 and 11.12). For EOR, a recording is considered as "normal" if the EOR duration is less than 1 s, and as "distracted" if its value is more than 3 s. For LEOR, a recording is considered as "normal" if the LEOR duration is less than 1 s. Otherwise, it is considered as "distracted."

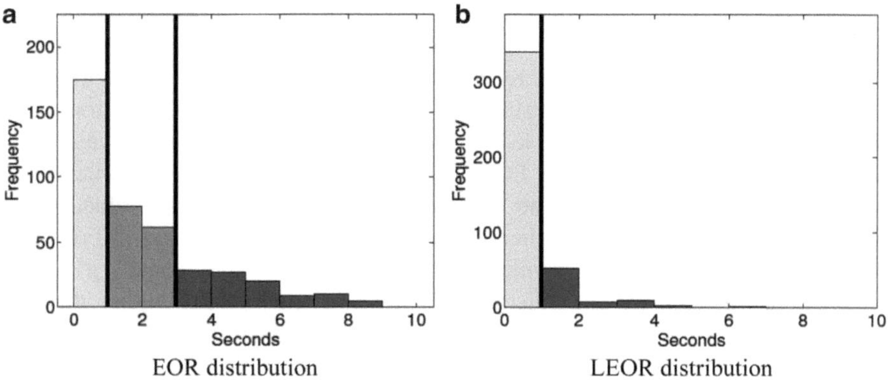

Fig. 11.10 Distributions for EOR and LEOR. The *vertical lines* give the thresholds defining the classes "normal" (*light gray*) and "distracted" (*dark gray*)

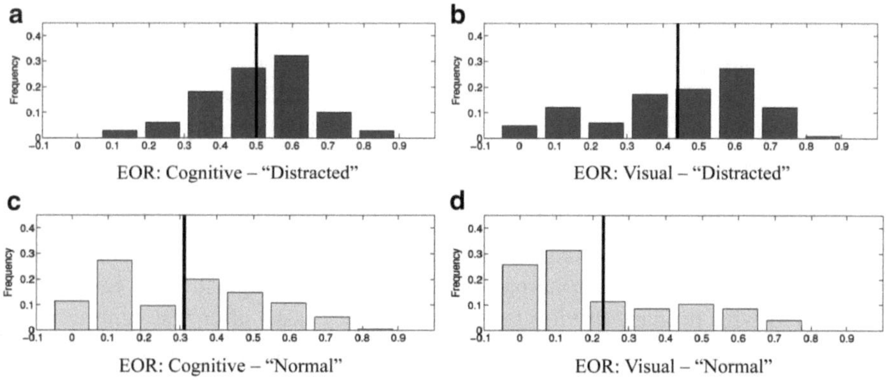

Fig. 11.11 Distribution of perceptual evaluation for extreme EOR values. (**a**) and (**b**) correspond to "distracted" class and (**c**) and (**d**) to "normal" class

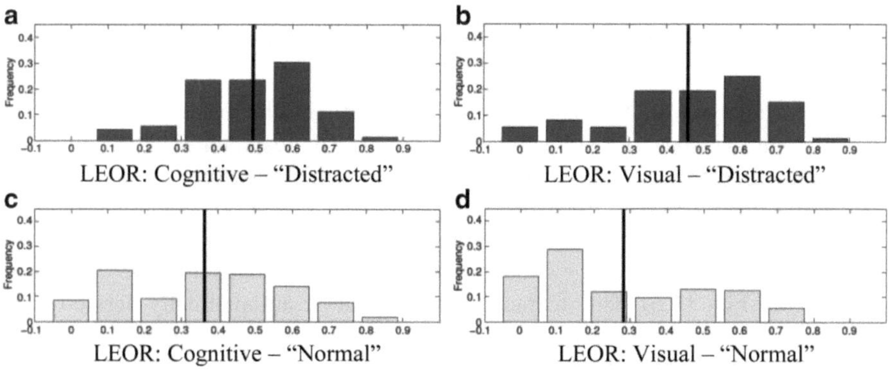

Fig. 11.12 Distribution of perceptual evaluation for extreme LEOR values. (**a**) and (**b**) correspond to "distracted" class and (**c**) and (**d**) to "normal" class

The analysis aims to identity whether the subjective evaluations capture the differences between the extreme video groups. We expect that the recordings with high EOR or LEOR values are perceived with higher distraction levels. We address this question by studying the distributions of visual and cognitive distraction scores assigned to the recordings labeled as "normal" and "distracted." Figures 11.11 and 11.12 report the results for EOR and LEOR, respectively. The vertical lines represent the means values. The distributions for the subjective evaluation are consistently skewed toward higher values for the "distracted" classes. For EOR, the mean values for both cognitive ($\mu^c{}_{distracted} = 0.50$) and visual ($\mu^v{}_{distracted} = 0.44$) distractions for the "distracted" class are significantly higher than the corresponding values for "normal" class ($\mu^c{}_{normal} = 0.23$, and $\mu^v{}_{normal} = 0.31$, respectively). The same results are observed for LEOR values.

Figure 11.11b presents a peak at 0.1. This peak may correspond to eye off the road actions associated to the primary driving task. While the EOR duration is above 3 s, the external observers may recognize that these actions do not represent distractions. Figures 11.11c and 11.12c show peaks at 0.4. These results suggest that the evaluators assigned moderate cognitive scores to recordings in which the drivers were looking at the road. These results may indicate that eye glance behaviors may provide an incomplete description of driver behaviors. As mentioned in Sect. 11.2.4, cognitive distracted drivers may have reduced peripheral visual awareness [27, 29, 30]. External observers may recognize the lack of eye glance movements as a signal of distraction.

The results reveal that subjective evaluations and eye glance behavior metrics provide consistent assessments of driver distractions (especially for visual distractions). Notice that certain eye off the road actions do not represent distractions (e.g., checking mirrors). External observers can distinguish between actions associated with primary driving tasks or secondary tasks after watching multiple cues in the road and driver videos. In these cases, the proposed two-dimensional space for perceptual evaluations can give a better representation of driver distractions.

11.6 Distraction Modes Defined by Subjective Evaluation

The final analysis in this chapter aims to highlight the benefits of using the visual-cognitive space for subjective evaluations. The results in Sect. 11.4.1 show important differences in the visual and cognitive distraction scores for certain tasks. An active safety system focusing only on visual distraction cannot provide a complete picture of the driver behaviors. These differences are captured by the proposed two-dimensional evaluation space, which defines natural distraction modes (see Fig. 11.6). The distraction modes can be automatically derived from the data by clustering the evaluations scores. The resulting modes can give a useful representation of driver distractions.

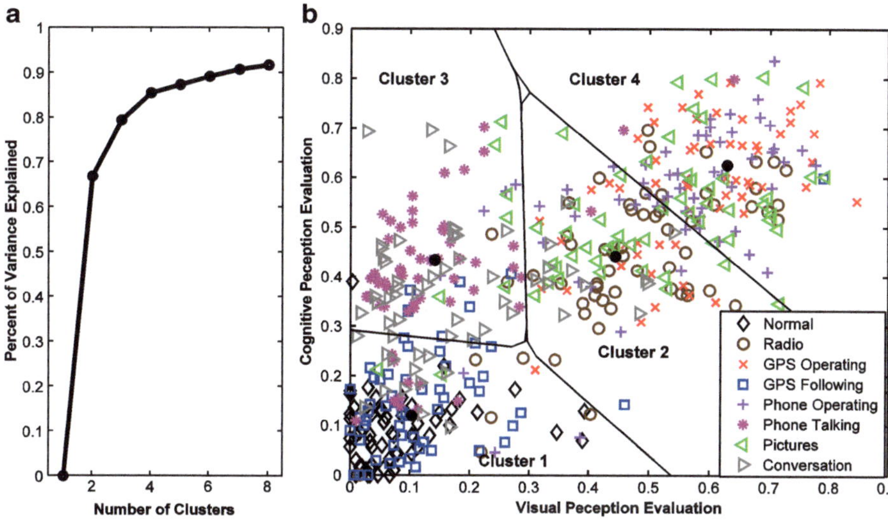

Fig. 11.13 Distraction modes defined by subjective evaluations (**a**) number of cluster defined by elbow analysis and (**b**) K-Means clusters in visual cognitive space

The clustering analysis relies on the K-means algorithm. An important aspect of the algorithm is the number of clusters, which is defined with the elbow criterion. In this approach, the number of clusters is increased, recording the percentage of variance explained by the corresponding clustering. Figure 11.13a shows that increasing the number of cluster above four does not reduce significantly the percentage of variance. Therefore, we set the number of clusters accordingly. Figure 11.13b shows the resulting clustering. The locations of the centroids suggest that drivers' distractions can be divided into (the most representative secondary tasks are given in brackets):

- *Cluster 1*—low visual and low cognitive distractions (*Normal* and *GPS—Following*).
- *Cluster 2*—medium visual and medium cognitive distractions (*Radio* and *Picture*);
- *Cluster 3*—low visual and medium cognitive distractions (*Phone—Talking* and *Conversation*);
- *Cluster 4*—high visual and high cognitive distractions (*GPS—Operating*, and *Phone—Operating*);

The proposed modes provide a new, useful representation space to characterize driving behaviors. It can be argued that clusters 3 and 4 are the most dangerous distraction modes. When a new IVIS is evaluated, multimodal features from the car and from the driver can be estimated to determine the underlying distraction mode. We are currently studying multiclass recognition problems (four class problem) and binary classification problems (one cluster versus the rest). Our preliminary analysis shows promising results in this area.

11.7 Discussion and Conclusions

This study explored the use of subjective evaluations from external observers to characterize driver behaviors. The goal is to define reference labels that can be used to train human-centric active safety systems. We conducted subjective evaluations to assess the perceived visual and cognitive distractions in randomly selected videos showing the driver and road. The analysis suggests that this two-dimensional space captures the multidimensional nature of distractions. The inter-evaluator agreement analysis shows very strong correlation for visual and cognitive assessments. The scores from external evaluators are consistent with self-evaluations collected from the drivers, and with eye glance metrics (videos with higher EOR and LEOR values are perceived more distracted).

The study suggests that perceptual evaluations from external observers have important advantages over other alternative approaches. First, multiple evaluators can provide reliable scores over short video recordings. This approach facilitates the study of relevant multimodal cues describing cognitive and visual distractions. Second, external evaluators can perceive important actions or cues that may be ignored by the drivers. For example, previous studies show the detrimental effects of the task *Phone—Talking* on the primary driving task [32, 33]. Drivers using cellphone may experience inattention blindness or selective withdrawal of attention, failing to see objects even though they are in front of them [32]. While this task was identified as the least distracted task by the self-evaluations, the external observers assigned higher scores. Third, external observers can capture the underlying driving dynamics, providing more reliable insights than metrics describing eye glance behaviors. For example, cognitive tasks reduce the drivers' peripheral visual awareness [27, 29, 30]. Therefore, lack of eye glances can signal cognitive distraction. While metrics such as eye off the road duration fail to capture these cues, external evaluators can complement their judgment by looking the driver's facial expressions.

The analysis suggests natural distraction modes to describe driver behaviors. These modes are estimated by clustering the evaluations in the visual-cognitive space. Some of these distraction modes can have a higher detrimental effect on the primary driving task (e.g., *clusters* 3 and 4). Our current research direction is to use these labels to build machine-learning algorithms to recognize the corresponding clusters. We are also planning to extend our database to include other secondary tasks, providing a better coverage of common distractions observed in real scenarios. The intended driver behavior monitoring system will provide feedbacks to inattentive drivers, preventing accidents, and increasing the security on the roads.

Acknowledgment The authors would like to thank Dr. John Hansen for his support with the UTDrive Platform. We want to thank the *Machine Perception Lab* (MPLab) at The University of California, San Diego, for providing the CERT software. The authors are also thankful to Ms. Rosarita Khadij M Lubag for her support and efforts with the data collection.

References

1. M. C. F. Aguilo, Development of guidelines for in-vehicle information presentation: text vs. speech, Master's thesis, Virginia Polytechnic Institute and State University, Blacksburg, Virginia, May 2004
2. P. Angkititrakul, D. Kwak, S. Choi, J. Kim, A. Phucphan, A. Sathyanarayana, and J.H.L. Hansen, Getting start with UTDrive: driver-behavior modeling and assessment of distraction for in-vehicle speech systems, in *Interspeech 2007*, Antwerp, Belgium, August 2007, pp. 1334–1337
3. A. Azman, Q. Meng, E. Edirisinghe, Non intrusive physiological measurement for driver cognitive distraction detection: Eye and mouth movements. In *International Conference on Advanced Computer Theory and Engineering (ICACTE 2010)*, vol. 3, Chengdu, China, August 2010
4. K. M. Bach, M.G. Jaeger, M.B. Skov, and N.G. Thomassen, Interacting with in-vehicle systems: understanding, measuring, and evaluating attention. In *Proceedings of the 23rd British HCI Group Annual Conference on People and Computers: Celebrating People and Technology*, Cambridge, United Kingdom, September 2009
5. M.S. Bartlett, G.C. Littlewort, M.G. Frank, C. Lainscsek, I. Fasel, J.R. Movellan, Automatic recognition of facial actions in spontaneous expressions. J. Multimedia **1**, 22–35 (2006)
6. C. Busso, J. Jain, Advances in multimodal tracking of driver distraction, in *Digital Signal Processing for In-Vehicle Systems and Safety*, ed. by J. Hansen, P. Boyraz, K. Takeda, H. Abut (Springer, New York, NY, 2011), pp. 253–270
7. Y. Dong, Z. Hu, K. Uchimura, N. Murayama, Driver inattention monitoring system for intelligent vehicles: a review. IEEE Trans. Intel. Trans. Syst. **12**(2), 596–614 (2011)
8. F.A. Drews, M. Pasupathi, D.L. Strayer, Passenger and cell phone conversations in simulated driving. J. Exp. Psychol. Appl. **14**(4), 392–400 (2008)
9. J. Engström, E. Johansson, J. Östlund, Effects of visual and cognitive load in real and simulated motorway driving. Trans. Res. Part F Traffic Psychol. Behav. **8**(2), 97–120 (2005)
10. T. Ersal, H.J.A. Fuller, O. Tsimhoni, J.L. Stein, H.K. Fathy, Model-based analysis and classification of driver distraction under secondary tasks. IEEE Trans. Intel. Trans. Syst. **11** (3), 692–701 (2010)
11. J.P. Foley, Now you see it, now you dont: visual occlusion as a surrogate distraction measurement technique, in *Driver Distraction: Theory Effects, and Mitigation*, ed. by M.A. Regan, J.D. Lee, K.L. Young (CRC Press, Boca Raton, FL, 2008), pp. 123–134
12. A. L. Glaze, J.M. Ellis, Pilot study of distracted drivers. Technical report, Transportation and Safety Training Center, Virginia Commonwealth University, Richmond, VA, USA, January 2003
13. P. Green, The 15-second rule for driver information systems. In *Intelligent Transportation Society (ITS) America Ninth Annual Meeting*, Washington, DC, USA, April 1999
14. J.L. Harbluk, Y.I. Noy, P.L. Trbovich, M. Eizenman, An on-road assessment of cognitive distraction: impacts on drivers' visual behavior and braking performance. Accid. Anal. Prev. **39**(2), 372–379 (2007)
15. J. Jain, C. Busso. Analysis of driver behaviors during common tasks using frontal video camera and CAN-Bus information. In *IEEE International Conference on Multi media and Expo (ICME 2011)*, Barcelona, Spain, July 2011
16. J.J. Jain, C. Busso. Assessment of driver's distraction using perceptual evaluations, self assessments and multimodal feature analysis. In *5th Biennial Workshop on DSP for In-Vehicle Systems*, Kiel, Germany, September 2011
17. S.G. Klauer, T.A. Dingus, V.L. Neale, J.D. Sudweeks, D.J. Ramsey, The impact of driver inattention on near-crash/crash risk: an analysis using the 100-car naturalistic driving study data. Technical Report DOT HS 810 594, National Highway Traffic Safety Administration, Blacksburg, VA, USA, April 2006

18. J.D. Lee, B. Caven, S. Haake, T.L. Brown, Speech-based interaction with in-vehicle computers: the effect of speech-based e-mail on drivers' attention to the road-way. Hum. Factors **43**(4), 631–640 (Winter 2001)
19. J.D. Lee, D.V. McGehee, T.L. Brown, M.L. Reyes, Collision warning timing, driver distraction, and driver response to imminent rear-end collisions in a high-fidelity driving simulator. Hum. Factors **44**, 314–334 (Summer 2002)
20. Y. Liang, M.L. Reyes, J.D. Lee, Real-time detection of driver cognitive distraction using support vector machines. IEEE Trans. Intel. Trans. Syst. **8**(2), 340–350 (2007)
21. S. Mattes, A. Hallén, Surrogate distraction measurement techniques: the lane change test, in *Driver Distraction: Theory, Effects, and Mitigation*, ed. by M.A. Regan, J.D. Lee, K.L. Young (CRC Press, Boca Raton, FL, 2008), pp. 107–122
22. J.C. McCall, M.M. Trivedi, Driver behavior and situation aware brake assistance for intelligent vehicles. Proc. IEEE **95**(2), 374–387 (2007)
23. B. Mehler, B. Reimer, J.F. Coughlin, J.A. Dusek, Impact of incremental increases in cognitive workload on physiological arousal and performance in young adult drivers. Trans. Res. Record **2138**, 6–12 (2009)
24. V. Neale, T. Dingus, S. Klauer, J. Sudweeks, M. Goodman, An overview of the 100-car naturalistic study and findings. Technical Report Paper No. 05-0400, National Highway Traffic Safety Administration, June 2005
25. W. Piechulla, C. Mayser, H. Gehrke, W. Koenig, Reducing drivers' mental work-load by means of an adaptive man–machine interface. Trans. Res. Part F Traffic Psychol. Behav. **6**(4), 233–248 (2003)
26. F. Putze, J.-P. Jarvis, T. Schultz, Multimodal recognition of cognitive workload for multitasking in the car. In *International Conference on Pattern Recognition (ICPR 2010)*, Istanbul, Turkey, August 2010
27. T.A. Ranney, Driver distraction: a review of the current state-of-knowledge. Technical Report DOT HS 810 787, National Highway Traffic Safety Administration, April 2008
28. T.A. Ranney, W.R. Garrott, M.J. Goodman, NHTSA driver distraction research: past, present, and future. Technical Report Paper No. 2001-06-0177, National High- way Traffic Safety Administration, June 2001
29. E.M. Rantanen, J.H. Goldberg, The effect of mental workload on the visual field size and shape. Ergonomics **42**(6), 816–834 (1999)
30. M.A. Recarte, L.M. Nunes, Mental workload while driving: effects on visual search, discrimination, and decision making. J. Exp. Psychol. Appl. **9**(2), 119–137 (2003)
31. A. Sathyanarayana, S. Nageswaren, H. Ghasemzadeh, R. Jafari, J.H.L. Hansen, Body sensor networks for driver distraction identification. In *IEEE International Conference on Vehicular Electronics and Safety (ICVES 2008)*, Columbus, OH, USA, September 2008
32. D.L. Strayer, J.M. Cooper, F.A. Drews. What do drivers fail to see when conversing on a cell phone? In *Proceedings of Human Factors and Ergonomics Society Annual Meeting*, volume 48, New Orleans, LA, USA, September 2004
33. D.L. Strayer, J.M. Watson, F.A. Drews, Cognitive distraction while multitasking in the automobile, in *The Psychology of Learning and Motivation*, ed. by B.H. Ross, vol. 54 (Academic, Burlington, MA, 2011), pp. 29–58
34. J.C. Stutts, D.W. Reinfurt, L. Staplin, E.A. Rodgman, The role of driver distraction in traffic crashes. Technical report, AAA Foundation for Traffic Safety, Washington, DC, USA, May 2001
35. F. Tango, M. Botta, Evaluation of distraction in a driver-vehicle-environment framework: an application of different data-mining techniques, in *Advances in Data Mining. Applications and Theoretical Aspects*, ed. by P. Perner. Lecture Notes in Computer Science, vol. 5633 (Springer, Berlin, 2009), pp. 176–190
36. T.W. Victor, J. Engstroem, J.L. Harbluk, Distraction assessment methods based on visual behavior and event detection, in *Driver Distraction: Theory, Effects, and Mitigation*, ed. by M.A. Regan, J.D. Lee, K.L. Young (CRC Press, Boca Raton, FL, 2008), pp. 135–165

37. W. Wierwille, L. Tijerina, S. Kiger, T. Rockwell, E. Lauber, A. Bittner Jr, Final report supplement—task 4: review of workload and related research. Technical Report DOT HS 808 467, U.S. Department of Transportation, National Highway Traffic Safety Administration, Washington, DC, USA, October 1996
38. Q. Wu, An overview of driving distraction measure methods. In *IEEE 10th International Conference on Computer-Aided Industrial Design & Conceptual Design (CAID CD 2009)*, Wenzhou, China, November 2009
39. K.L. Young, M.A. Regan, J.D. Lee, Measuring the effects of driver distraction: direct driving performance methods and measures, in *Driver Distraction: Theory Effects, and Mitigation*, ed. by M.A. Regan, J.D. Lee, K.L. Young (CRC Press, Boca Raton, FL, 2008), pp. 85–105
40. Y. Zhang, Y. Owechko, J. Zhang. Driver cognitive workload estimation: a data-driven perspective. In *IEEE Intelligent Transportation Systems*, Washington, DC, USA, October 2004, pp. 642–647

Part IV
Driving Behavior and User Profiling

Chapter 12
Evaluation Method for Safe Driving Skill Based on Driving Behavior Analysis and Situational Information at Intersections

Yosuke Yoshida, Matti Pouke, Masahiro Tada, Haruo Noma, and Masaru Noda

Abstract The overall number of traffic accidents is on the decrease, however still at high level. To reduce accidents, we propose an automated method for evaluating driving skill by measuring and analyzing safe driving behavior. Earlier, we have proposed an evaluation method HAS for safe driving skill using the sight distance generated by geometrical features of intersections. For this study, we have performed an experiment with 38 subjects on public road to confirm the effectiveness of our method and compared its results with an evaluation given by a driving instructor. As a result of experiment, correlation coefficient between the evaluation results calculated by HAS and the ones evaluated by a driving instructor is 0.71. Therefore, our measurement system shows effectiveness. In addition, we propose more understandable result presentation method that employs probability theory.

Keywords Evaluation method • Safe driving skill • Situational information at intersection • Wearable sensors

12.1 Introduction

To reduce the amount of traffic accidents, a lot of effort has been made for the improvement of vehicle and road safety equipment. However, generally the number of traffic accidents is still at high level. In addition to improving vehicles and

Y. Yoshida (✉) • M. Noda
Nara Institute of Science and Technology, Nara, Japan
e-mail: yosuke-y@is.naist.jp

M. Pouke
Oulu University, Oulu, Finland
e-mail: matti.pouke@oulu.fi

M. Tada • H. Noma
Advanced Telecommunications Research Institute International, Kyoto, Japan
e-mail: mtada@atr.jp

G. Schmidt et al. (eds.), *Smart Mobile In-Vehicle Systems: Next Generation Advancements*, DOI 10.1007/978-1-4614-9120-0_12,

roadside safety, driving behavior should also be considered for traffic accident reduction. The driver can reduce traffic risks significantly with his/her own behavior so we aim to promote safe driving by retraining licensed drivers. To achieve this, we propose an automated method for evaluating driving skill by measuring safe driving behavior using wearable sensors [1].

Many traffic accidents in intersections are caused by a combination of factors like visibility reduced by geometrical features and the amount and speed of other traffic, which we refer to as traffic flux. Therefore, when we evaluate safe driving skill, we should consider not only the driver's behavior itself but also its relation to these other factors. Based on the idea, we have earlier proposed Highest Admitted Speed (HAS) as an evaluation method of driving skill [2]. By using HAS, we calculated the risk of collision with other vehicles using sight distance and found out it can be used to display the skill of the driver objectively.

However, HAS has two major shortcomings, the first is that the effectiveness of the evaluation by HAS has not yet been confirmed. Thus, in this paper, we confirmed its effectiveness by comparing evaluation results by our method to subjective evaluation results given by a driving instructor. The other shortcoming is that the evaluation by HAS is not easy to understand. Therefore, we propose a more understandable result presentation method based on probability theory by considering the speed distribution of other vehicles around the intersection which we define as traffic flux. Considering the traffic flux in relation to driving behavior allows quantitatively evaluating the risk of collision in probability form and will give drivers' evaluation results in a more easy and intuitive way.

12.2 Definition of HAS

The sight distance of an intersection deeply depends on its geometrical features. To pass through the intersection safely, the drivers should adjust their driving behavior according to these features. Therefore, we should consider the sight distance as a factor when evaluating driving behavior. Also the vehicle speed, scanning behavior, and pedal operation are used to calculate HAS. The scanning behavior should be performed early enough for the driver to be able to avoid a possible collision. Therefore, HAS includes only the scanning behavior which is done at a distance from where the driver can still apply the brakes and stop his vehicle before entering the intersection. Each scanning behavior is judged by stopping distance based on such as idle running time or coefficient of friction [3]. We calculate the sight distance using these scanning behavior positions and the geometrical features. If a driver performs scanning behavior to the right side at the position which is shown in Fig. 12.1, the relation between triangle A and triangle B makes equation (12.1):

$$\frac{D - a - W/4}{b + c} = \frac{a}{x - c - L} \tag{12.1}$$

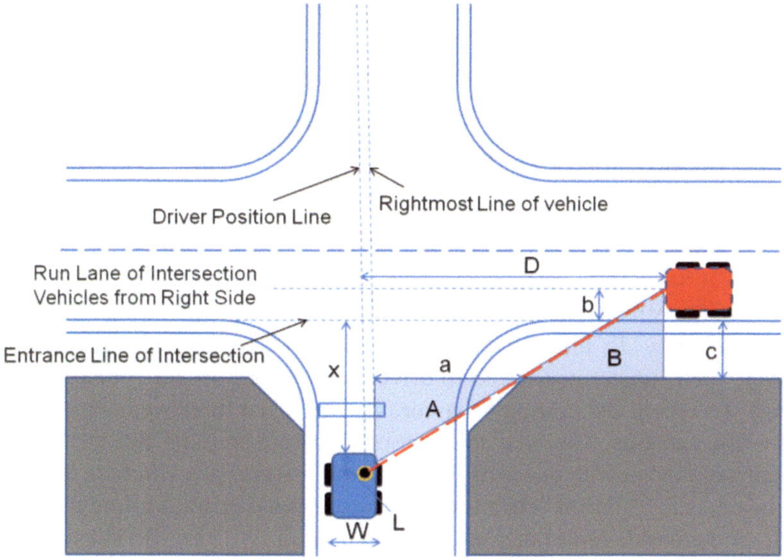

Fig. 12.1 Calculation of sight distance

Therefore, the sight distance D is calculated by (12.2):

$$D = \frac{a}{x - c - L} \cdot (b + c) + a + \frac{W}{4} \tag{12.2}$$

At this equation, W is the width of the vehicle. Since Japan adopted keep-left rule, we considered the driver sits at the foremost right quarter of the vehicle. L is the distance from the tip of the vehicle to the driver.

As discussed above, the sight distance from the driver's viewpoint is estimated by the detected scanning behavior and geometrical features around the intersection. Using the sight distance, HAS is given by (12.3):

$$HAS = D/t \tag{12.3}$$

Here, t is the time required to travel from the position where scanning behavior was performed to the collision area which could be calculated from vehicle average speed. Therefore, HAS resulted as the critical speed of an approaching vehicle which barely allows the driver to avoid risk of collision by slamming on the brakes. If a driver got low HAS result, it means that the driver could have collided with the other vehicles even under legal speed, i.e., the driver's behavior is dangerous. On the other hand, if a driver got high HAS result, it means that the driver could pass the intersection safely. In this way, HAS allows us to score each driver's safety driving skill. Note that HAS is individually defined for each approaching vehicles or bicycles from both direction using the same way.

In this paper, we calculate four kinds of HAS (vehicles/bicycles from right/left side).

12.3 Validation Study of HAS

12.3.1 Experiment on Public Road

To confirm the effectiveness of HAS, we performed an experiment on public road with 38 subjects. In this experiment, to observe the subjects' regular driving behavior, each subject was asked to drive a predefined 40 km course (almost 1 h driving). To record the driving behavior, we employed three wearable sensors and three video cameras. The wearable sensors have a sampling rate of 25 Hz. The subjects wore a cap with one sensor to measure drivers' head motion. Another one was placed on the subjects' right foot to measure pedal operation, and the other was placed on the dashboard to measure vehicle motion. We also use video cameras to record drivers' head/foot motion. In addition, the position and speed data of the vehicle was obtained by GPS with a frequency of 1 Hz. As our experimental scenario, we focused on the scene that subjects passed through the non-priority road of one unsignalized intersection (Fig. 12.2).

Figure 12.3 shows the geometrical features of the intersection, and Fig. 12.4 shows a relation between the position of the subject vehicle and the sight distance to

Fig. 12.2 Target intersection

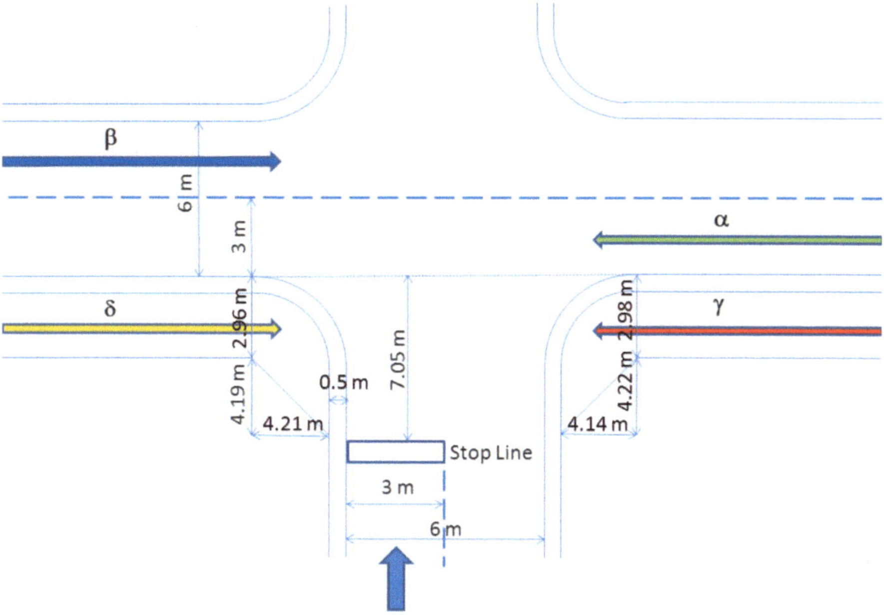

Fig. 12.3 Geometrical features of target intersection

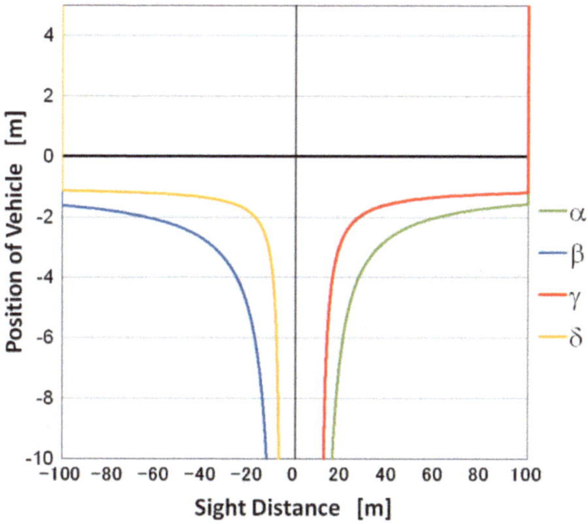

Fig. 12.4 Sight distance depends on the position of vehicle

other vehicles and bicycles. As this intersection has a low visibility, it is notorious for its traffic accidents.

The subjects' personal profiles are shown in Table 12.1. In this paper, we manually extracted the drivers' scanning behavior and pedal operation by using

Table 12.1 Subjects' personal profile

	Range	Average
Age	30–78	56.9
Driving career (year)	10–57	34.7
Driving opportunity (day/week)	2–7	5.9

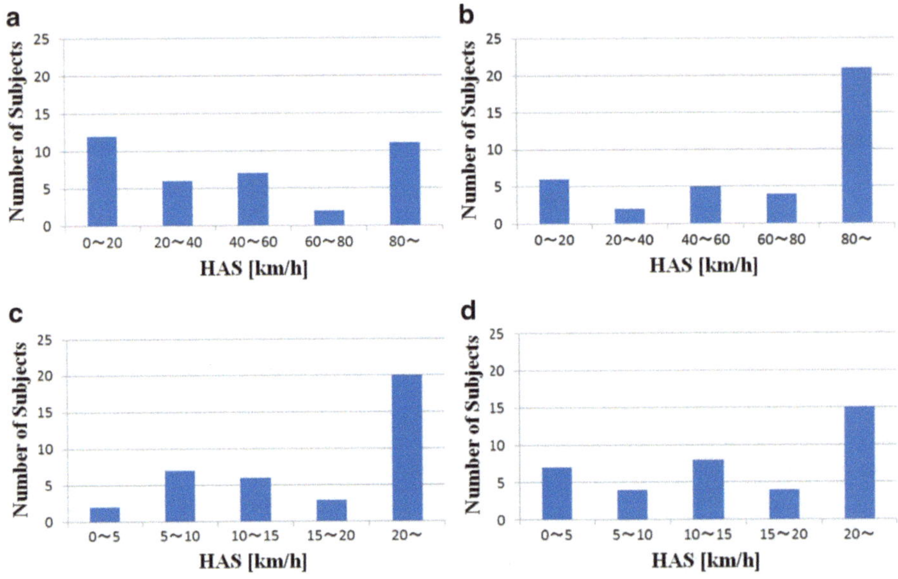

Fig. 12.5 Histograms of HAS results. (**a**) HAS for vehicle from the right side. (**b**) HAS for vehicle from the left side. (**c**) HAS for bicycle from the right side. (**d**) HAS for bicycle from the left side

video data by 0.1 s precision. Then, we calculated HAS for each approaching vehicle/bicycle from both sides as shown in Fig. 12.5.

When focusing on the HAS results for vehicles, the results from the left side are better than ones from the right side. We can suppose this difference is caused by different driving lanes. Since Japan adopted keep-left rule, the driving lane of vehicles approaching from the left side is 3 m behind of those from the right side. Therefore, subjects had a longer time to scan traffics to the left side than to the right side. When focusing on the HAS results for bicycles, the results for the right side are better than those for the left side. This is supposed to happen because the subjects can get a longer sight distance to the right side than to the left side from the same position.

As shown in Fig. 12.6, the cross-correlation among HAS results for each approaching vehicle/bicycle is not so strong. From the results, we can consider that subjects who pay enough attention for one side do not always pay enough attention for the other.

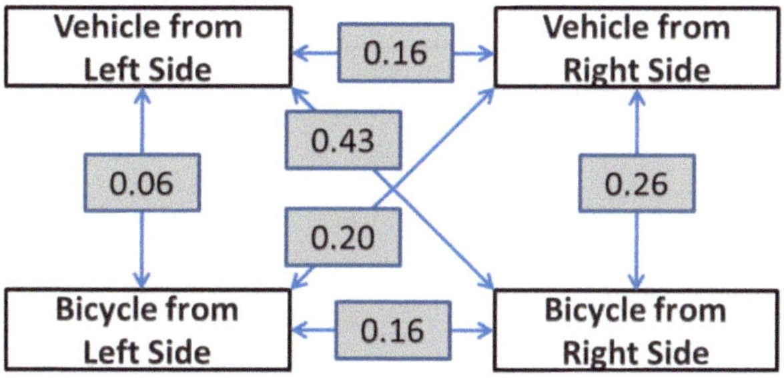

Fig. 12.6 Cross-correlation among HAS results

Fig. 12.7 An example of recorded video data

12.3.2 Evaluation Given by Driving Instructor

In this section we discuss the effectiveness of HAS in comparison with the professional driving instructor's judgment. We asked the driving instructor to give subjective evaluation of our subjects' safe driving skills. The driving instructor scored each subjects' driving skill with four levels: "worst," "bad," "good," and "best." In the subjective evaluation procedure, we showed the driving instructor our experiment video data and never showed HAS results. The video data consisted of the front view, driver's face, and driver's feet as shown in Fig. 12.7.

It should be noted that the instructor gave an overall score only, i.e., he didn't evaluate individual parts of the driving behavior such as scanning behavior to the left side. Instead, to clarify the instructor's judgment aspects, we asked him to leave comments of certain points and recorded these comments with a voice recorder.

Table 12.2 Instructor's judgment result

	Evaluation score	Number of subjects
Bad	1	1
	2	10
	3	15
Good	4	12

The driving instructor took about 5 min in average to evaluate each subject. Table 12.2 shows subjective evaluation result. Each figure in Table 12.2 represents the number of subjects at each level. About one-third of the subjects were judged as bad score "worst" or "bad."

12.3.3 Comparing HAS Results to the Driving Instructor's Judgment

Figure 12.8 shows histograms that itemize Fig. 12.5 according to the driving instructor's judgment result. When focusing on the evaluation results for vehicles from the right side, most of the subjects who were given high evaluation results by HAS were also evaluated as "good" or "best" by the driving instructor. The coefficient of correlation of both evaluation results is 0.64 ($p < 0.01$). However, when focusing on the evaluation results for other approaching vehicles and bicycles, they showed relatively weak correlations with the instructor's judgment results as shown in Fig. 12.6. Each coefficient of correlation between instructor's judgment and HAS result for vehicles from the left side, bicycles from the right side, and bicycles from the left side is 0.36 ($p < 0.05$), 0.32 ($p < 0.05$), and 0.37 ($p < 0.05$), respectively. A reason for this is probably because of the difference in evaluation procedure. As discussed above, since the driving instructor gave an overall score only, he gave a bad score when a subject performs poorly with one approaching vehicle or bicycle even if the subject behaved well with the others. In contrast, HAS did not provide overall score but a set of scores for the individual target (vehicles/bicycles from right/left side). Accordingly we integrated four kinds of HAS by using (12.4).

$$HAS_s = \frac{HAS_{rv} + HAS_{lv} + 4(HAS_{rb} + HAS_{lb})}{4} \quad (12.4)$$

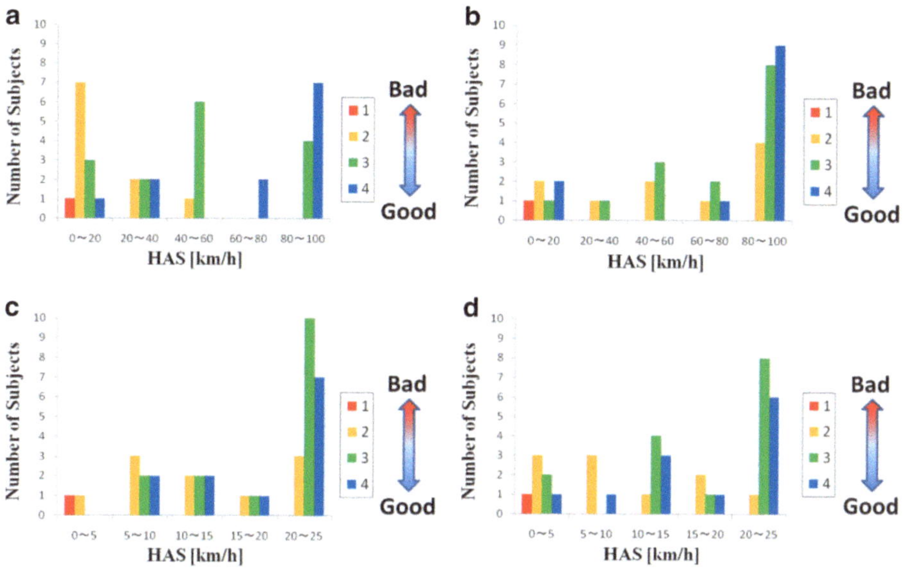

Fig. 12.8 HAS results itemized according the instructor's judgment. (**a**) HAS for vehicle from the right side. (**b**) HAS for vehicle from the left side. (**c**) HAS for bicycle from the right side. (**d**) HAS for bicycle from the left side

Here, HAS_{rv}, HAS_{lv}, HAS_{rb}, HAS_{lb} are HAS for vehicles from the left side, vehicles from the right side, bicycles from the left side, and bicycles from the right side, respectively. Here, we set limit of HAS for vehicles and bicycles as 100 km/h and 25 km/h, respectively. Therefore, we multiplied the HAS for bicycles by four to calculate a normalized average. By using HASs given this way, we made a comparison to the evaluation results given by the driving instructor again.

Figure 12.9 shows the relations between HAS results and driving instructor's judgment. By this integration the coefficient of correlation between HASs and the driving instructor's judgment became 0.71. This indicates the effectiveness of HAS as the evaluation method for safe driving.

12.4 More Understandable Result Presentation Method

As discussed in Sect. 12.3, we confirmed that HAS results showed similar pattern to the driving instructor's judgment results. In this section, we discuss applying HAS into safe driving lecture given by a driving school.

As a result of an interview for the driving instructor, we found out the fact that one of major problems in safe driving lecture is how to quantitatively show drivers their shortcomings in their safe driving skills in an easy understandable way.

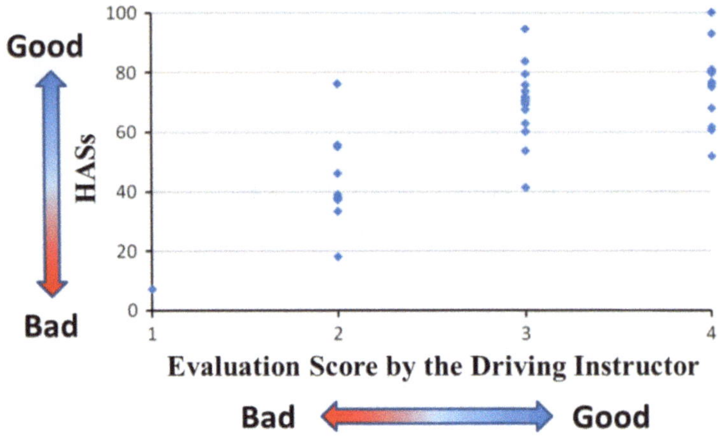

Fig. 12.9 Comparing HASs to instructor's judgment

From this viewpoint, HAS has some shortcomings. Firstly, HAS does not consider the traffic flux of intersections. Let us assume that a driver's HAS result for vehicles is 30 km/h. If there are few vehicles driving through the intersection over 30 km/h, the HAS result "30 km/h" means the driving behavior is a safe one. In contrast, if there are many vehicles driving through the intersection driving over 30 km/h, the HAS result "30 km/h" means the driving behavior is a dangerous one. In this way, without preliminary knowledge of traffic flux, it is difficult to interpret HAS result. Secondly, since HAS does not consider the traffic flux, trainees of the lecture might think HAS result is not realistic.

Accordingly, to resolve these problems we propose a more understandable result presentation method for HAS which shows the risk of collision in a probability form by considering characteristics of traffic flux. In this paper, to represent characteristics of traffic flux, we use speed distribution of other vehicles at the intersection. Figure 12.10 shows our basic idea. By overlaying HAS results on to the speed distribution, we could quantitatively evaluate risk of collision in probability form. Our method allows presenting evaluation results in a more intuitive way.

To estimate the speed distribution of approaching vehicles, by using a radar speed gun, we measured the speed of the approaching vehicles at the intersection at daytime of a working day for 2 h and obtained 121 vehicle data. Figure 12.11 shows speed distribution of the intersection. Because of the limitation of the radar speed gun's function, we could not measure vehicle speed under 16 km/h. Therefore, we gathered under 16 km/h data records into one range. By comparing the speed distribution to the histogram of HAS results, we can show the risk of collision in a probability form as shown in Fig. 12.11.

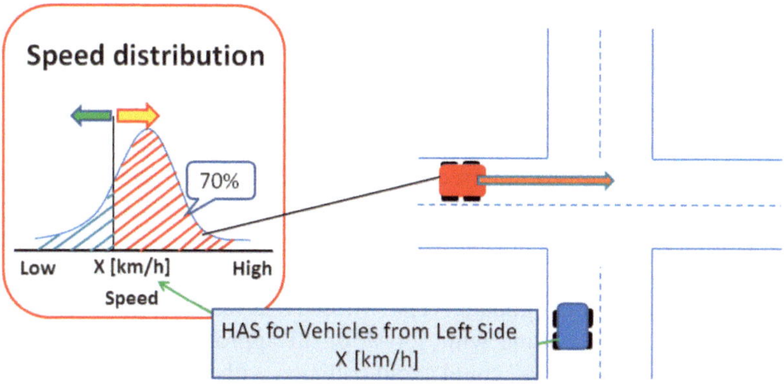

Fig. 12.10 Basic idea of understandable result presentation method for HAS

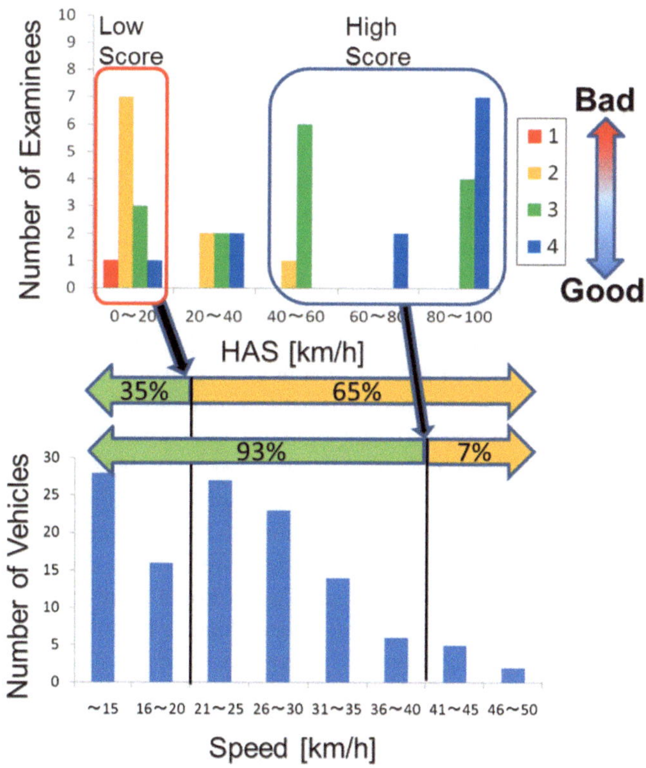

Fig. 12.11 Understandable result presentation method for HAS

Here, let us focus on Fig. 12.8 again. Figure 12.8 shows that almost 70 % of the subjects who were judged as a bad driver by the driving instructor obtained HAS result of under 20 km/h. In contrast, almost 80 % of the subjects who were judged as a good driver obtained HAS of over 40 km/h.

Let us now return to Fig. 12.11. Speed distribution shows that 65 % of all vehicles passing through the intersection drove over 20 km/h. This result indicates that 70 % of "bad drivers" whose HAS results marked under 20 km/h have potential risks to collide with 65 % of all approaching vehicles. In contrast, speed distribution also shows that almost 7.4 % of all vehicles passing through the intersection drove over 40 km/h, i.e., 80 % of "good drivers" whose HAS results marked over 40 km/h have potential risks to collide with 7.4 % of all approaching vehicles. In other words, 80 % of "good drivers" could avoid collision with 92.6 % of all approaching vehicles.

In this way, our method allows to quantitatively show the difference between "good drivers" and "bad drivers" as the difference of potential risks to collide with approaching vehicle in an easy understandable probabilistic form.

12.5 Discussion

In this paper, we confirmed that the coefficient of correlation between the evaluations computed by HAS and those given by an actual driving instructor marked 0.71. As the evaluation given by a driving instructor is the current definite method for evaluating one's driving skill, the result shows the possibility that our method could be used for evaluating the drivers' safe driving skill.

In this paper, to integrate four kinds of HAS for vehicles/bicycles from the right/left side, we simply averaged HAS results assuming that each HAS has the same importance. However, intersections having different kinds of geographical features would require drivers to pay different kinds of attention. Therefore, as a future work, we have to estimate the priority order of each HAS using the driving instructors' safety driving knowledge to develop weighted average method for HAS integration. Additionally, in calculation of HAS, approaching objects are considered to keep the same speed, but there were many drivers who slowed down as they approached the intersection. To develop a more realistic representation of the traffic flux by modeling these speed changes is also our future work.

From the viewpoint of educational effectiveness, the evaluation results should be given to the subjects as quickly as possible. Currently, we manually checked the video data to detect scanning/pedal operation behavior. However, in our other project, we have already developed the automated method to detect driving behavior from wearable sensor data. Therefore, by applying the method, we can automate detection procedure of scanning/pedal operation behavior. This would be a great help for quick HAS calculation.

12.6 Conclusion

In this study, we confirmed the effectiveness of our evaluation method HAS for safe driving skill based on sight distance. The effectiveness of HAS was validated by comparing the evaluation results with the ones given by driving instructors. Through experiments on public road with 38 subjects, the effectiveness of HAS was confirmed by achieving high similarity with the driving instructor's judgment.

In addition, to give drivers an impression of their own driving behavior, we proposed understandable result presentation method for HAS based on probability theory by adding information of other vehicle speed distribution. By overlaying HAS results on to the speed distribution, our method allows to quantitatively show the difference between "good drivers" and "bad drivers": while 70 % of "bad drivers" have potential risks to collide with 65 % of all approaching vehicles, 80 % of "good drivers" could avoid collision with 92.6 % of all approaching vehicles.

As the next step, we will plan to apply our method to a driving school and to study the educational effects at safe driving lectures initiated in October 2011.

References

1. M. Tada, H. Noma, A. Utsumi, M. Okada, K. Renge, Automatic evaluation system of driving skill using wearable sensors for personalized safe driving lecture, mobile learning 2012, pp. 173–180, Berlin, Germany, 13 Mar 2012
2. T. Kamon, S. Umehara, H. Kosaka, M. Noda, H. Nishitani, Y. Mizoguti, M. Obana, K. Sasaki, Evaluating risk levels of driver behaviors on basis of vehicle speed and driver eye-movement and pedal operation. Rev. Automotive Eng. **2**, 144–149 (2009)
3. M. Green, How long does it take to stop methodological analysis of driver perception-brake times. Transport Hum Factors **1**(3), 195–216 (2000)

Chapter 13
Pre- and Postaccident Emotion Analysis on Driving Behavior

Abdul Wahab, Norhaslinda Kamaruddin, Norzaliza M. Nor, and Hüseyin Abut

Abstract There are many contributing factors that result in high number of traffic accidents on the roads and highways today. Globally, the human (operator) error is observed to be the leading cause. These errors may be transpired by the driver's emotional state that leads to his/her uncontrolled driving behavior. It has been reported in a number of recent studies that emotion has direct influence on the driver behavior. In this chapter, the pre- and postaccident emotion of the driver is studied in order to better understand the behavior of the driver. A two-dimensional Affective Space Model (ASM) is used to determine the correlation between the driver behavior and the driver emotion. A 2-D ASM developed in this study consists of the valance and arousal values extracted from electroencephalogram (EEG) signals of ten subjects while driving a simulator under three different conditions consisting of initialization, pre-accident, and postaccident. The initialization condition refers to the subject's brain signals during the initial period where he/she is asked to open and close his/her eyes. In order to elicit appropriate precursor emotion for the driver, the selected picture stimuli for three basic emotions, namely, happiness, fear, and sadness are used. The brain signals of the drivers are captured and labeled as the EEG reference signals for each driver. The Mel frequency cepstral coefficient (MFCC) feature extraction method is then

A. Wahab (✉) • N.M. Nor
Kuliyyah of Information & Communication Technology, Department of Computer Science, International Islamic University, Malaysia, Jalan Gombak, Kuala Lumpur 53100, Malaysia
e-mail: abdulwahab@iium.edu.my

N. Kamaruddin
Universiti Teknologi Mara, Selangor Darul Ehsan, Malaysia
e-mail: norhaslinda@fskm.uitm.edu.my

H. Abut
ECE Department (Emeritus), San Diego State University, San Diego, CA 92182, USA

EEE Department, Boğaziçi University, Istanbul, Turkey
e-mail: abut@anadolu.sdsu.edu

G. Schmidt et al. (eds.), *Smart Mobile In-Vehicle Systems: Next Generation Advancements*, DOI 10.1007/978-1-4614-9120-0_13,

employed to extract relevant features to be used by the multilayer perceptron (MLP) classifier to verify emotion. Experimental results show an acceptable accuracy for emotion verification and subject identification. Subsequently, a two-dimensional Affective Space Model (ASM) is employed to determine the correlation between the emotion and the behavior of drivers. The analysis using the 2-D ASM provides a visualization tool to facilitate better understanding of the pre- and postaccident driver emotion.

Keywords Driver behavior • Valance • Arousal • Pre- and postaccident emotion • Mel frequency cepstral coefficients (MFCC) • Multilayer perceptron (MLP)

13.1 Introduction

Driving requires making critical decisions in very short period of time, and oftentimes, such decision is needed under extraneous circumstances. To make an informed maneuvering decision, drivers rely on input from a number of sources including the road condition, traffic volume, other road users, the condition of the vehicle, duration of the trip, and the environment [1]. Drivers are observed to lack prudent decision-making ability (a) if their secondary tasks like mobile phone usage and texting and/or (b) if they are under the influence of alcohol, drugs, stress, fatigue, and excessive emotion. These are known to distract drivers' concentration and often cause accidents [2]. Hence, the understanding of drivers' emotion, in particular, on the pre- and postaccident instances is important to give us cues the way emotion impacts the driving activity. The following three conjectures are prevalent in the research community:

1. Emotion influences driver's behavior.
2. Individual emotional brain signal can be measured.
3. There are variations among brain signals of people for a particular emotion, and it could even be unique for each person.

Emotion is a very important factor in human life to interact and even to control his/her own behavior. Moreover, the uncontrolled emotion—i.e., negative emotion—during driving is observed to result in poor judgment calls and causing accidents with serious consequences.

It is often observed that anger impairs the driver's decision-making ability leading him/her to drive dangerously, taking unnecessary risks, or even force other drives to slow down or swerve, i.e., road rage.

Fear is observed to introduce overconscious behavior that makes drivers to hesitate or act in a non-confident way. For instance, if a driver has fear of speed (tachophobia syndrome), he/she does not want to drive higher than the speed that he/she is comfortable with, which poses concerns and even results in accidents at roads with a minimum speed limit. These drivers are observed to suffer from breathlessness, palpitations, and even full-blown anxiety attack.

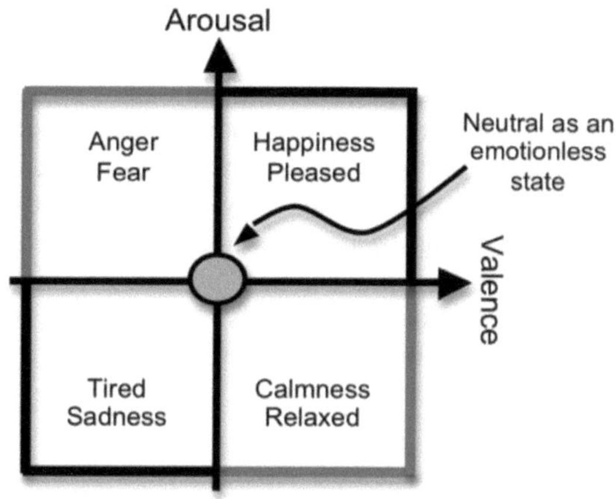

Fig. 13.1 Affective space model with axis valance (*vertical*) and arousal (*horizontal*)

Sadness can also influence driver behavior by making him/her lose concentration and long delays in reacting. This condition poses high risks in congested traffic and highways.

On the other hand, positive emotion can as well distract drivers. Often, drivers are observed to focus on his/her happiness and ignore the world around him/her. He/she could miss critical driving cues, such as, indication of empty fuel tank, blinking warning signs in the vehicle, and even road signs on construction and activities on the shoulder of the highway.

In a number studies, the emotional effect has been conceptualized in terms of emotion primitives of valance and arousal values [3, 4]. Valance (v) refers to the impact of the emotion on oneself ranging from a positive to a negative effect, i.e., the extent of pleasure or sadness. It can be described as a bipolar continuum of positive and negative value of hedonic level [5]. The arousal (A) ranges from calm to excited. These two values can be used to generate an Affective Space Model (ASM) to illustrate different emotion boundaries. Figure 13.1 shows the affective space model with several labeled emotions and neutral as a black dot in the middle of the model according to Russel [6].

13.2 Related Work

In recent years a number of research teams have focused on capturing emotions from EEG recordings [7]. Chanel et al. have tried to recognize only the arousal dimension of emotion from their EEG database and other physiological measures [8]. Classification rates were around 60 % when using two emotional classes, and if an additional class is added, that number dropped to 50 %. Most studies in the literature are based on a two-dimensional model of emotions, valance (positive–negative) and arousal (calm–exciting). Emotions are then thought to be a point in a two-dimensional plane of valance vs. arousal as depicted in Fig. 13.1.

In a study at Israel Institute of Technology (Technion), the driving data has been collected by an in-vehicle data recorder (IVDR) called Drive Diagnostics. This IVDR has been designed to monitor and analyze driver behavior not only in crash or pre-crash events, but also in normal driving situations. It records the movement of the vehicle and uses this information to indicate overall trip safety. Their findings show strong correlations between the two datasets, suggesting that the driving risk indices can be used as indicators of the risk involvement in car crashes [9]. This connection has enabled our study on the potential impact of the system on driving behavior and on safety. Access to the feedback provided by the system has further impact on driver performance.

13.3 Methodology

13.3.1 Electrode Positions and Preprocessing

Five EEG electrodes were pasted on the scalp of subjects (C3, C4, T3, and T4) according to the "International 10–20" Standards and a Cz as reference. Figure 13.2 shows the block diagram of the procedure that was used to analyze the brain signals and their spectrum from the subjects under investigation (drivers).

Brain waves obtained from each channel are then decimated in order to decrease the sampling rate and to filter the data. As expected, the decimation process filters the input data with a low-pass filter and then resamples the resulting smoothed signal at a lower rate. The matlab code below reduces the sampling rate by a factor of 3:

input = decimate (input, 3).

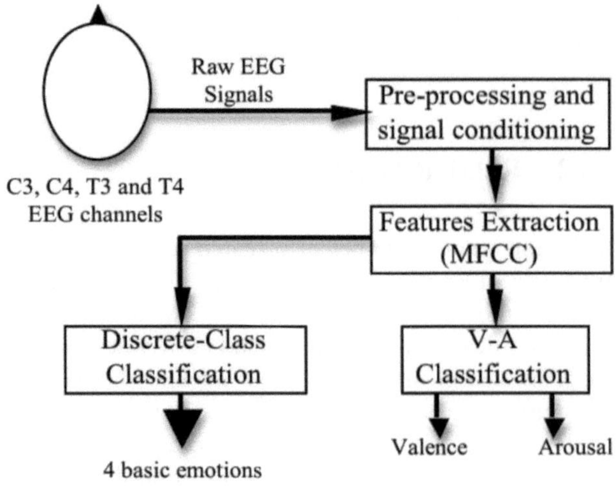

Fig. 13.2 Proposed procedure for analysis

13.3.2 Feature Extraction

Mel frequency cepstral coefficients (MFCC) were used as features in this study, which is used frequently in dimension reduction applications for waveforms. *melcepst* tool is utilized to calculate the cepstrum of the signal. We have used ten MFCC coefficients for capturing the relevant feature of the EEG. The final combined dataset from four channels gives a total of 40 features for classification. *enframe*, *rfft*, *melbankm*, and *rdct* are also utilized during processing.

13.3.3 Classification

The last step in this process is the classification of the features with a meaningful and yet computationally efficient classifier. Multilayer perceptron (MLP) method has been chosen to classify the extracted features so that it can derive the pre-emotion of the driver which relates to the driving behavior. Multilayer perceptron with a feedforward artificial neural network architecture maps sets of input data onto a set of appropriate output. Optimal model selection for the number of layers and the neurons needed for the best MLP architecture is essential to ensure a respectable performance. Data fed into the input layer are the 40 features obtained from the previous MFCC stage. Each of the data is processed by the network by multiplying it with assigned weights in the hidden layers synapses. In this research study, the mean square error (mse) goal was set to 0.1 with a single hidden layer consisting of ten neurons. In addition, we have used *tan-sig* as the activation function for the hidden layer and *purelin* as the output layer with 0.01 learning rate.

In experiments, *eyes open* data was used for the *calm emotion* since the subjects have not been burdened by any task at this stage. The data obtained from the subjects (drivers) are tested against the emotion data of affective state of *happy*, *calm*, *fear*, and *sad*.

In order to get reliable results with a high percentage of accuracy, we have used the k-fold cross-validation for our global validation. K-fold cross-validation builds on the idea of holdout testing in a clever way by rotating data through the process 0. Data is again divided randomly into groups, but now k equal-sized groups are used. The train-test process is repeated k times, each time leaving a different segment of the data out as the test set. The dataset and its desired result are randomized and sliced into fivefolds which mean that the process is repeated five times. This is required to eliminate any biases towards the data [10]. The slicing process enables to have different training and testing datasets. Each dataset consists of 440 instances by which 352 (80 %) instances are used for training and the remaining 88 (20 %) for testing.

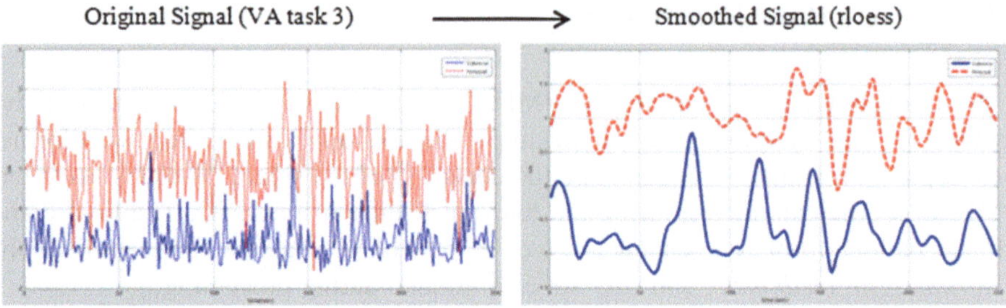

Fig. 13.3 Signal smoothing

13.3.4 Smoothing

This function basically smoothens the original data that has been selected. The process of smoothing the original dataset produces a new dataset containing smoothed response values. The smoothed signal is displayed in Fig. 13.3.

13.4 Experiment Design and Stimuli

Subjects are briefed on the experimental procedures and were asked to sign an informed consent form for participating in experiments. Then, subjects are seated in a lighted, quiet, and temperature-controlled room. Before the data collection, each subject is made familiar with the driving simulators. Next, the electrodes are placed on the scalp of each subject. The acquisition of signals is achieved by a module called BMC Acquisition. Initially, subjects are instructed to open their eyes for 1 min and then close for 1 min. Afterwards, the movie clips with three basic emotions are shown to them for 1 min per movie clip. Finally, they were asked to drive according to the three tasks given to them, and the recorded brain waves are then saved for offline processing.

13.4.1 Stimuli

In this study, we have used the movie clips with scenarios depicting emotions to obtain emotional responses and a driving simulator platform to simulate the driving framework. The drivers were exposed to three basic emotions by using (1) the International Affective Picture (IAPS), (2) Bernard Bouchard's synthesized musical clips, and (3) movie clips of Gross and Levenson which can be used to elicit emotional responses [8]. Then, they were asked to drive in three different types of conditions: Task 1—easy driving, where they were subject to noisy sounds that could disturb them while driving. Task 2—bulked driving, where they were interviewed by the experimenter deliberately to see their behavior while answering cognitive questions. Finally, Task 3—heavy driving, subjects had to deal with traffic congestion where their driving skills were challenged at this stage.

13.5 Results and Discussions

Since the correlation between the pre-emotional state of the subject and driving behavior is the primary interest, the accuracy of the acquired data gains importance to demonstrate that the results will be more robust from the proposed valance analysis (VA). To achieve that a memory test was performed for all subjects to see the level of accuracy, either it can be accepted or rejected. Next, we have performed a fivefold validation test to obtain the intensity of the selected emotions for each subject.

13.5.1 Pre-emotion (Memory Test and Fivefold)

Here, the accuracy is calculated from the valance analysis instead of directly getting the accuracy from MLP. There are two VA analyses in the identification of a particular; first: the memory test which consists of 100 % test data (Fig. 13.4) and second: the fivefold validation (Fig. 13.5). As it is clearly seen, four basic emotions can be identified, and higher than 80 % of accuracy can be achieved. The best accuracy value was at 0.1 of mean square error goal. Consequently, the emotions data can be used as the base for the subsequent driving task analysis. Furthermore, the highest intensity of emotion for each subject is shown in Fig. 13.5. Here the plots are according to the average k-fold percentage for each subject.

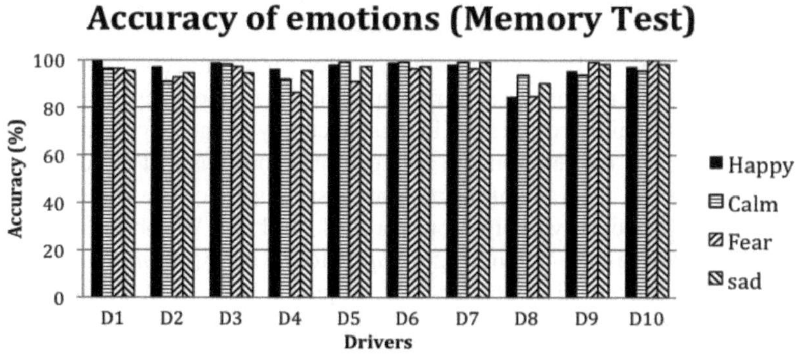

Fig. 13.4 Accuracy of emotion based on the memory test

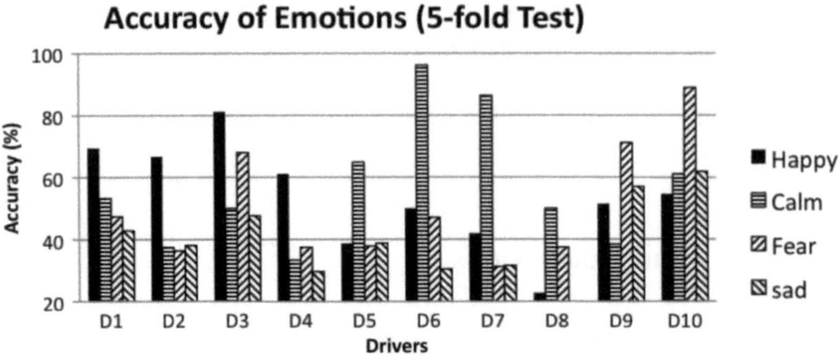

Fig. 13.5 Emotion intensity based on the average accuracy of k-fold test

Table 13.1 Confusion matrix, average emotion accuracy (memory test), subject 1 (happy)

		Expected emotion (%)			
		Happy	Calm	Fear	Sad
Output emotion (%)	Happy	100	0	3.6	0
	Calm	0	96.3	0	1.8
	Fear	0	0	96.3	2.7
	Sad	0	3.6	0	95.4

It is worth noting that when the emotion that gains the highest intensity in the k-fold yields the same result in the memory tests. In addition, there are four subjects with *happy* as the highest intensity of emotions, another four subjects fall into *calm*, and two subjects has *fear* as their pre-emotion. None of them is *sad*. The results indicate that each subject has their own pre-emotion which may affect their driving behavior.

13.5.2 Valance (V) and Arousal (A) Analysis

The results above point to four critical emotions relevant to driving, which are *happy*, *calm*, *fear*, and *sad*. With these findings, we extend the analysis to the valance and arousal analysis where emotions for three subjects studied in detail. The subjects have been chosen randomly based on the memory test results, and they show the emotion which exhibits the highest accuracy in the k-fold analysis. Here, we would like to see the relationship between pre-emotions and the driving tasks. Table 13.1 shows the accuracy of the memory test for subject 1, whereas Table 13.2 shows the average accuracy of emotions for the k-fold test. We observe that subject 1 has exhibited the highest accuracy for *happy* followed by *calm*, *fear*, and *sad*.

In Fig. 13.6, the emotion clusters for subject 1 have been plotted from the memory test experiment in order to see if each emotion falls into its own quadrant based on valance and arousal. As it is clearly seen that none of the *happy* emotion clusters falls into other three quadrants, whereas the other three emotions have reciprocally spread to each other.

Table 13.2 Confusion matrix, average emotion accuracy (VA), subject 1 (happy)

		Expected emotion (%)			
		Happy	Calm	Fear	Sad
Output emotion (%)	Happy	69.45	14.83	6.53	5.71
	Calm	23.71	53.29	8.44	9.76
	Fear	5.59	13.93	47.40	41.67
	Sad	1.25	17.95	37.63	42.86

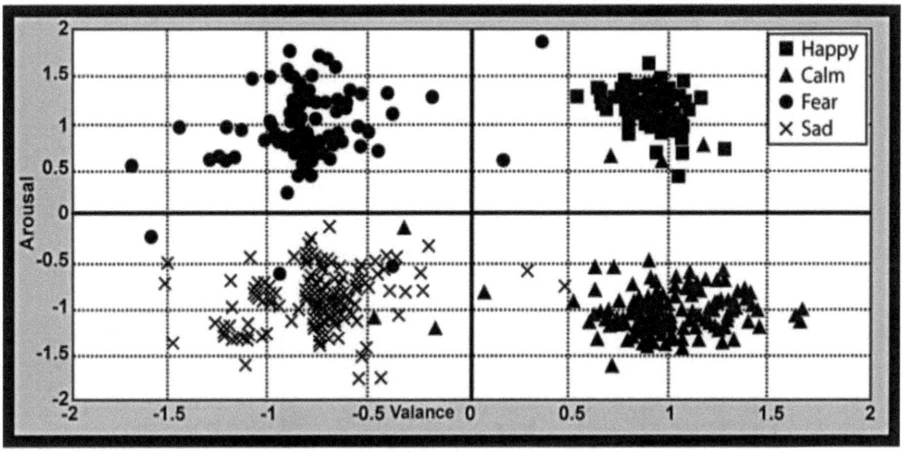

Fig. 13.6 Emotion clustering for subject 1

Table 13.3 Confusion matrix, average emotion accuracy (VA) for fivefold test, subject 5 (calm)

		Expected emotion (%)			
		Happy	Calm	Fear	Sad
Output emotion (%)	Happy	38.7	18.5	29.1	19.1
	Calm	9.6	64.9	7.3	12.6
	Fear	30.9	1.6	37.9	29.5
	Sad	20.7	15.1	25.7	38.8

Table 13.4 Confusion matrix, average emotion accuracy (VA), subject 9 (fear)

		Expected emotion (%)			
		Happy	Calm	Fear	Sad
Output emotion (%)	Happy	51.4	31.5	4.0	9.2
	Calm	27.2	38.4	5.7	9.3
	Fear	12.1	12.3	71.2	24.6
	Sad	9.2	17.8	19.0	57.0

Table 13.3 shows the emotion accuracy for subject 5 with the highest being *calm*. Finally for this VA analysis, subject 9 exhibits *fear* as the highest intensity of emotion as shown in Table 13.4.

For these three subjects, we can conclude that each subject has their own highest intensity for the pre-emotion, which could be due to their different backgrounds, cultures, or experiences of having positive or negative emotions that already exist in each subject, i.e., preexisting conditions.

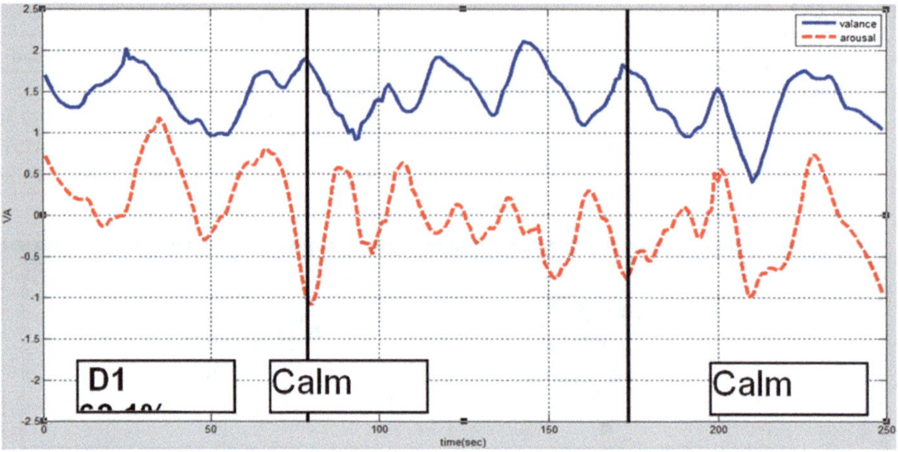

Fig. 13.7 Dynamic movement task 1, subject 1 (driving and sounds)

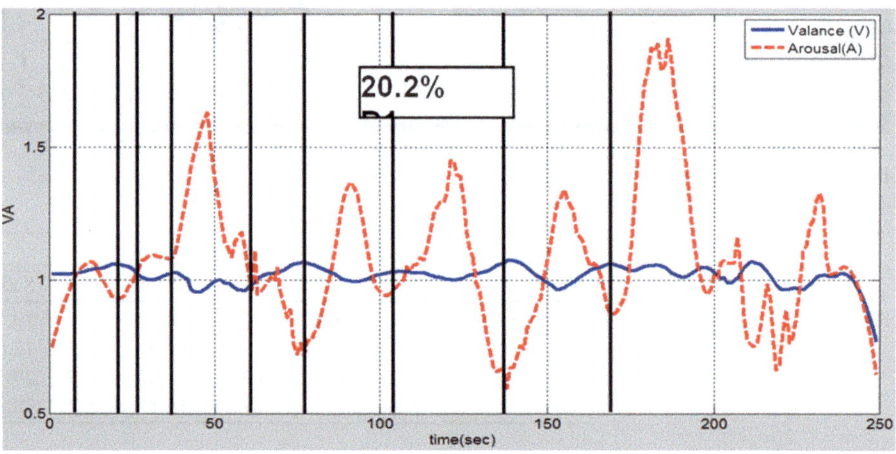

Fig. 13.8 Trajectory task 2, subject 1 (driving and interview)

13.5.3 Trajectory of the Driving Task

Each task that was given to the driver is expected to invoke some stress during driving, so we can observe the driver behavior while under duress. From these experiments, we see that each driver has a diverse driving behavior. For subject 1, who has not been affected by an accident, the emotions remain positive beginning with the first task until the one as they can be seen in Figs. 13.7, 13.8, and 13.9. Besides, this particular driver (female) has changed her emotion from *happy* to *calm* when the accident occurred during the first task. This result implies that this driver has a very high intensity of *happy* emotion.

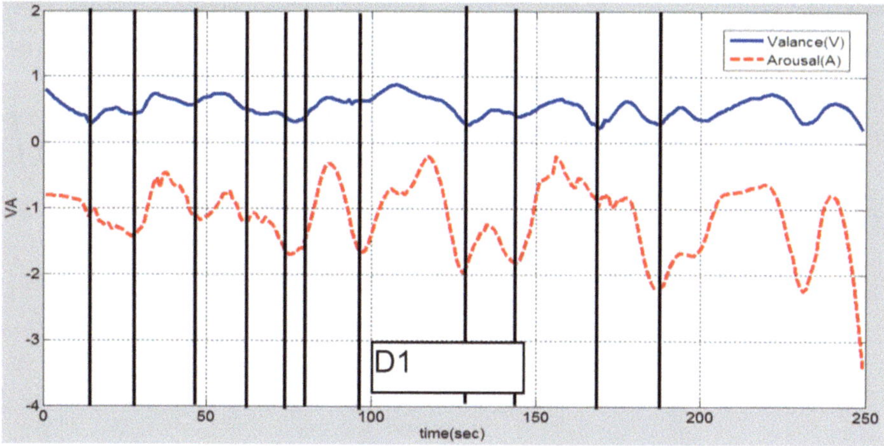

Fig. 13.9 Dynamic movement task 3, subject 1 (driving and congested traffic)

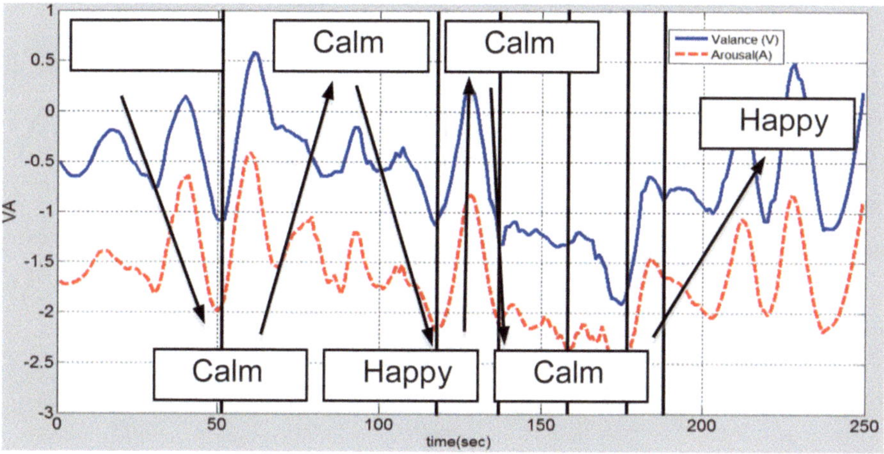

Fig. 13.10 Dynamic movement task 1, subject 5 (driving and sounds)

In contrast to subject 1, subject 5 has the same pre-emotion since the beginning, but his emotional state has changed when the accident occurred. However, he has managed to complete the driving task with *calm* emotion. Furthermore, this driver took some time to get back to the pre-emotion that he had. As we can see from Figs. 13.10, 13.11, and 13.12, the vertical solid black lines represent that accidents occurred while the subject was driving the vehicle. The trajectory of task 1 and 3 (Figs. 13.10 and 13.12, respectively) is mostly from *calm* to *sad* and vice versa, whereas for task 2 (Fig. 13.11), he was *sad* for the whole task. This is the indicator of giving up, as he had sighed a lot in order to maneuver the car while answering the question from the experimenter.

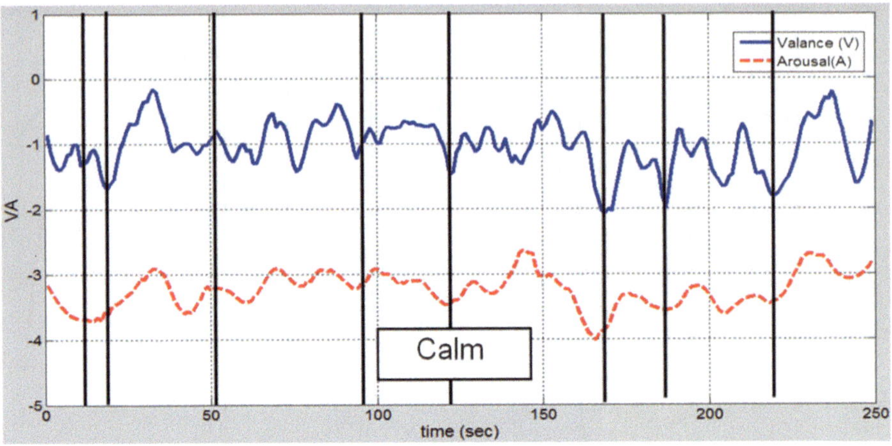

Fig. 13.11 Trajectory task 2, subject 5 (driving and interview)

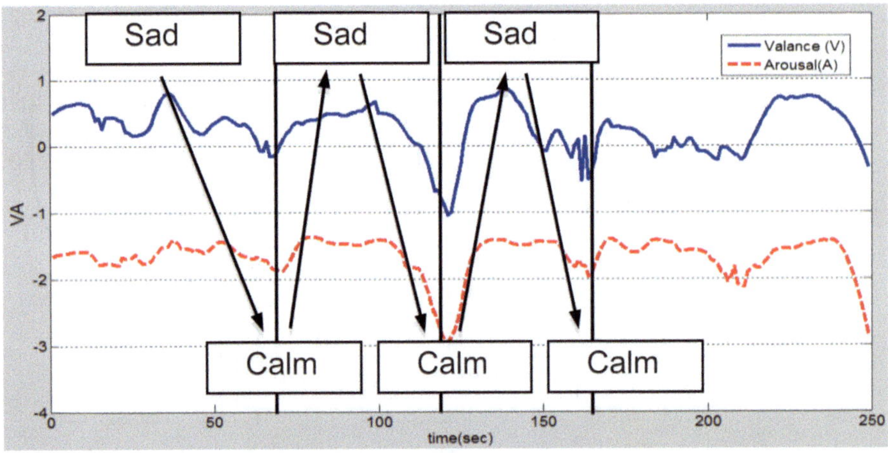

Fig. 13.12 Dynamic movement task 3, subject 5 (driving and congested traffic)

Finally, we include Figs. 13.13, 13.14, and 13.15 to illustrate the trajectories for subject 9. As we can see from Fig. 13.13, the trajectory has turned from *fear* to *sad* for task 1, whereas for task 2, he has just stayed *sad*. This could be interpreted as this subject was nervous behind the wheel at the beginning of the first task given to him, but after several accidents, he had just given up and became *sad* as it is obvious from Fig. 13.14. Finally, he has started with *happy* emotion for task 3 and turned to *fear* when accident occurred. After the accident he became *happy* again. Therefore, he demonstrated willingness to drive after a long period of driving but still manages to come back to the pre-emotion state even if there is an accident. This could be interpreted as the subject fears easily regardless of his pre-emotion state, i.e., before driving tasks.

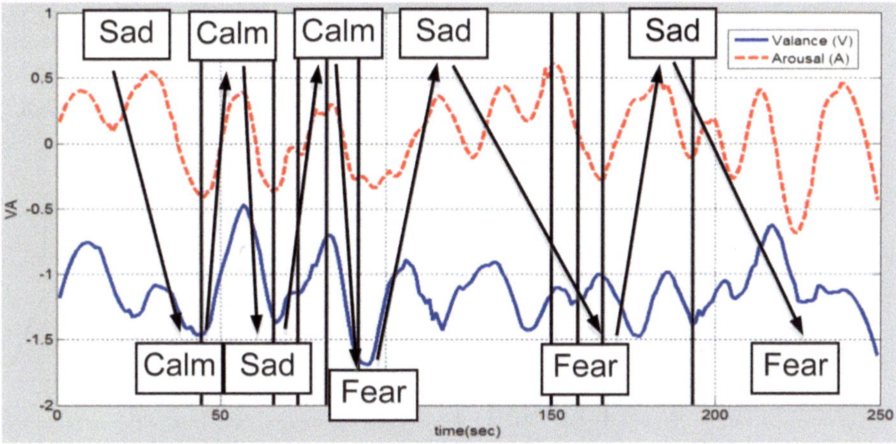

Fig. 13.13 Dynamic movement task 1, subject 5 (driving and sounds)

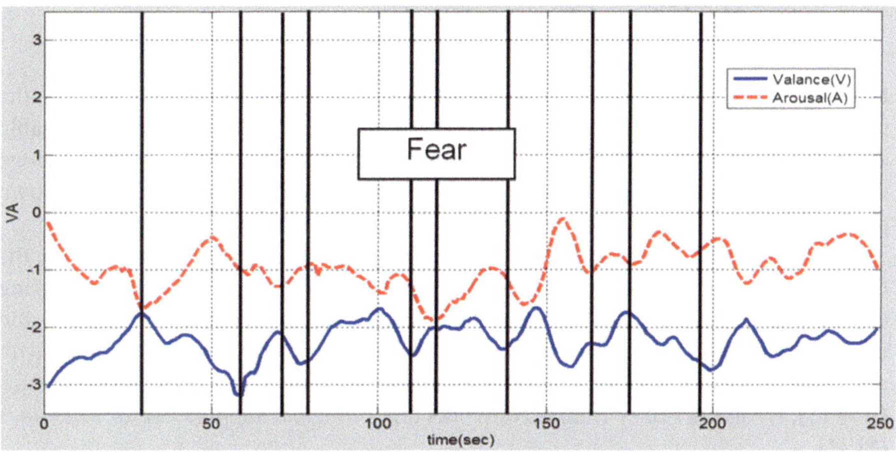

Fig. 13.14 Dynamic movement task 2, subject 5 (driving and interview)

From these results, we can conclude that each subject has his/her own pre-emotion state and a precursor emotion that impacts the driving behavior. The pre-emotion is the emotional state that the subject was in before coming for the experiment; while precursor is the emotion that was already in the mind of the subject caused by previous experiences or emotions that he/she already has experienced in the past. Therefore, these two emotional states have a strong relationship between each other since they affect the subject during the driving tasks.

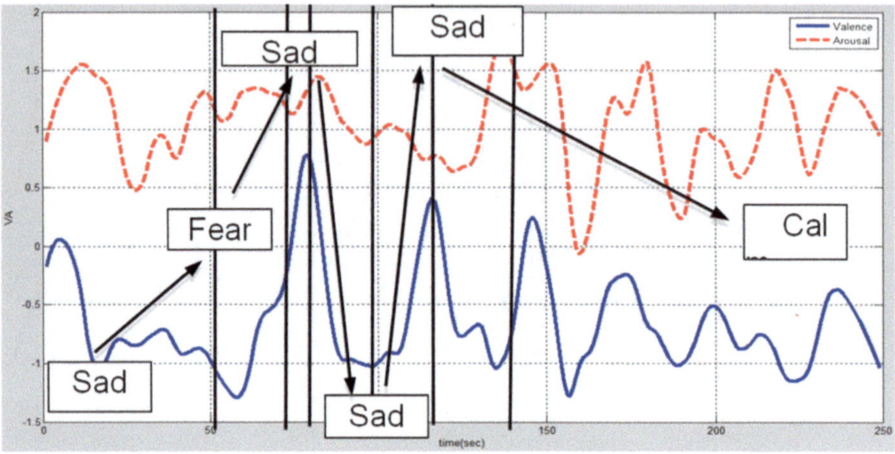

Fig. 13.15 Dynamic movement task 3 (driving and congested traffic)

13.6 Conclusion and Future Work

From these findings it can be deduced that there is a strong correlation between the pre-emotional state of drivers and their driving behavior. In addition, unstable emotion could potentially lead to accidents, and the drivers could easily change their positive emotion to a negative one. We also see that each driver has their own pre-emotion that could impact their driving behavior since the beginning.

In future work, we are planning to enlarge the driver database and to explore the behavior of a larger number of subjects from the same background and the driving culture to better understand the driving experience under different road, traffic, and environmental conditions. In addition, we would like to study the performance with a number of other classifiers including Adaptive Neuro-Fuzzy Inference Systems (*ANFIS*), Evolving Fuzzy Neural Networks (*eFuNN*), and *Support Vector Machines (SVM)*.

Acknowledgment This study is supported in part by the IIUM Endowment Fund (EDW B10-108-0447). The authors would like to thank all families who supported in this study and Bjorn Cruts from Biometrisch Centrum for sponsoring our EEG machine.

References

1. N. Kamaruddin, A. Wahab, Driver behavior analysis through speech emotion understanding. IEEE International Symposium on Intelligent Vehicle, 2010 (IV 2010), pp. 238–243, San Diego, California, USA, 21–24 June 2010
2. M.R. Othman, Z. Zhang, T. Imamura, T. Miyake, A study of analysis method for driver features extraction. IEEE International Conference on Systems, Man and Cybernetics, 2008 (SMC 2008), pp. 1501–1505, Singapore, 12–15 Oct 2008

3. J.A. Russell, A circumplex model of affect. J. Pers. Soc. Psychol. **39**, 1161–1178 (1980)
4. H. Schlosberg, Three dimensions of emotion. Psychol. Rev. **61**(2), 81–88 (1954)
5. P.A. Lewis, H.D. Critchley, P. Rotshtein, R.J. Dolan, Neural correlates of processing valence and arousal in affective words. Cereb. Cortex **17**, 742–748 (2007). Advance Access publication, 2006
6. J.A. Russell, Culture and the categorization of emotions. Psychol. Bull. **110**, 426–450 (1991)
7. W. Heller, J. Nitschike, D. Lindsay, Neuro psychological correlates arousal in self-reported emotion. Neurosci. Lett. **11**(4), 383–402 (1997)
8. G. Chanel, J. Kronegg, G. Grandjean, P. Pun, Emotion assessment: arousal evaluation using EEG's and peripheral physiological signals. Computer Vision Group, Computing Science Center, University of Geneva, Tech. Rep. 5 Feb 2005
9. L. Tsippy, T. Toledo, In-Vehicle Data Recorder for evaluation of Driving Behavior and Safety. Israel Institute of technology, pp 122–119, 2006
10. S.M. Weiss, C.A. Kulikowski, *Computer Systems That Learn: Classification and Prediction Methods from Statistics, Neural Nets, Machine Learning and Expert Systems* (Morgan Kaufmann, San Francisco, 1990)

Part V
Driving Scene Analysis

Chapter 14
Content-Based Driving Scene Retrieval Using Driving Behavior and Environmental Driving Signals

Yiyang Li, Ryo Nakagawa, Chiyomi Miyajima, Norihide Kitaoka, and Kazuya Takeda

Abstract With the increasing presence of drive recorders and advances in their technology, a large variety of driving data, including video images and sensor signals such as vehicle velocity and acceleration, can be continuously recorded and stored. Although these advances may contribute to traffic safety, the increasing amount of driving data complicates retrieval of desired information from large databases. One of our previous research projects focused on a browsing and retrieval system for driving scenes using driving behavior signals. In order to further its development, in this chapter we propose two driving scene retrieval systems. The first system also measures similarities between driving behavior signals. Experimental results show that a retrieval accuracy of more than 95 % is achieved for driving scenes involving stops, starts, and right and left turns. However, the accuracy is relatively lower for driving scenes of right and left lane changes and going up and down hills. The second system measures similarities between environmental driving signals, focusing on surrounding vehicles and driving road configuration. A subjective score from 1 to 5 is used to indicate retrieval performance, where a score of 1 means that the retrieved scene is completely dissimilar from the query scene and a score of 5 means that they are exactly the same. In a driving scene retrieval experiment, an average score of more than 3.21 is achieved for queries of driving scenes categorized as straight, curve, lane change, and traffic jam, when data from both road configuration and surroundings are employed.

Keywords Content-based retrieval • Driving data • Drive recorder • Similarity measure • Surrounding environment

Y. Li (✉) • R. Nakagawa • C. Miyajima • N. Kitaoka • K. Takeda
Graduate School of Information Science, Nagoya University, Nagoya, Japan
e-mail: yiyang.li@g.sp.m.is.nagoya-u.ac.jp

G. Schmidt et al. (eds.), *Smart Mobile In-Vehicle Systems: Next Generation Advancements*, DOI 10.1007/978-1-4614-9120-0_14,

14.1 Introduction

Drive recorders are used to investigate the causes of traffic accidents and to improve drivers' safety awareness. With the increasing presence of more advanced drive recorders, a large variety of driving data, including video images and sensor signals such as vehicle velocity and acceleration, can be continuously recorded and stored. Although these advances may contribute to traffic safety, the increasing amount of driving data complicates retrieval of desired information from large databases. Some researchers have studied methods for recognizing driving events, such as lane changing and passing, using HMM-based dynamic models [1–3]. In our previous work, a similarity-based retrieval system for finding driving data was proposed [4]. However, since our method used differences in histograms of driving behavior signals as the similarity measurement, it did not efficiently use dynamic information from driving scenes for retrieval. In this chapter, we study two driving scene retrieval systems that utilize dynamic information from driving scenes.

In the first study, we focus on driving behavior signals. The first retrieval system captures dynamic information from driving scenes by directly using sequences of driving behavior signals and utilizes changes in these signals over time. Six kinds of driving behavior signals (velocity, longitudinal and lateral acceleration, gas and brake pedal pressures, and steering angle) are used for calculating similarity between driving scenes. We compared the use of both early and late integration to integrate these signals.

In the second study, we focus on environmental driving data that is collected from the road and surrounding vehicles. The second retrieval system uses a similarity measure to compare the road configuration and motion of surrounding vehicles. Positions of surrounding vehicles and roadside barriers are detected with laser scanners mounted on the front and back of an instrumented vehicle, and the velocities of surrounding vehicles are estimated from their relative positions to the vehicle. Each scanned frame of a driving scene is categorized based on three general features, i.e., road type, congestion level, and the positions of surrounding objects. Also, the motion of each surrounding vehicle is tracked to obtain its motion features, so we can measure the similarity between vehicles. Categorization results and detected vehicle path are integrated to measure similarity between driving scenes.

14.2 Data Collection

The driving data used in our study was collected on real roads and was recorded using the instrumented vehicle shown in Fig. 14.1. The collected signals included velocity [km/h], longitudinal and lateral acceleration [G], gas and brake pedal pressures [N], and steering wheel angle [deg]. Two laser scanners were mounted on the front and back of the vehicle to detect surrounding objects. The laser scanners covered 80° arcs at both the front and back of the vehicle, to an effective

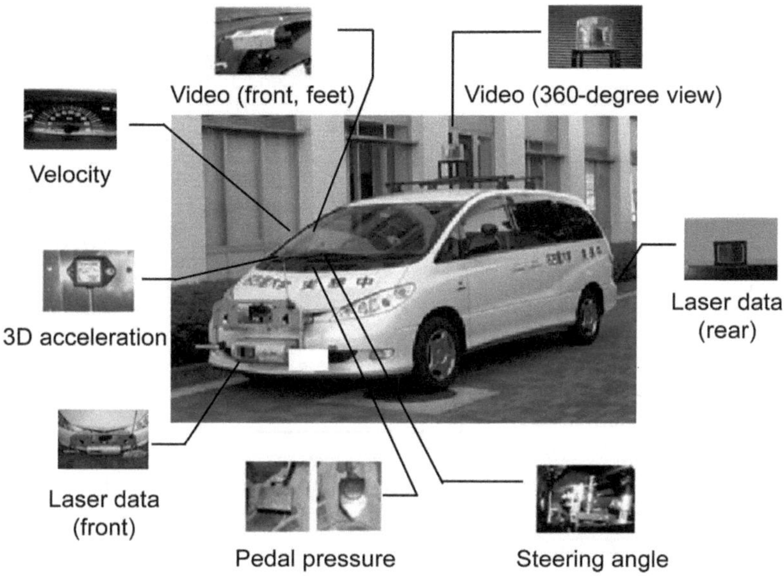

Fig. 14.1 Instrumented vehicle used for driving data collection [8]

range of about 100 m to the front and 55 m to the rear. A Kalman filter was employed to predict the motions of objects in blind areas. In order to assist in the subjective confirmation of retrieved scenes, synchronously recorded front and driver's feet scenes, as well as a 360° panoramic scene of the surroundings from an omnidirectional camera, were also available for every retrieved scene.

14.3 Driving Scene Retrieval Using Driving Behavior Signals

In this section, we describe the first similarity-based driving scene retrieval system, which uses similarity of driving behavior signals. Six driving signals (velocity, longitudinal and lateral acceleration, gas and brake pedal pressures, and steering angle) were used for calculating similarity between driving scenes. We compared the use of early and late integration to integrate these signals.

14.3.1 Integration Methods for Driving Behavior Signals

14.3.1.1 Method 1: Early Integration

We retrieved similar driving scenes using two methods, early and late integration. Figure 14.2 shows the procedure for early integration. The six kinds of signals mentioned above were extracted from the scene to be retrieved, and each signal

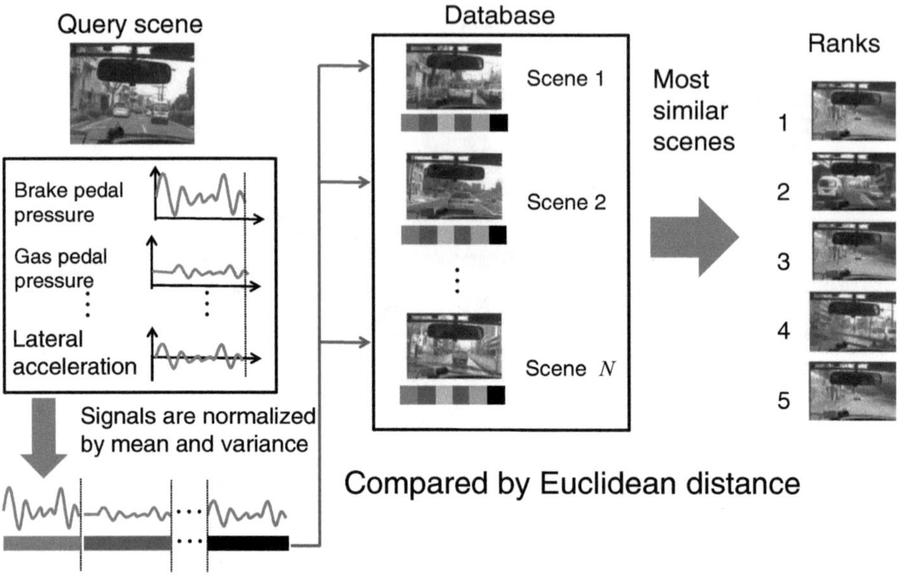

Fig. 14.2 Driving scene retrieval using driving behavior signals (early integration)

was normalized by mean and variance using all the data for all the drivers. The normalized signals of the query scene were represented as a vector, and the Euclidean distance between the vectors of the query scene and every scene in the database was measured. The database for the search consisted of about 200,000 vectors, one for each recorded scene. A fast retrieval technique was used to reduce retrieval time. The top five scenes with the smallest distances were chosen as similar scenes.

14.3.1.2 Method 2: Late Integration

The other retrieval method used was late integration, shown in Fig. 14.3. Each of the six kinds of signals of a scene was represented as a vector, and the Euclidean distance between the vectors of the query scene and those of all the other scenes was calculated for each signal. The sum of the ranks of the six signals was calculated, and the five scenes that had the lowest summation were retrieved as similar scenes.

14.3.2 Retrieval Performance Evaluation

To evaluate these methods, we conducted a driving scene retrieval experiment using driving data collected on city roads from 74 drivers (35 males and 39 females). There was about 45 min of recorded driving data per driver, for a total of about 54 h of driving data. The sampling rate of the driving signals was 10 Hz.

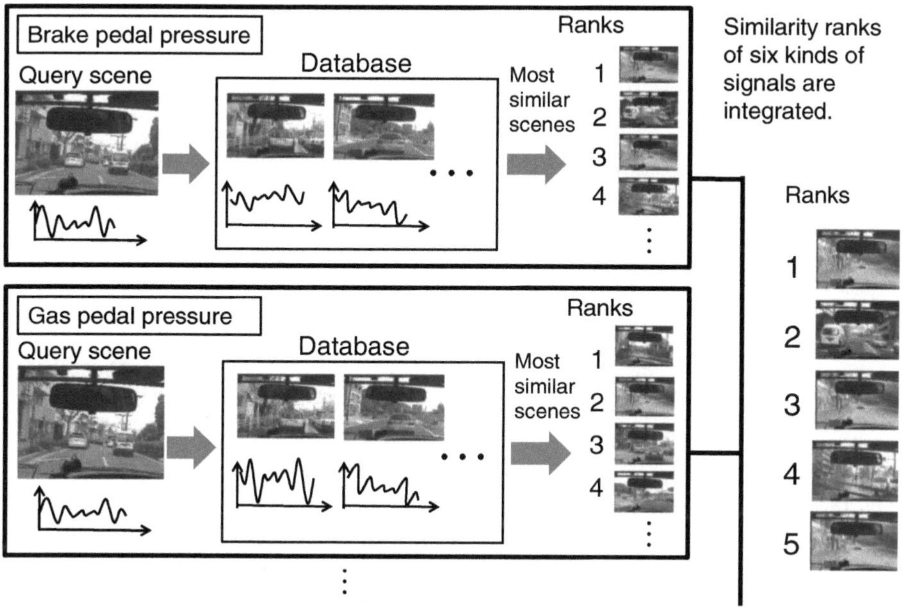

Fig. 14.3 Driving scene retrieval using driving behavior signals (late integration)

14.3.2.1 Experimental Condition

Eight kinds of driving events (stops, starts, right and left turns, right and left lane changes, and going up and down hills) were selected as query scenes, and similar scenes were retrieved using the two techniques described in Sect. 14.3.1. Scenes occurring less than 2 s before or after the query scene, and scenes which had already been retrieved, were excluded from being candidates for retrieval. We chose a total of 80 query scenes, which included about 10 scenes for each type of event.

Retrieval performance was evaluated in terms of retrieval accuracy, i.e., the percentage of correctly retrieved scenes in proportion to the total number of retrieved scenes. Whether or not a scene was correctly retrieved was determined subjectively by human validation.

14.3.2.2 Results

Experimental results are shown in Fig. 14.4. Retrieval accuracy averaging more than 95 % was achieved for driving scenes of stops, starts, and right and left turns, while accuracy was relatively lower for scenes of right and left lane changes, and going up and down hills. Retrieval accuracy of situations involving right turns was higher using the early integration method, but for scenes going down hills, the late integration method was more accurate. On average, the early integration method gave slightly better performance.

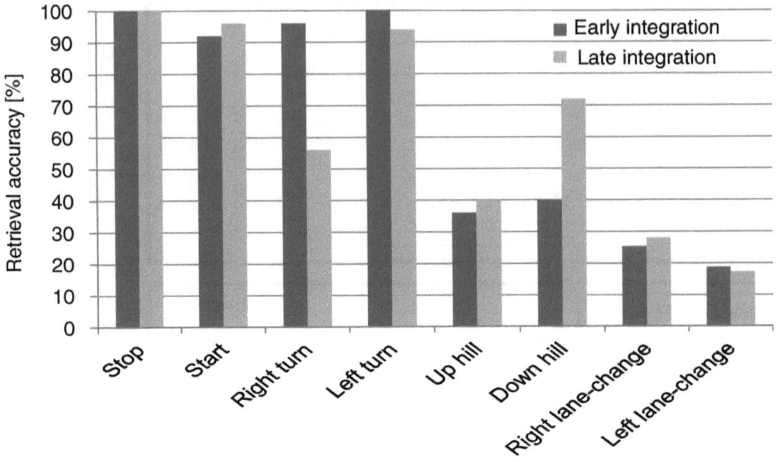

Fig. 14.4 Retrieval accuracy for driving behavior signals

14.4 Driving Scene Retrieval Using Environmental Driving Signals

In contrast to the first study, which employed in-vehicle driving signals, in this section we measured the similarity between scenes by comparing driving environments as detected by laser scanners.

14.4.1 Laser Data Preprocessing

14.4.1.1 Clustering of Laser Data and Tracking of Vehicles

The first step towards automatic scene retrieval was the clustering of discrete laser dots obtained with laser scanners from surrounding driving environments. Each cluster was a set of distance measurements in a plane, grouped closely to each other, and thus probably belonging to a single object. While many approaches have been used to calculate such physical distances [5], we simply used Euclidean distance here. Due to laser dot detection errors, not every cluster actually represented a separate object, i.e., sometimes more than one cluster could belong to a single object. Since all of the laser data were recorded on expressways in this study, in most cases a laser dot must belong to either a vehicle or a roadside barrier, so it was not difficult to integrate some clusters with our prior knowledge of the shapes of these objects [6]. Then, each surrounding vehicle was modeled as a rigid box, characterized by its orientation, position, and velocity. By tracking the vehicles with a Kalman filter, we estimated their dynamic features, even if they were outside the range of the laser scanners.

14.4.1.2 Frame Categorization

A frame categorization method was used to categorize laser-acquired driving frames based on three general features, in order to reduce the number of candidates and facilitate fast retrieval. The scenes were categorized based on road type, congestion level, and the relative positions of surroundings objects. The three features were defined as follows:

- Road type was divided into three classes: left curve, straight line, and right curve. Since two laser scanners were used, one on the front of the vehicle and one on the back bumper, they collected information about road types separately. Their combined data was used to define the road type for each frame of a driving scene, for example, "left curve, straight."
- Road congestion level was divided into two classes: "free flow" and "traffic jam." A Greenshields model [7] was employed to estimate the congestion level for each lane. The road congestion level of a driving frame was designated "traffic jam" if any lane in the frame was estimated as "traffic jam"; otherwise, the frame was designated "free flow."
- Relative positions of surrounding vehicles were classified into 450 situations based on whether there was another vehicle in each of eight surrounding directions and whether there was a roadside barrier on the left or right of the driver's vehicle.

For example, a frame could be represented as "(left curve, straight)," "traffic jam," and "21."

14.4.2 Similarity Measure for Surrounding Environment

Here, we measured the similarity between driving scenes based on the surrounding environment, using three procedures: first, their frame categories (given in Sect. 14.4.1.2) were compared; second, the relative positions of the surrounding vehicles were calculated; and finally, their motion features were compared.

14.4.2.1 Comparison of Frame Categories

In this study, each driving scene consisted of 100 frames (10 s), so each scene could be represented as a vector with 400 dimensions. We then calculated the difference between scenes using Hamming distance to measure how similar the frame categories of two scenes were. Hamming distance between two elements of the vectors took 0 only when the compared features were exactly the same. If the two features

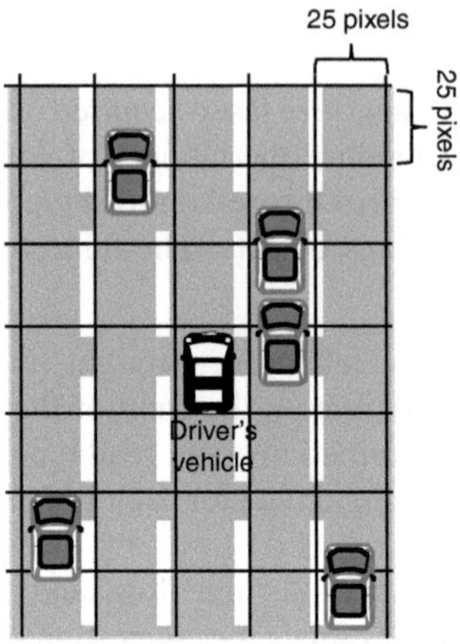

$$G = \begin{bmatrix} 0 & 0 & 0 & 0 & 0 \\ 0 & 1 & 0 & 0 & 0 \\ 0 & 0 & 0 & 2 & 0 \\ 0 & 0 & 1 & 0 & 0 \\ 0 & 0 & 0 & 0 & 0 \\ 1 & 0 & 0 & 0 & 0 \\ 0 & 0 & 0 & 0 & 1 \end{bmatrix}$$

Fig. 14.5 Example of a frame and its matrix. *Left*: Each cell of the grid is composed of 25×25 pixels. The grid is centered on the host vehicle. *Right*: The value of each element of matrix represents the number of vehicles in the corresponding cell

were different, the value was 1. So, if the total Hamming distance was 0, two scenes were identical, and if the total value was 400, they were completely dissimilar. The scenes with a Hamming distance below a threshold of 150 were extracted as candidates for further processing.

14.4.2.2 Comparison of Surrounding Vehicle Positions

The second step was to compare the positions of vehicles in key frames of two scenes. We assumed here that the first frames of scenes were key frames because people generally focus on the first frames of scenes. As shown in Fig. 14.5, a key frame was divided into a grid, and the frame was represented as a matrix G. Each cell of the matrix shows the number of vehicles in the corresponding cell of the grid.

Assume that frames F_1 and F_2 are represented by symbolized matrices G_1 and G_2. To compute the similarity of the two matrices, we first matched all cells in the two frames:

$$\Delta G(F_1, F_2) = \sum_i \sum_j \left| g_{i,j}^{(1)} - g_{i,j}^{(2)} \right|, \tag{14.1}$$

where $g_{i,j}^{(1)}$ and $g_{i,j}^{(2)}$ denote the number of vehicles in cell (i,j) in G_1 and G_2, respectively, and the value of ΔG represents the distance between them. For instance, we can say frames F_1 and F_2 match perfectly if and only if the value of ΔG equals zero. However, this rarely happens because even if two frames are almost identical, this symbolization method sometimes puts vehicles with the similar positions into different cells. To decrease errors caused by such problems, we also allowed soft matching. We assumed vehicles in two frames matched if there were the same numbers of vehicles in the cells at the same position in two matrices. In addition, we also considered vehicles to match if there were an equal number of vehicles in nearby cells, using a cost function. Thus, the final distance between frames F_1 and F_2 is defined as

$$d(F_1, F_2) = \Delta G'(F_1, F_2) + \frac{k}{K}, \tag{14.2}$$

where $\Delta G'$ is the value of ΔG after soft matching; k is the number of soft matches in $\Delta G'$, and K is an empirically defined normalization factor for the penalty of soft matching.

After that, distance $d(F_1,F_2)$ was used to calculate the similarity between F_1 and F_2:

$$s(F_1, F_2) = \frac{d(F_1, F_2)}{n_1 + n_2}, \tag{14.3}$$

where n_1 and n_2 denote the numbers of vehicles in frames F_1 and F_2, respectively. Frames with a distance below 0.5 from the first frame of a query scene, as well as between their preceding and following frames within 2 s, were selected as key frames for the next step in processing.

14.4.2.3 Comparison of Surrounding Vehicle Motion

If surrounding vehicles have nearly the same positions in the first frames of scenes, as well as similar trajectories and velocities, we believe there is a high probability that these are matching scenes. Also, comparing the motion of surrounding vehicles overcomes problems caused by grid division and achieves a faster search than with frame-to-frame matching between scenes.

Assume that scenes S_1 and S_2 are represented by their vehicle sets (excluding the host vehicle), $V_1 = \left\{v_1^{(1)}, v_2^{(1)}, \ldots, v_M^{(1)}\right\}$ and $V_2 = \left\{v_1^{(2)}, v_2^{(2)}, \ldots, v_N^{(2)}\right\}$, where M and N are total numbers of surrounding vehicles observed in S_1 and S_2. At point in time, t, each surrounding vehicle, $v_i^{(1)}$ or $v_j^{(2)}$, is represented as a sequence of vehicle motion feature vectors, consisting of longitudinal position y_i and lateral position x_i with their first-order dynamics Δy_i and Δx_i:

$$(y_i(t), x_i(t), \Delta y_i(t), \Delta x_i(t))^{\mathrm{T}}. \tag{14.4}$$

Dynamic features were calculated by the following equation:

$$\Delta y_i(t) = \frac{\sum_{l=-L}^{L} l \cdot y_i(t+l)}{\sum_{l=-L}^{L} l^2}, \tag{14.5}$$

in which $y_i(t)$ is the ith vehicle's driving signal at point in time t, and L is window size for linear regression. $\Delta x_i(t)$ was calculated in the same way. The distance between vehicles $v_i^{(1)}$ and $v_j^{(2)}$ in two scenes S_1 and S_2, respectively, were calculated as a Mahalanobis distance:

$$d^2\left(v_i^{(1)}, v_j^{(2)}\right) = \left(\mu_{v_i^{(1)}} - \mu_{v_j^{(2)}}\right)^{\mathrm{T}} \Sigma^{-1}_{v_i^{(1)}, v_j^{(2)}} \left(\mu_{v_i^{(1)}} - \mu_{v_j^{(2)}}\right), \tag{14.6}$$

where μ_v represents a four-dimensional vector (including the means of longitudinal position, lateral position, as well as their first-order dynamics) of a vehicle v, and $\sum_{v_i^{(1)}, v_j^{(2)}}$ is a four-by-four covariance matrix of the four features for vehicle $v_i^{(1)}$ and $v_j^{(2)}$. This calculates the distance between a pair of vehicles by comparing the distribution of their four-dimensional features. Based on our preliminary experiment, a pair of vehicles with a Mahalanobis distance below a threshold of 15.0 was believed to be similar to each other.

To acquire a vehicle-to-vehicle match, we calculated $d(v_i, v_j')$ for all i and j between scenes and selected them from smallest to largest. We considered scenes to be similar to each other if there were enough similar vehicles in both scenes. Similarity p between S_1 and S_2 was defined as the summation of the weights of similar vehicles divided by the summation of the weights of all the vehicles in the scenes:

$$p(S_1, S_2) = \frac{\sum\limits_{i \in X_1} \sum\limits_{t \in T_i^{(1)}} w_t^{(i)} + \sum\limits_{i \in X_2} \sum\limits_{t \in T_i^{(2)}} w_t^{(i)}}{\sum\limits_{i \in Y_1} \sum\limits_{t \in T_i^{(1)}} w_t^{(i)} + \sum\limits_{i \in Y_2} \sum\limits_{t \in T_i^{(2)}} w_t^{(i)}}, \tag{14.7}$$

where X_1 and X_2 denote the sets of similar vehicles, and Y_1 and Y_2 denote the sets of all vehicles in S_1 and S_2, respectively. $w_t^{i)}$ denotes the weight of vehicle v_i at time t. $T_i^{(1)}$ and $T_i^{(2)}$ are the sets of frame numbers where $v_i^{(1)}$ or $v_i^{(2)}$ was observed in S_1 or S_2, respectively. Here, "weight" means the importance of a surrounding vehicle, which was represented as a value of a modified Gaussian distribution as illustrated in Fig. 14.6. The reason we used a modified Gaussian distribution which was stretched towards the front value as a similarity metric is that, generally, a driver is more aware of nearby leading vehicles while driving. For example, the

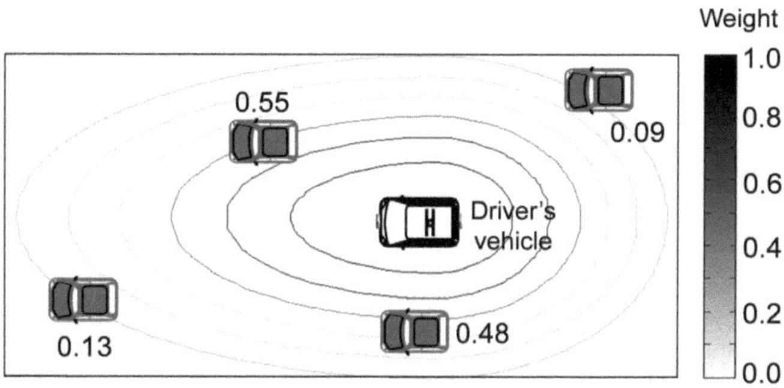

Fig. 14.6 A modified two-dimensional Gaussian distribution, centered on the driver's vehicle, where surrounding vehicles with higher values denote greater importance

surrounding vehicles in front of a driver's vehicle are more important than those on either side of or behind the driver's vehicle. It can be inferred that a pair of similar vehicles near the driver's vehicle makes scenes more similar than pairs located farther away.

14.4.3 Retrieval Performance Evaluation

The proposed driving scene retrieval system was evaluated using database-containing expressway scenes from 57 drivers (28 males and 29 females) recorded with the instrumented vehicle shown in Fig. 14.1. The database contained approximately 140,000 driving frames. All of the driving data were sampled at 10 Hz. We compared retrieval accuracy and speed for different types of scenes under various retrieval conditions, by using subjective scores and by measuring retrieval speed in CPU time. Here, "retrieval conditions" mean some combinations of the similarity measures presented in Sect. 14.4.2:

(a) Based on frame category
(b) Based on surrounding vehicle position
(c) Based on surrounding vehicle motion

The combinations are represented as a, c, $a + c$, $b + c$, and $a + b + c$. We did not use b or $a + b$, since b only considered the first frame of a scene and would not be accurate if used alone.

The experiment was conducted as follows:

- Five driving scenes each, for straight road, curve, traffic jam, and lane change, were randomly selected as queries.
- For each query scene, we evaluated retrieval accuracy and retrieval speed for each retrieval condition. For each condition, the top five similar scenes were retrieved, and they were used for the evaluation.

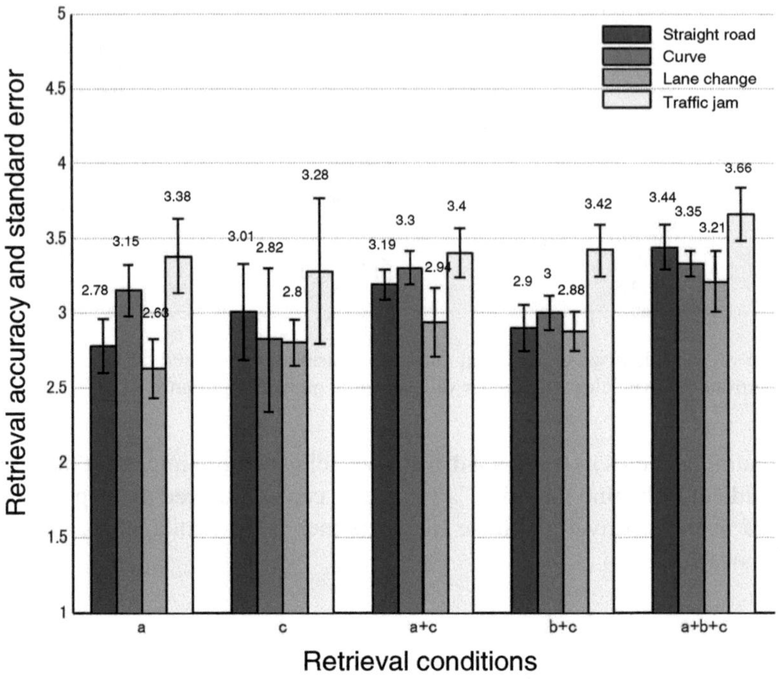

Fig. 14.7 Comparison of retrieval accuracy

14.4.3.1 Comparison of Retrieval Accuracy Using Subjective Scores

In this comparison, the subjective scores of five volunteers were used to judge which retrieval condition, or combination of retrieval conditions, was able to select scenes with the highest similarity to a query scene for a given driving situation. Each volunteer gave scores, from 1 (lowest) to 5 (highest), to the top five retrieved scenes for each query under each retrieval condition. Scenes with a score of 3 or higher were considered to be similar. A score of 5 indicated perfect similarity, while a score of 1 indicated complete dissimilarity. The retrieval accuracy of a given scene under a given retrieval condition was estimated as the average of the scores from the five volunteers.

The experimental results are shown in Fig. 14.7, which indicate that condition $a + b + c$ demonstrated much higher accuracy than the other conditions, in various driving situations.

14.4.3.2 Comparison of Retrieval Speed Using CPU Time

In order to compare processing speed, the proposed driving scene retrieval system was installed on a Core i5 CPU 650@3.20 GHz PC using the Windows 7 operating system. The CPU time for each query process was recorded for each retrieval condition.

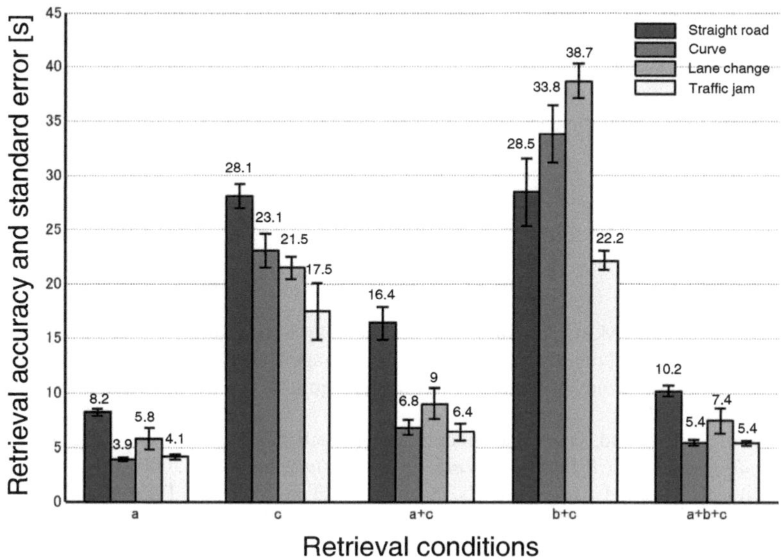

Fig. 14.8 Comparison of retrieval speed

The average retrieval time for top five driving scenes was calculated. This was considered to represent system speed performance under a given retrieval condition for each scene. Figure 14.8 shows the average retrieval time taken to retrieve scenes from the 140,000-frame database. On average, retrieval condition a took the least time, and condition $a + b + c$ was the next fastest.

14.5 Conclusions

In this chapter, we developed two systems for retrieving recorded driving scenes based on measuring the similarity of driving behavior and environmental driving signals. In the first study, similar scenes were retrieved using driving behavior signals, and they were integrated using two methods, early and late integration. Experimental results showed that an average of more than 95 % retrieval accuracy was achieved for driving scenes of stops, starts, and right and left turns. In most situations, the early integration method achieved better performance than the late integration method. In the second study, we used environmental driving signals with the idea that similar driving scenes could be retrieved by measuring similarity in surrounding environments. Experimental results showed that the integrated use of information from surrounding vehicles and road conditions achieved higher retrieval accuracy than the use of either type of information alone.

Currently, we are working to integrate these two systems, to see if retrieval accuracy can be further improved.

Acknowledgement This work was partially supported by the Strategic Information and Communications R & D Promotion Programme (SCOPE) of the Ministry of Internal Affairs and Communications of Japan under No. 082006002, by Grant-in-Aid for Scientific Research (C) from the Japan Society for the Promotion of Science (JSPS) under No. 24500200, and by the Core Research of Evolutional Science and Technology (CREST) of the Japan Science and Technology Agency (JST).

References

1. S.Y. Cheng, S. Park, M.M. Trivedi, Multispectral and multi-perspective video arrays for driver body tracking and activity analysis. Comput. Vis. Image Understand. **106**, 245–247 (2007)
2. D. Mitrovic, Reliable method for driving events recognition. IEEE Trans. Intell. Transp. Syst. **6** (2), 198–205 (2005)
3. N. Oliver, A. Pentland, Graphical models for driver behavior recognition in a SmartCar, in *Proceedings of the IEEE Intelligent Vehicles Symposium,* pp. 7–12, 2000
4. M. Naito, A. Ozaki, C. Miyajima, N. Kitaoka, R. Terashima, K. Takeda, A browsing and retrieval system for driving data, in *Proceedings of the IEEE Intelligent Vehicles Symposium,* pp. 1159–1165, June 2010
5. J.Z. Wang, J. Li, G. Wiederhold, SIMPLIcity: semantics-sensitive integrated matching for picture libraries. IEEE Trans. Pattern Anal. Mach. Intell. **23**(9), 947–963 (2001)
6. N. Kaempchen, Feature-level fusion of laser scanner and video data scanner and video for advanced driver assistance systems, Ph.D. dissertation, University of Ulm, Germany, 2007
7. H. Rakha, B. Crowther, Comparison of Greenshields, pipes, and Van aerde car-following and traffic stream models. J. Transp. Res. Board **1802**, 248–262 (2007)
8. K. Takeda, J. Hansen, P. Boyraz, L. Malta, C. Miyajima, H. Abut, International large-scale vehicle corpora for research on driver behavior on the road. IEEE Trans. Intell. Transp. Syst. **12**, 1609–1623 (2011)

Chapter 15
Driving Event Detection by Low-Complexity Analysis of Video-Encoding Features

Elias S.G. Carotti and Enrico Masala

Abstract All standard video-encoding algorithms rely on differential encoding with motion compensation to improve the compression. When a video from a front-facing camera onboard a vehicle is compressed, the information computed for compression purposes, in particular motion vectors, can be effectively used to gain some understanding of the driving dynamics and eventually to support driver decisions and improve driving safety. In this chapter an algorithm that can use such side information to detect a number of driving events is presented. Numerous potential applications are envisaged. Since video-encoding software and hardware are usually strongly optimized, it is possible to implement the proposed algorithms in battery-powered embedded devices with strict limits on processing capabilities such as camera-equipped mobile phones mounted on the car dashboard and consequently allow different types of low cost vehicles, which in most cases do not include cameras as a standard equipment, to be fitted with at least a warning device with very low cost. If the video is captured in the context of a video surveillance scenario, differentiating the events could be used to automatically decide which portion of the video should be transferred to a remote monitoring center thus optimizing network resources usage and costs.

Keywords Driving event detection • Event classification • Support vector machines • Video analysis • Video coding

---

E.S.G. Carotti (✉) • E. Masala
Control and Computer Engineering Department, Politecnico di Torino, Torino, Italy
e-mail: carotti@polito.it; masala@polito.it

G. Schmidt et al. (eds.), *Smart Mobile In-Vehicle Systems: Next Generation Advancements*, DOI 10.1007/978-1-4614-9120-0_15,

15.1 Introduction

The availability of cheap video cameras is constantly increasing due to their low cost and usefulness in a number of contexts. In the vehicular environment, for instance, applications include courtesy cameras to facilitate maneuvering, video logging to facilitate reconstruction of events in case of, e.g., an accident happened, and remote surveillance. In order to compress the huge amount of data produced by video capturing, video encoders perform a number of computationally heavy operations, the most complex one being motion compensation. This chapter shows that such processing operations, when performed on a video captured from a vehicle onboard camera oriented towards the travel direction, could be used to gain some insight in the semantics of the scene and eventually improve driving safety.

For the vehicular environment, several context-understanding techniques have been developed during the recent years [1]. However, most of them require computationally heavy operations, identification of single elements inside the scene, etc. The novelty of our approach resides in the extremely low complexity of the proposed techniques that make them suitable for battery-powered devices as well as devices with limited processing power. The key idea is to exploit the side information produced by the encoder to gain some understanding of the driving dynamics going on. Examples of side information include macroblock (MB) size, mean squared error (MSE) with respect to the uncompressed content, type of macroblock chosen during encoding, and, most importantly, motion vectors (MV) for each differentially encoded frame. Although motion vectors do not always exactly represent the motion actually happening in the real scene, as the optical flow is expected to do, we argue that nevertheless useful insight on the scene might be extracted with very low complexity, as opposed to the optical flow approach that presents high complexity.

The possibility to access part of the huge database of driving behavior signals collected by the University of Nagoya project [2] allows to perform a statistical analysis of the side information normally produced by a video encoder in various driving conditions. The analysis is used to build a model that can identify, given the features induced by the side information, and with no additional driving signals such as those available from vehicle sensors, the driving conditions and potentially act to increase safety by, e.g., activating warning signals. The possibility to deploy such a system in a battery-powered device with strict limits on processing capabilities would have a number of applications. For instance, it could be run on embedded devices, such as camera-equipped mobile phones mounted on the car dashboard, and consequently allow different types of low-cost vehicles, which in most cases do not include cameras as standard equipment, to be fitted with a safety warning device with very low cost.

15.2 Feature Analysis

This section presents a preliminary analysis of the characteristics of very simple features that can be extracted as side information from a state-of-the-art video encoder at extremely low cost, such as size and MSE of each frame. For instance, those two features alone could be exploited to decide, without any other input, if the vehicle is moving or not. Intuitively, the scene captured by a camera in front of the driver is much more static when the vehicle is not moving. Thus, an H.264 encoder [3] operating at constant quality would approximately use the same amount of bits and introduce almost the same distortion for a number of consecutive intercoded frames.

As seen in Fig. 15.1, the distribution of the size and distortion of each video frame presents two different clusters, corresponding to the *moving* and *still* vehicle conditions.

Motivated by the fact that those results seem quite promising, a more systematic approach is employed in order to compute the relevance of each feature that can be easily extracted from the compressed video stream, so that classification complexity could be reduced. Feature selection is a very important problem in pattern analysis and classification, and many different algorithms have been proposed in literature, based on diverse criteria, among them mutual information [4] between features and classes (which correspond to events in our case).

The mutual information of two random variables is a quantity that measures the statistical dependence between the two variables, that is, the reduction in the uncertainty (as measured by Shannon's Entropy) about one random variable yielded by knowledge of the other one [5]. Given two discrete random variables X, Y, $X \in \mathrm{X}$, $Y \in \mathrm{Y}$ and the corresponding probability mass functions (p.m.f.) $p_X(x)$, $p_Y(y)$, along with their joint p.m.f. $p_{XY}(x,y)$, their mutual information $I(X;Y)$ is defined as

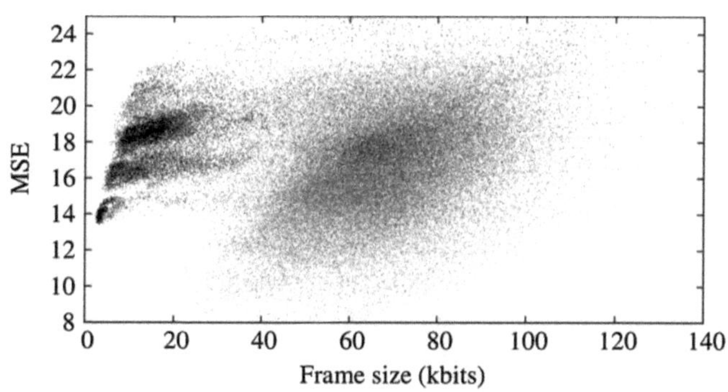

Fig. 15.1 Scatter plot of MSE vs. frame size of compressed video frames when the vehicle is moving (*grey*) or still (*black*)

Table 15.1 Video features considered in this work

Level	Name	Value type
Frame	Size	Integer
	MSE w.r.t. original	Float
Macroblock	Size	Integer
	MSE w.r.t. original	Float
	Motion vectors (MVx, MVy)	Fractional (1/4-pel precision)
	Number of MVs	Integer (1, 2, or 4)

$$I(X;Y) = \sum_{x \in X} \sum_{y \in Y} p_{XY}(x,y) \log \frac{p_{XY}(x,y)}{p_X(x)p_Y(y)}. \tag{15.1}$$

In principle, given a feature set $F = (X_1, X_2, \ldots, X_n)$ with n features and a target class C, one would like to find the subset $S \subset F$ for a given $m \leq n$ which bears the most information about the class C, i.e., which has the largest dependency on the target class (Max-Dependency):

$$\arg \max_{S \subset F} I(S = (X_{k_1}, X_{k_2}, \ldots, X_{k_m}); C). \tag{15.2}$$

When $m = 1$ the solution is almost trivial and is equivalent to finding the feature whose mutual information with the class is maximal. When more features are involved, i.e., $m > 1$, the mutual information between the feature should be considered as well because the information they bear to the reduction of the uncertainty on the class can be partially overlapping. Directly maximizing (15.2) in this case can be hardly done in practice, especially if m is large; thus, usually Max-Dependency (15.2) is approximated with the simpler Max-Relevance:

$$\arg \max_{S \subset F} \frac{1}{|S|} \sum_{X_i \in S} I(X_i; C), \tag{15.3}$$

which amounts to taking the subset of m features individually maximizing the mutual information exchanged with the class. Note that mutual information among the features is ignored in (15.3), i.e., the joint mutual information with the class may be less than the sum of the individual mutual informations between each feature and the class. However, the Max-Relevance criterion is simple to implement and proved to be quite effective for our purposes.

To compute mutual information we estimated all the probability mass functions by frequency counts on a training set. The considered features at frame and macroblock level are listed in Table 15.1, where MVx and MVy represent the horizontal and vertical components of the MV, respectively. The events of interest considered here are listed in Table 15.2.

Figure 15.2 represents typical frames captured in the highway scenario, urban scenario, and dual carriageway, respectively. Macroblock-level features can also

Table 15.2 Events of interest

Event	Description	Possible values
Moving	Car is moving	Yes/no
Scenario	Type of road	Urban/highway
Dual_carriageway	Opposite car flows are separated	Yes/no

Fig. 15.2 Example of a highway, urban, and dual carriageway conditions (from *top* to *bottom*)

consider more specific and numerous features such as motion vectors. Table 15.1 reports a sample list of macroblock-level features that can be easily extracted as side information during the encoding process. Their individual contribution towards identifying a number of situations is shown in Figs. 15.3, 15.4, 15.5, 15.6, 15.7, and 15.8. Brighter colors represent higher mutual information values.

Figure 15.3 shows that the contribution of the size of each macroblock to the detection of the *motion* event is almost uniform throughout the picture. In Fig. 15.4 the macroblocks in the position corresponding to the converging horizontal white lines depicted on the road surface are shown to be important for the detection of the driving scenario. Concerning motion vectors, their horizontal component seems to

Fig. 15.3 Mutual information between macroblock size and *moving* event

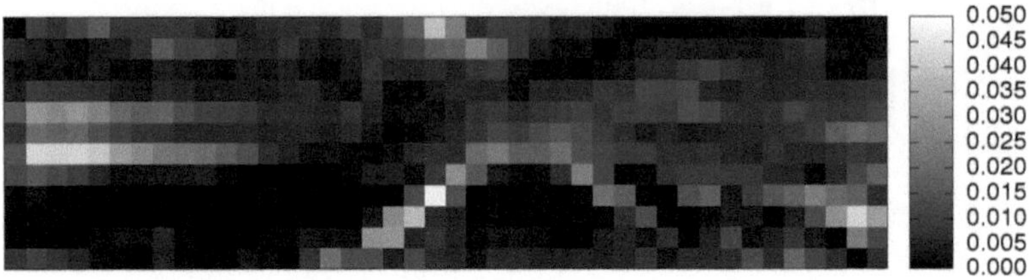

Fig. 15.4 Mutual information between number of MVs in the macroblock and *scenario* event

Fig. 15.5 Mutual information between MVx and *moving* event

Fig. 15.6 Mutual information between MVx and *scenario* event

Fig. 15.7 Mutual information between MVx and *dual carriageway* event

Fig. 15.8 Mutual information between MVy and *scenario* event

be one of the most important features to consider for detecting various events. For instance, in Fig. 15.5 the contribution of all macroblocks is important to detect the movement of the vehicle, except for the macroblocks which are immediately in front of the vehicle (bottom center part of the figure). This is probably due to the fact that the area represented by them is very often a uniform road surface where movements cannot be clearly detected by simply using motion compensation due to the uniformity. The driving scenario seems to be detected mainly by analyzing the converging horizontal white lines depicted on the road surface (Fig. 15.6), as well as some macroblocks in the left and right parts of the image. The left part probably includes the highly textured area at the side of the road in the urban scenario, while the right part might indicate the presence of vehicles quickly moving in the opposite direction. The *dual_carriageway* event can be detected by focusing on the left part of the image, as shown in Fig. 15.7, again probably due to the highly textured area at the side of the road. Finally, Fig. 15.8 shows the contribution of the vertical components of the motion vectors to the detection of the scenario. The contribution seems to be very concentrated in the area of the converging horizontal white lines depicted on the road surface, suggesting that their movement can be used to discriminate among different scenarios.

For events such as *scenario*, which has been classified for the purpose of these experiments using a binary value (urban/highway), it can be easily seen that the macroblocks that are in the position corresponding to the horizontal white stripes on the road surface, which are aimed at separating the lanes, give the highest contribution compared to the rest of the image. This is expected since the database contained a test path in which, for almost all roads, horizontal road signs are clearly marked. In addition, the left part of the image also gives a large contribution since that part of the image is expected to contain visual information coming from the roadside. This moves quite uniformly when the vehicle changes direction, for instance, due to a turn or a lane change. Moreover, the intensity and direction of motion are also highly related to the vehicle direction of movement. It is more difficult to get this insight from the opposite side of the image because the road surface is more uniform, and, often, the movements represented by the motion vectors do not closely represent the real motion in the video scene. Note also that this result is influenced by the fact that the video sequences used for the experiments have been acquired in Japan, where cars are required to keep the left side. The opposite consideration would hold in the case of right-side driving.

The mutual information approach is useful since it provides quantitative values for each input feature. Since the number of macroblock for each frame is quite high, and the video frame rate adds complexity to the problem, the mutual information values can be used as a guidance in order to reduce the number of features needed to detect the current driving situation. The underlying idea is that many macroblocks are located in parts of the image which have low significance in terms of helping to determine the current driving situation. For instance, the upper macroblocks generally present low utility values since they might represent a portion of the sky which might not present particular patterns; therefore, motion vectors in that area are not very useful. The ordering that mutual information creates between the macroblocks will be employed to reduce the complexity of the detection algorithm, hopefully without a significant impact on the performance of the algorithm.

15.3 Detection Algorithm

We experimented with different detection algorithms relying on either a generative model or a more efficient discriminative model. In the first case, a Gaussian Mixture Model (GMM) can be trained for each of the two conditions. Then, given certain values for the input parameters, the output of each model is the likelihood of the corresponding possible outcome, e.g., *moving* or *still*, and a simple comparison of the likelihoods can be used to determine the maximum and which of the two conditions apply accordingly.

We also employed two algorithms based on a discriminative model. The first one is a binary linear Support Vector Machine (SVM) [6, 7]. For each binary event we try to detect, we map the two possible outcomes on the labels $\{-1, +1\}$, and given a set of instance-label pairs $(\underline{x}_i, y_i), i = 1, \ldots, m, \underline{x}_i \in \Re^m, y_i \in \{-1, +1\}$, the aim is to solve the unconstrained optimization problem:

$$\min_{\underline{w}} \frac{1}{2} \underline{w}^T \underline{w} + \sum_{i=1}^{m} \max\left(1 - y_i \underline{w}^T \underline{x}_i, 0\right)^2. \tag{15.4}$$

Once the $\underline{w}$ vector is learned, binary classification proceeds as follows: a new instance is assigned the $+1$ label if $\underline{w}^T \underline{x} > 0$ and -1 otherwise.

A second algorithm relies on an SVM whose kernel is the radial basis function. This is more complex than the linear SVM approach; hence, it will be used to present reference results and for comparison purposes. In order to perform experiments with this type of SVM, we relied on the software described in [8].

15.4 Results

We tested our algorithm by means of tenfold stratified cross-validation over a set of 77,060 frames. Performance was assessed in terms of accuracy, precision, and recall. Accuracy measures how many times the classifier is right on average, i.e., how many times it correctly identifies if an event is occurring or not (e.g., moving or not moving), and is given by:

$$\text{Accuracy} = (\text{True\_positives} + \text{True\_negatives})/\text{Number\_of\_instances}$$

While accuracy gives an indication of the average performance of the classifier, corresponding to the average identification rate, precision, and recall complements this information. Precision is, in fact, defined as:

$$\text{Precision} = \text{True\_positives}/(\text{True\_positives} + \text{False\_positives}),$$

and, as such, measures the fraction of times an event is identified (e.g., the classifier outputs *moving* and the event is actually occurring). Recall, instead, measures the fraction of occurring events actually identified by the classifier and is defined as:

$$\text{Recall} = \text{True\_positives}/(\text{True\_positives} + \text{False\_negatives}).$$

High recall implies that if an event occurs it is very likely the classifier will spot it, while high precision indicates that most of the time the classifier is right when it identifies an event.

15.4.1 Gaussian Mixture Model

First, the GMM algorithm has been tested. Average detection results are presented in Table 15.3 which shows that, even with very few components for each GMM, correct decisions can be taken successfully on more than 90 % of the analyzed frames. It is noteworthy to say that this simple algorithm does not jointly consider consecutive frames nor it considers other features available as side information at the encoder.

Further investigation shows that there are a few cases where the algorithm fails to correctly identify the *still* condition. This usually happens on frames characterized by a mostly still image but with few and localized movements in the scene,

Table 15.3 GMM algorithm accuracy

Number of components	Identification rate
2	90.72 %
5	91.36 %

Fig. 15.9 Vehicle crossing the intersection at a traffic light stop. Superimposed motion vectors present a relatively high intensity in the direction of apparent motion when they are located in the car area while they are nearly zero in the rest of the image

such as vehicles crossing an intersection, while the vehicle is waiting at a traffic light. Figure 15.9 shows such a condition, where the vehicle at the center of the scene is rapidly moving in the direction which is perpendicular to the camera axis, while the rest of the scene is completely static. Therefore, we expect that incorporating the information coming from the motion estimation algorithm could improve the prediction accuracy.

Moreover, also note that the ground truth (i.e., the decision between *moving* and *still* conditions as well as for the other events) has been determined by manually annotating the video, and it is not always easy to determine if the vehicle is moving or not by looking at the video when the speed is low. Therefore, such low-speed conditions could be weighted differently when evaluating the performance of the detection algorithm.

15.4.2 Support Vector Machine

The following results incorporate the features available at the macroblock level. The linear SVM algorithm has been employed in this case, since it can easily cope with the high number of features available for each frame. Table 15.4 shows the performance results in terms of accuracy, precision, and recall when all the motion vector components (both horizontal and vertical) of all macroblocks in each frame are used, as well as the total frame size and MSE value. The results vary depending on the event to be detected; however, all values are significant.

Improving the performance shown in Table 15.4 can be achieved by simply considering that it is extremely unlikely that the value of the event rapidly changes over time, especially for each frame. Therefore, Table 15.5 shows the results

Table 15.4 Performance of the SVM algorithm considering each frame individually

Event	Accuracy	Precision	Recall
Moving	91.44 %	92.85 %	96.00 %
Scenario	87.02 %	82.87 %	68.34 %
Dual carriageway	66.79 %	66.60 %	82.49 %

Table 15.5 Performance of the SVM algorithm considering five frames at a time

Event	Accuracy	Precision	Recall
Moving	91.84 %	93.64 %	95.62 %
Scenario	89.85 %	85.67 %	77.12 %
Dual carriageway	67.94 %	68.09 %	81.24 %

obtained by grouping, as the SVM input, the features of five frames at a time. The performance significantly improves, up to 3 % for both accuracy and precision.

The previous approach provides a performance improvement at the cost of an additional complexity in terms of the number of input features. Therefore, to reduce complexity, for each feature and event to be detected, macroblocks have been sorted using the mutual information value computed as described in Sect. 15.2, and only the ones with the highest value have been given as input to the SVM algorithm.

Figure 15.10 reports the results as a function of the considered number of macroblocks (the first 20, 50, 100). The all macroblocks case (504) has been considered for reference purpose. Note that in the case of five frames, to further reduce the complexity, only the horizontal component of the MVs has been considered in addition to the other features. The performance gap between the case of 20 macroblocks and the full case is limited; therefore, it is possible to effectively reduce the complexity of the algorithm without significantly affecting the performance. Note also that when 100 macroblocks are employed for the five-frame case, the complexity, in terms of input features, is equivalent to the case of considering all macroblocks of each frame for the one-frame case.

Table 15.6 shows the results obtained by means of the SVM approach based on the radial basis function described at the end of Sect. 15.3. Almost the same accuracy is achieved with respect to the linear SVM approach; however, precision and recall may significantly change. For instance, for the moving event, accuracy slightly increases (0.3 %), whereas precision increase is in the order of 5 %, but the recall is reduced by 4–5 %. Consequently, when the event is identified by the classifier, it is more likely to be actually happening; however, there are some more events which are not spotted by the classifier. For the dual carriageway case, a moderate decrease in accuracy (0.5–2 %) is accompanied by a strong increase in precision (about 29–30 %), whereas the recall decrease is about 22 %. In this case, there is a much higher probability that when the classifier identifies the event, it is actually happening, while some of them are missed due to the lower recall. With the scenario event, accuracy does not change significantly, while precision and recall move in the opposite direction compared with the previous case.

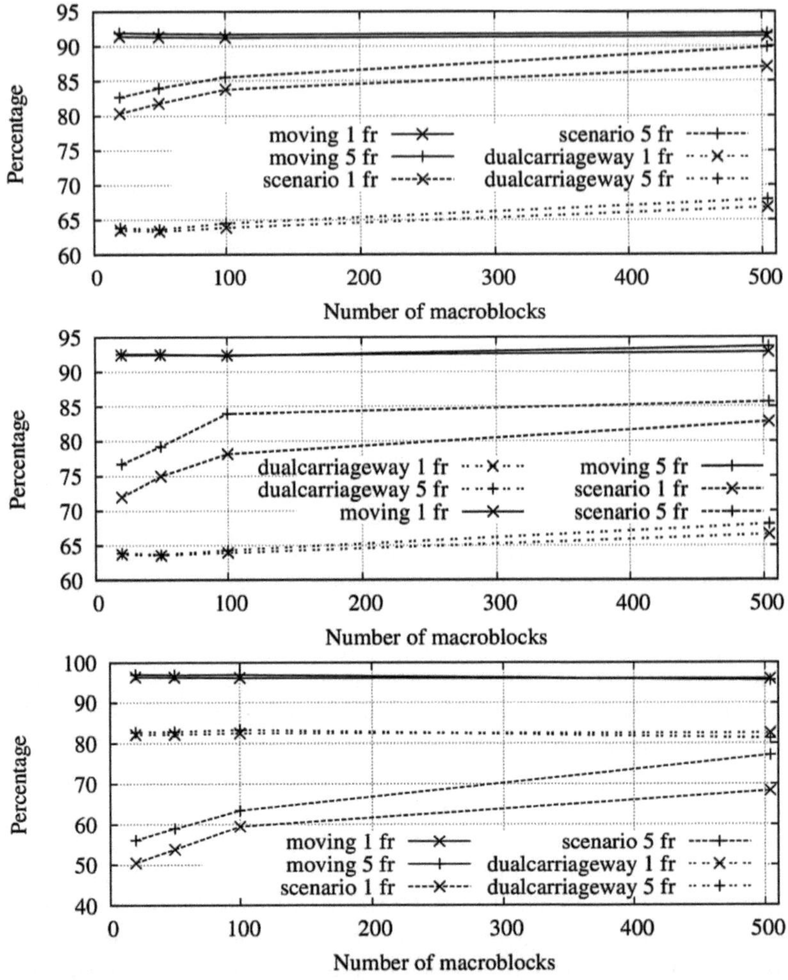

Fig. 15.10 Accuracy, precision, and recall (from *top* to *bottom*) as a function of the number of considered macroblocks, considering one or five frames at a time

15.5 Conclusions

This chapter presented driving event detection algorithms based on the side information available from a video compression procedure. The algorithms achieve good identification rate by employing both a GMM-based and SVM-based approach, using numerous input features such as the ones provided by the motion estimation algorithm of a video encoder. The results showed that the algorithms are able to identify, with good reliability and low complexity, a number of driving conditions. The comparison between the linear SVM approach and the SVM approach based on

Table 15.6 Performance increase/decrease of the nonlinear SVM algorithm, considering five frames at a time

Event	Number of macroblocks	20	50	100
Moving	Accuracy	+0.35 %	+0.32 %	+0.33 %
	Precision	+5.69 %	+5.44 %	+5.40 %
	Recall	−5.08 %	−4.91 %	−4.87 %
Scenario	Accuracy	+1.13 %	−0.10 %	−0.70 %
	Precision	−20.29 %	−24.07 %	−27.37 %
	Recall	+25.27 %	+22.84 %	+23.21 %
Dual carriageway	Accuracy	−0.53 %	−1.75 %	−2.61 %
	Precision	+29.24 %	+29.37 %	+29.56 %
	Recall	−21.47 %	−22.18 %	−22.13 %

the radial basis function shows that the performance of the linear approach is close to the one provided by the other, more complex, approach.

Further work will be devoted to investigate the possibility to identify more interesting events such as lane changes and vehicles unexpectedly entering in the scene. In case of a video surveillance scenario, the detection of different types of events could be used to automatically decide which portion of the video should be uploaded to a remote monitoring center thus reducing communication costs. Finally, other possible applications besides driving assistance could be investigated, such as a simple helper program to automatically produce a draft annotation for all video signals captured from a front-facing camera in a signal collection campaign. Such a draft annotation would then be refined by human annotators more quickly than starting from scratch. Since during a signal collection campaign, the vehicle usually follows roughly the same path in each recording session; it is expected that such an approach could be quite accurate and could speed up the annotation of the huge amount of collected video especially if the model is trained over a careful human annotated recording session.

References

1. D. Alonso, L. Salgado, M. Nieto, Robust vehicle detection through multidimensional classification for on board video based systems. Proc. Intl. Conf. Image Processing **4**, 321–324 (2007)
2. Driving behavior signal processing based on large scale real world database (2008), http://www.sp.m.is.nagoya-u.ac.jp/NEDO
3. ITU-T Rec. H.264 & ISO/IEC 14496-10 AVC, Advanced video coding for generic audiovisual services, May 2003
4. H. Peng, F. Long, C. Ding, Feature selection based on mutual information criteria of max-dependency, max-relevance, and min-redundancy. IEEE Trans. Pattern Anal. Mach. Intell. **27**(8), 1226–1238 (2005)
5. T.M. Cover, J.A. Thomas, *Entropy, Relative Entropy and Mutual Information* (John Wiley & Sons Inc., New York, 2001)
6. R.-E. Fan, K.-W. Chang, C.-J. Hsieh, X.-R. Wang, C.-J. Lin, Liblinear: a library for large linear classification. J. Mach. Learn. Res. **9**, 1871–1874 (2008)

7. C.-J. Hsieh, K.-W. Chang, C.-J. Lin, S. S. Keerthi, S. Sundararajan, A dual coordinate descent method for large-scale linear SVM, in *Proceedings of the 25th International Conference on Machine Learning*, ser (ICML '08) (ACM, New York, NY, USA, 2008), pp. 408–415
8. C.-C. Chang, C.-J. Lin, LIBSVM: a library for support vector machines. ACM Trans. Intell. Syst. Technol. **2**(3), 1–27 (2011)

Chapter 16
Target Shape Estimation Using an Automotive Radar

Florian Engels

Abstract Estimating the shape of vehicles is crucial for radar-based advanced driver assistance and safety systems. However, conventional radar processing is unable to resolve the different parts of a vehicle as required for this task. To address this issue a two-stage approach is considered which employs high-resolution techniques in combination with conventional Fourier-based methods. Single- and two-dimensional high-resolution estimation is discussed, which includes range and range-rate estimation in the temporal dimensions of the radar data. A novel technique referred to as cell interpolation is proposed, which can employ range and range-rate estimates in combination with Fourier-domain data for direction-of-arrival estimation. Two-stage processing has been implemented in the case of two-dimensional high-resolution estimation using the spectral RELAX algorithm and experimental results are shown.

Keywords Automotive radar • Target shape estimation • Complex target model • High-resolution frequency estimation

16.1 Introduction

Advanced driver assistance systems (ADA), as well as active and passive safety systems need precise knowledge of their environment. Radar sensors are commonly applied in such systems as they can operate in inclement weather and provide direct and reliable measurements of range and range-rate [15]. The drawback of today's automotive radar sensors is their limited angular and therefore lateral resolution [14]. Estimating the lateral extents of a target with respect to the driving direction is

F. Engels (✉)
A.D.C. Automotive Distance Control Systems GmbH, Peter-Dornier-Str. 10, D-88232 Lindau, Germany
e-mail: florian.engels@continental-corporation.com

G. Schmidt et al. (eds.), *Smart Mobile In-Vehicle Systems: Next Generation Advancements*, DOI 10.1007/978-1-4614-9120-0_16,

however of utmost importance for ADA and safety systems. For example adaptive cruise control (ACC), lane change assistance (LCA), as well as forward vehicle collision mitigation system (FVCMS) need to report the highway lane in which a target vehicle is located. This is surely not possible without knowledge of the target vehicle's width. For future collision avoidance systems [20], the task is generalized to estimating the free space in front of the subject vehicle (the car equipped with the radar sensor) in order to adjust the driving path to avoid a crash. A precise estimate of the lateral obstacle extensions is therefore crucial. When a crash becomes unavoidable, precrash systems need to classify the crash to determine which airbags have to be fired or if belt pretensioners have to be activated. The crash class depends on the overlap of the subject car and the obstacle and therefore on the obstacle width [6].

The limited angular resolution of automotive radar sensors is a consequence of commonly employed limited aperture antenna arrays in combination with digital beamforming. As a small sensor size is imperative for vehicle integration, increasing the array aperture is not an option. Therefore the focus of this chapter is signal processing strategies which improve the angular resolution. Existing work in the automotive context [7, 17] aims at separating multiple vehicles based on a point target or far-range assumption. The far-range scattering of a target can be locally modeled by a single plane wave, which causes a sinusoidal variation over the antenna array with a frequency proportional to the sine of the target angle. For multiple targets the array response becomes a sum of complex sinusoids, where each frequency corresponds to one target angle. Therefore parametric methods for line spectra estimation such as the ML method, MUSIC, or ESPRIT can be applied [7, 17]. In the context of array processing, such methods are also known as high-resolution direction of arrival (DOA) techniques.

If the far-range assumption is dropped (as in Sect. 16.3), the array response for a single target can also be modeled as a sum of complex sinusoids, where each frequency is associated with the angle of different target parts. Therefore high-resolution DOA estimation techniques can be employed to separate scattering contributions from different parts of a target. As the sum of complex sinusoids model holds also in the temporal dimensions of the radar data (see Sect. 16.3), line spectra methods can be used for high-resolution range and range-rate estimation. The angle corresponding to high-resolution range or range-rate estimates can be obtained by applying a new technique introduced in Sect. 16.4, which is called cell interpolation. This enables one to choose the data dimension in which high-resolution estimation is applied according to the radar sensor design.

In particular automotive safety systems require high angular resolution only for targets in the driving direction of the subject vehicle. Therefore rough position estimates of relevant targets can be used to limit the use of high-resolution estimation to a subset of the radar data. This enables a two-stage approach to high-resolution estimation where the first stage provides coarse target estimates which are refined in the second stage using line spectra estimation techniques. Two-stage processing is discussed in Sect. 16.4 and applied to experimental data in Sect. 16.6.

16.2 Scattering Model

This section gives a short overview of the dominant mechanisms involved in radar scattering from typical automotive targets, such as cars, trucks, or motorcycles. The extent of such targets is large with respect to wavelengths of 3.7 mm (79 GHz), 3.9 mm (76 GHz), and 12.4 mm (24 GHz), which are commonly used in automotive radar systems [20]. Therefore the principles of high-frequency scattering apply [12] and a targets' scattered field can be described as the superposition of local fields originating from different parts of the target. The complex target shape can thus be understood as a collection of simple shapes, such as curved surfaces, edges, or corners, each attributed with a characteristic high-frequency scattering mechanism [3, 10], which is illustrated in Fig. 16.1.

The sides of a car give rise to geometrical optics (GO) type or specular scattering, which is very strong when the incident wave's propagation direction is normal to the car's side. The same holds for a car's back and in principle for all singly or doubly curved surfaces of a car. The scattered or reflected GO field propagates in the direction of the so-called GO ray, a straight path in homogeneous media. Along the GO ray the reflected field is locally plane [12]. The direction of the reflected GO ray depends on the direction of incidence with respect to the surface normal and is determined by the law of reflection, which is also known as Snell's law [19]. Note that the direction of interest is the direction back to the radar sensor. Straight or curved edges, such as the radiator grill or the wheelhouse edges, lead to diffracted fields which are described by the geometrical theory diffraction (GTD) or the uniform geometrical theory of diffraction (UTD) [12]. Diffracted waves propagate in the direction of the so-called diffracted ray, which is formally equivalent to the GO ray. As in the GO case, the diffracted field is locally plane along the diffracted ray. The direction of the diffracted ray depends on the direction of incidence with respect to the edge and is determined by the law of edge diffraction [12]. The wheelhouses of a car produce a multi bounce GO return, see ([2], Chap. 13) and references therein. This means a ray entering the wheelhouse is reflected multiple times before eventually leaving the wheelhouse. As the field scattered from the wheelhouse can be modeled as GO field, it is also locally plane along the reflected ray. The direction of the reflected ray depends on the direction of incidence and the inner geometry of the wheelhouse. Scattering from the underbody of the car can reach the radar indirectly over the street surface, which acts as a

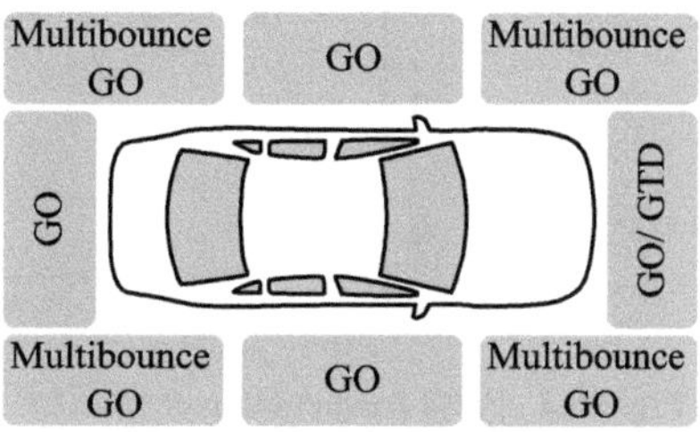

Fig. 16.1 Parts of a car attributed with dominant scattering mechanisms

mirror at 79, 76, and 24 GHz [16]. Being also of GO type, the scattered fields from the car's underbody are also locally plane along their respective rays.

All mentioned high-frequency scattering contributions lead to reflected fields, which are locally plane along the direction of propagation. Where multiple reflected rays in the direction of the radar sensor exist, the local field at the radar sensor becomes a superposition of plane waves with each corresponding to a different part of a target. This is an important result and will be used to derive the signal model presented in the following section. Experimental results showing the high-frequency radar scattering of a car can be found in [1].

16.3 Automotive Radar Sensors

Automotive radar sensors differ in their employed waveform and their applied angular measurement principle. Commonly chosen waveforms are frequency-modulated continuous wave (FMCW), stepped frequency, and coherent linear frequency-modulated (LFM) pulse train signals [20]. FMCW and stepped frequency signals are popular due to their low complexity. The main drawback of either waveform is that range and range-rate cannot be measured independently. This is particularly unfavorable in multi-target situations [20] and equivalently for estimating the extent of a target. In contrast, coherent LFM pulse trains enable almost independent range and range-rate measurements [11]. Therefore the remainder of this chapter focuses on coherent LFM pulse trains, and the other waveforms will not be discussed further.

Figure 16.2 shows the frequency of transmitted (solid line) and received (dashed line) LFM pulses for one coherent processing interval (CPI) and the time interval for transmitting and receiving all M pulses. The frequency axis is normalized to the center frequency f_c and each pulse has a bandwidth B. The pulses are transmitted at multiples of the pulse repetition time t_r. The n-th pulse is reflected from a target and arrives at the radar sensor delayed by t_n. If the target is not moving relative to the radar sensor, τ_n will be equal for all N_p pulses. On the other hand, if the target is

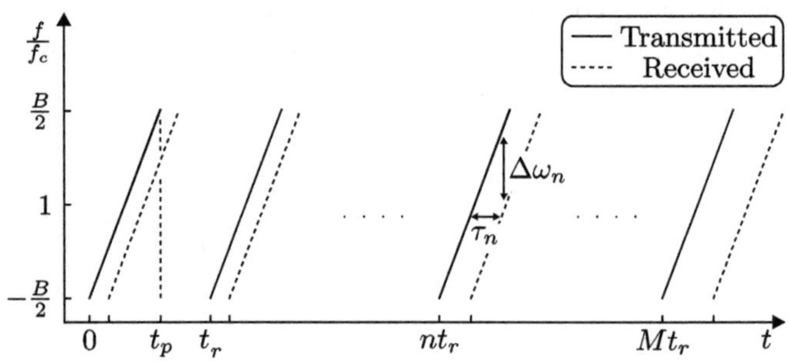

Fig. 16.2 Pulse frequency over time for one coherent processing interval

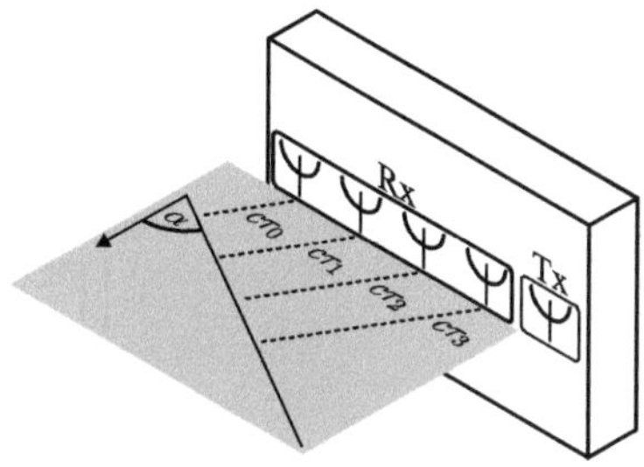

Fig. 16.3 Plane wave impinging on a four-element ULA

moving, τ_n becomes a function of the target's range at time nt_r. Thus the delay change over all pulses in one CPI can be utilized to measure the range-rate of a target, which will be shown in Sect. 16.3. The target's range can also be measured by exploiting the fact that τ_n is proportional to the difference frequency $\Delta\omega_n$. Mixing each transmitted with the corresponding received pulse followed by band pass filtering yields a sinusoidal signal with a frequency of $\Delta\omega_n$, which can be used for range measurement. This concept is known as stretch pulse compression [11].

Two popular angular measurement principles in automotive radar systems are monopulse and array processing. Monopulse techniques have no angular resolution capability and are not considered further. On the other hand, array techniques offer angular resolution via multiple spatially distributed antennas in combination with frequency estimation methods. A simple and often used antenna geometry is the uniform linear array (ULA), which consists of equidistant antennas in horizontal or vertical direction.

Figure 16.3 shows an azimuthal projection of a locally plane wave front impinging on a four element ULA and the outline of the radar sensor, and transmit antenna is provided for reference. The wave propagates with velocity c and forms an azimuth angle α with the array normal. The wave front arrives at the n-th antenna with a delay τ_n, which is a function of sin α and the antenna position. The delay differences between the antennas can be used to estimate α. Note that the exemplary ULA has no measurement capability in elevation.

A radar system employing a coherent LFM pulse train and array processing is depicted in Fig. 16.4. In this system the generated LFM pulses are successively transmitted and directly mixed with the received pulses. After filtering and amplification the resulting signal is sampled with a sampling frequency f_s. As observed from Fig. 16.4 this is done for each antenna in parallel. Thus the data $y(\boldsymbol{n})$ for one CPI is three dimensional and is indexed by a vector

$$\boldsymbol{n} = (n_0 n_1 n_2)^{\mathrm{T}} \in \mathbb{N}_L \times \mathbb{N}_M \times \mathbb{N}_N, \tag{16.1}$$

where L denotes the number of samples per pulse, M the number of pulses, N the number of antennas, and

$$\mathbb{N}_i = \{0, \cdots, i-1\}$$

are sets comprising all integer values from 0 to i. For example, $y(\boldsymbol{n}_1)$ with $\boldsymbol{n}_1 = (\,0 \quad 1 \quad 2\,)$ is the first sample of the second pulse received with the third antenna.

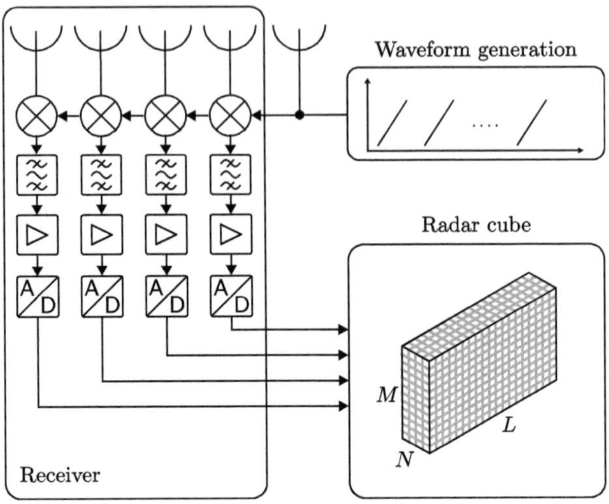

Fig. 16.4 Radar system employing a coherent LFM pulse train and array processing

Table 16.1 Parameters for a sample radar sensor

System parameter		
Center frequency	f_c	24 GHz
Pulse bandwidth	B	200 MHz
Pulse duration	t_p	8 μs
Number of samples	L	256
Pulse repetition time	t_γ	76.3 μs
Number of pulses	M	256
Antenna distance	d_a	6.207 mm
Number of antennas	N	7

As seen in Fig. 16.4, $y(\boldsymbol{n})$ can be visualized as a cube with the edges corresponding to the sample, pulse, and antenna data dimensions. Therefore $y(\boldsymbol{n})$ is referred to as the radar cube in the following.

Table 16.1 lists the parameters of a sample radar system employing a coherent LFM pulse train and array processing. These parameters will be used in the following sections, particularly in Sect. 16.6, where experimental results using such a radar system are discussed.

16.4 Signal Model

It was discussed in Sect. 16.2 that the scattered field of a complex target such as a vehicle can be locally modeled as a superposition of plane waves. Localized plane wave sources lead to harmonic variations in the sample, the pulse, and the antenna dimension of the radar cube when a radar system such as the one introduced in

Sect. 16.3 is employed [9, 15]. Thus the radar cube can be modeled as a sum of complex sinusoids

$$y(\boldsymbol{n}) = \sum_{k=0}^{K-1} a_k e^{j\boldsymbol{\omega}_k \boldsymbol{n}} + w(\boldsymbol{n}), \tag{16.2}$$

where each complex sinusoid is parameterized with a frequency vector

$$\boldsymbol{\omega}_k = (\rho_k \mu_k \nu_k)^{\mathrm{T}} \in \mathbb{R}^3 \tag{16.3}$$

and a complex amplitude a_k. The index k refers to one of K target parts. The index vector $\boldsymbol{n}$ is given by (16.1) and $w(\boldsymbol{n})$ denotes an additive noise term. The elements of (16.3) are normalized to the respective sampling frequencies f_s, $f_r = t_r^{-1}$, and $f_a = d_a^{-1}$ and are functions of range r, range-rate v, and azimuth angle α, respectively [9, 15]:

$$\begin{aligned} \rho_k &= c_0 r_k \\ \mu_k &= c_1 v_k \\ \nu_2^k &= c_2 \sin \alpha_k, \end{aligned} \tag{16.4}$$

where the constants

$$\begin{aligned} c_0 &= \frac{c t_p f_s}{4\pi B} \\ c_1 &= \frac{c f_r}{4\pi f_c} \\ c_2 &= \frac{c f_a}{2\pi f_c} \end{aligned} \tag{16.5}$$

depend on the radar system parameters given in Table 16.1. Estimating r, v, and α of different parts of a target is thus a frequency estimation problem which is in principle solved by minimizing

$$\sum_{n} \left| y(\boldsymbol{n}) - \sum_{k=0}^{K-1} a_k e^{j\boldsymbol{\omega}_k \boldsymbol{n}} \right|^2 \tag{16.6}$$

with respect to $(a_0 \cdots a_{K-1}, \boldsymbol{\omega}_0 \cdots \boldsymbol{\omega}_{K-1})$ [18].

Note that the number K will change with the target's position and orientation with respect to the radar. Even if K is equal for different orientations, the target parts contributing to the radar cube can be different. Figure 16.5 illustrates two possible

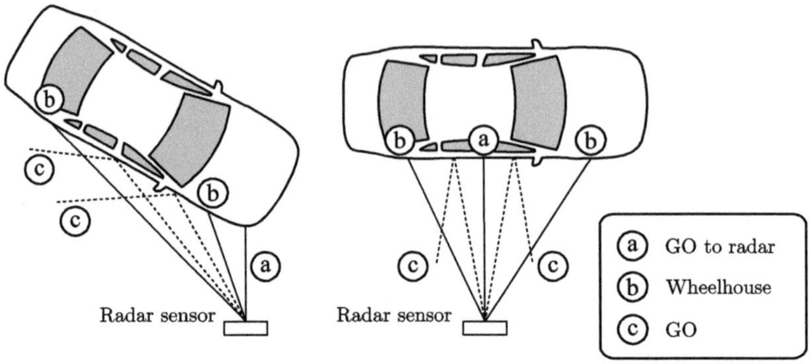

Fig. 16.5 Examples of rays directed away and back to the radar sensor

target orientations with $K = 3$. Here the solid lines represent ray paths leading back to the radar sensor, which are included in the sum of (16.2). The dashed lines denote ray paths which lead away from the radar and therefore do not contribute to the radar cube.

16.5 Signal Processing

The objective of radar signal processing is to extract information from the radar cube which is needed for the realization of a given ADA or safety function. As pointed out in Sect. 16.1, a common requirement is to know the positions, extents, and dynamic properties of function relevant targets. Thus a generic signal processing approach could be to provide a list of all targets detected by the radar sensor, and let the specific ADA function pick the relevant ones from this list.

Radar signal processing happens at two different time scales, one spanning a single CPI and the other multiple CPIs. Single CPI processing provides a list of detections based on the most recent radar cube. Using multiple CPIs enables target estimates which are tracked over time and which are updated with detections from the most recent CPI. The multi CPI processing is therefore commonly referred to as tracker. The single CPI detection list ideally contains multiple detections per target, where a detection is parameterized by range r, range-rate v, and azimuth angle α. Those detections are used by the tracker to estimate the target extent as well as the target orientation [5], which are crucial parameters for ADA and safety functions (see Sect. 16.1).

Here the focus will be single CPI processing, which is based on estimating the frequencies of the radar cube given in (16.2). Putting these frequency estimates in (16.4) results in range r, range-rate v, and azimuth angle α, which constitute a detection. In the following sections a two-stage approach to frequency estimation is discussed. The first stage employs the so-called periodogram to provide coarse estimates, which are used to select the region of interest for the second stage. The second stage then applies high-resolution estimation in one or two dimensions. Both stages are discussed in the following sections.

16.5.1 Periodogram-Based Processing

The periodogram is a frequency estimation method based on the discrete Fourier transform (DFT). It is popular in automotive radar systems because it can be calculated efficiently via the fast Fourier transform (FFT) algorithm. In the following the application of the periodogram to the radar cube is briefly discussed.

If it is assumed that either

$$\Delta\rho_R = \inf_{k \neq p} \left|\rho_k - \rho_p\right| > \frac{2\pi}{L}, \tag{16.7}$$

or

$$\Delta\mu_R = \inf_{k \neq p} \left|\mu_k - \mu_p\right| > \frac{2\pi}{M}, \tag{16.8}$$

or

$$\Delta\nu_R = \inf_{k \neq p} \left|\nu_k - \nu_p\right| > \frac{2\pi}{N}, \tag{16.9}$$

holds, then (16.6) is approximately minimized by the frequencies corresponding to the K largest maxima of the periodogram [18]

$$P(\boldsymbol{\omega}) = |Y(\boldsymbol{\omega})|^2, \tag{16.10}$$

where

$$Y(\boldsymbol{\omega}) = \frac{1}{\widetilde{N}} \sum_{\boldsymbol{n}} y(\boldsymbol{n}) e^{-j\boldsymbol{\omega}\boldsymbol{n}}, \tag{16.11}$$

with $\boldsymbol{\omega} \in \mathbb{R}^3$ and $\widetilde{N} = LMN$ (see Table 16.1). Equations (16.7), (16.8), and (16.9) are known as Rayleigh limits. Note that in (16.11) the sum over an index vector equals the sums over its elements. Therefore (16.11) represents the three-dimensional time discrete Fourier transform of the radar cube, which is commonly calculated at the DFT frequencies

$$\rho_s = \left\{ \frac{2\pi}{L} s \right\}_{s=0}^{L-1}, \quad \mu_s = \left\{ \frac{2\pi}{M} s \right\}_{s=0}^{M-1}, \quad \nu_s = \left\{ \frac{2\pi}{N} s \right\}_{s=0}^{N-1}$$

using the FFT algorithm.

Note that the K largest maxima of (16.10) do not have to correspond to a single target. For automotive applications the number of targets can be rather large as not only other vehicles, but also the road border, street signs, and other stationary targets contribute to the radar cube. Therefore no a priori knowledge of K is available, and the largest maxima are determined by finding all local maxima of (16.10). Owing to the limited sampling points obtained from the DFT, the frequency estimates are refined using local interpolation [8]. The frequencies corresponding to the local maxima which exceed a detection threshold constitute the periodogram frequency estimates.

16.5.2 Two-Stage Processing

Most ADA and safety systems require high angular resolution only in a narrow region around the driving path of the subject vehicle. Therefore the resolution capability of the periodogram is sufficient in regions which have a large lateral offset with respect to the subject vehicle. High-resolution estimation can thus be restricted to a small angular sector around the driving path. Furthermore the periodogram can be used to limit the region of interest to a longitudinal distance around detections in the driving path. This allows a two-stage approach where high-resolution estimation is applied only to preprocessed subsets of the radar cube, which will be referred to as cells. Cell preprocessing is done by applying (16.11) only in the so-called cell dimensions. The remaining dimensions of the radar cube which are not preprocessed are referred to as high-resolution dimensions. Note that the cell, as well as the high-resolution dimensions, is related to the respective parameter by (16.4) and (16.5).

One- and two-dimensional high-resolution estimation can be applied to the radar cube. This leads to six possible combinations of cell parameters and high-resolution parameters, which are listed in Table 16.2 together with the cell data to which high-resolution estimation is applied. Depending on the cell parameters, Fourier-based preprocessing is only applied in one or two dimensions. The frequency and index vectors $\boldsymbol{\omega}_c$, $\boldsymbol{n}$, and $\boldsymbol{m}$ in Table 16.2 are therefore two dimensional. For high-resolution range-rate estimation (high-resolution parameter v and cell parameters (r, α)), the complete processing chain is shown in Fig. 16.7. The processing starts with calculating samples of (16.11) using a three-dimensional DFT. This is followed by taking the absolute value squared to obtain (16.10). A local maxima search and applied thresholds result in frequency estimates corresponding to the L strongest maxima of the frequency domain radar cube. The corresponding frequencies are refined by local frequency interpolation and then used to obtain

Table 16.2 High-resolution (HR) and cell parameters for two-stage processing

HR parameter	Cell parameter	Preprocessed data
r	(r,α)	$Y_{(v,\alpha)}(n,\boldsymbol{\omega}_c) = \sum_{\boldsymbol{m}} y\left((n,m_0,m_1)^{\mathrm{T}}\right)e^{-j\boldsymbol{\omega}_c\boldsymbol{m}}$
v	(r,α)	$Y_{(r,\alpha)}(n,\boldsymbol{\omega}_c) = \sum_{\boldsymbol{m}} y\left((m_0,n,m_1)^{\mathrm{T}}\right)e^{-j\boldsymbol{\omega}_c\boldsymbol{m}}$
α	(r,v)	$Y_{(r,v)}(n,\boldsymbol{\omega}_c) = \sum_{\boldsymbol{m}} y\left((m_0,m_1,n)^{\mathrm{T}}\right)e^{-j\boldsymbol{\omega}_c\boldsymbol{m}}$
(r,v)	α	$Y_\alpha(\boldsymbol{n},\omega_c) = \sum_m y\left((n_0,n_1,m)^{\mathrm{T}}\right)e^{-j\omega_c m}$
(r,α)	v	$Y_v(\boldsymbol{n},\omega_c) = \sum_m y\left((n_0,m,n_1)^{\mathrm{T}}\right)e^{-j\omega_c m}$
(r,α)	r	$Y_r(\boldsymbol{n},\omega_c) = \sum_m y\left((m,n_0,n_1)^{\mathrm{T}}\right)e^{-j\omega_c m}$

the detection parameter estimates range, range-rate, and angle by employing (16.4). One detection of interest $(\hat{r}_0 \quad \hat{v}_0 \quad \hat{\alpha}_0)$ is extracted from the L detections, and the corresponding frequencies $(\hat{\rho}_0 \quad \hat{\mu}_0 \quad \hat{\nu}_0)$ are used for cell selection. The corresponding cell data $Y_{(r,\alpha)}\left(m,\left(\hat{\rho}_0 \quad \hat{\nu}_0\right)^T\right), m = 0,\cdots,M-1$ is then used for high-resolution estimation yielding K frequency estimates $\hat{\mu}_k$ for the range-rate dimension. By using cell interpolation the cell dimension frequency estimates $\hat{\rho}_k$ and $\hat{\nu}_k$ are obtained. By applying (16.4), the frequency estimates are converted to range, range-rate, and angle and are finally (together with the periodogram detections) passed to the tracker.

Which of the six possible choices in Table 16.2 is most beneficial depends on the radar sensor design. For example, an ultra-wideband short-range radar sensor might employ high-resolution range estimation, because the large bandwidth results in high-range resolution [15]. In contrast, long-range radar sensors with comparatively large antenna arrays might use high-resolution DOA estimation to increase the angular resolution. Whether to use one- or two-dimensional high-resolution estimation is a trade-off between computational complexity, signal-to-noise ratio (SNR), and resolution performance. Two-dimensional algorithms will in general be more computationally demanding than one-dimensional algorithms, but offer a higher resolution capability [13]. On the other hand, cell preprocessing via the DFT increases the SNR with each cell dimension. This means one-dimensional high-resolution estimation can benefit from a higher SNR compared to the two-dimensional case.

16.5.3 Cell Interpolation

Based on high-resolution estimates in one dimension of the radar cube, cell interpolation yields estimates in the remaining cell dimensions. This is shown in Fig. 16.6, where range and azimuth angle estimates based on high-resolution

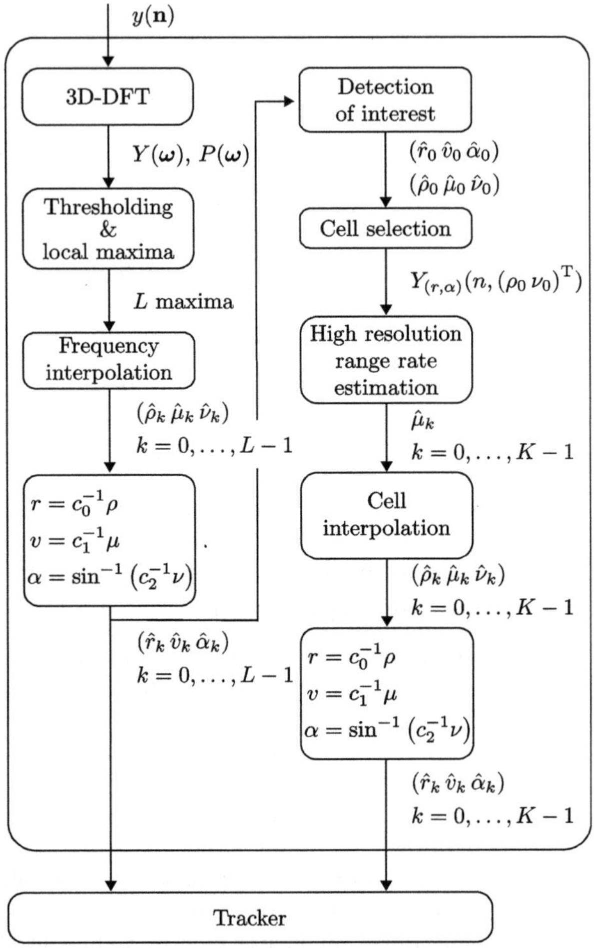

Fig. 16.6 Two-stage signal processing including high-resolution range-rate estimation

range-rate estimates are obtained. The idea behind cell interpolation is to estimate the complex amplitudes corresponding to high-resolution estimates in the cells adjacent to the so-called main cell. The main cell is the cell which is selected for high-resolution estimation, as in Fig. 16.6. Looking at frequency domain data as in (16.11), the complex amplitudes in the adjacent cells will be samples of the spectral response of each complex sinusoid in (16.2). Their absolute values can thus be used to estimate the maximizing frequency. This will be derived in the following for high-resolution range-rate estimation in combination with range-angle cells.

The frequency domain radar cube in (16.11) can be written as

$$Y(\boldsymbol{\omega}) = \sum_{k-0}^{K-1} a_k S(\boldsymbol{\omega}_k - \boldsymbol{\omega}) + W(\boldsymbol{\omega}), \tag{16.12}$$

where

$$S(\boldsymbol{\omega}) = S_L(\rho)S_M(\mu)S_N(\nu)$$

and

$$S_P(\omega) = e^{j\frac{P-1}{2}\omega}\frac{\sin \omega P/2}{\sin \omega/2}.$$

By dropping the noise term and fixing ρ and v to ρ_c and ν_c, respectively, (16.12) can be written as

$$Y(\boldsymbol{\omega}) = \sum_{k=0}^{K-i} c_k S_M(\mu_k - \mu), \tag{16.13}$$

where

$$c_k = a_k S_L(\rho_k - \rho_c) S_N(\nu_k - \nu_c). \tag{16.14}$$

The complex amplitudes (16.14) can be estimated from a set of N samples

$$\boldsymbol{y} = \left(Y\left(\left(\rho_c \mu_s^{(0)} \nu_c\right)^{\mathrm{T}}\right) \cdots Y\left(\left(\rho_c \mu_s^{(N-1)} \nu_c\right)^{\mathrm{T}}\right)\right)^{\mathrm{T}}$$

at the sampling points $\{\mu_s^{(0)}, \cdots, \mu_s^{(N-1)}\}$. Using (16.13) for each sample and gathering (16.14) in a vector

$$c = (c_0 \cdots c_{K-1})^{\mathrm{T}} \tag{16.15}$$

yields

$$\boldsymbol{y} = \boldsymbol{Sc}, \tag{16.16}$$

where

$$\boldsymbol{S} = \begin{bmatrix} S\left(\mu_0 - \mu_s^{(0)}\right) & \cdots & S\left(\mu_{K-1} - \mu_s^{(0)}\right) \\ \vdots & \ddots & \vdots \\ S\left(\mu_0 - \mu_s^{(N-1)}\right) & \cdots & S\left(\mu_{K-1} - \mu_s^{(N-1)}\right) \end{bmatrix}. \tag{16.17}$$

If the μ_k, $k = 0, \cdots, K - 1$ in (16.17) are known, (16.16) is solved by

$$\hat{\boldsymbol{c}} = \left(\boldsymbol{S}^H \boldsymbol{S}\right)^{-1} \boldsymbol{S}^H \boldsymbol{y} \tag{16.18}$$

in a least squares sense [18].

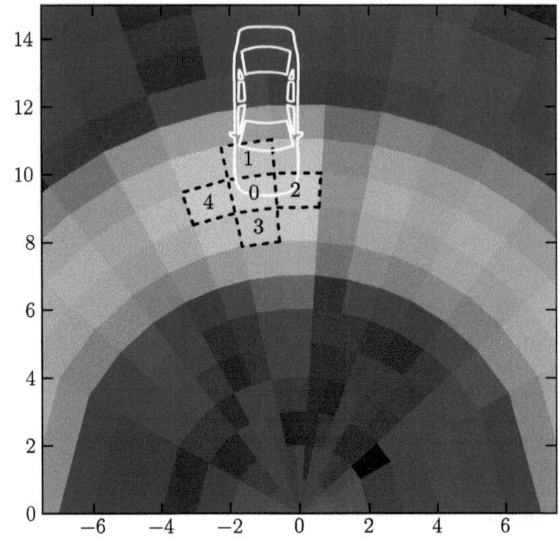

Fig. 16.7 Range-angle cells for a target vehicle

Equation (16.18) can be used to obtain angle and range based on high-resolution range-rate estimates and the frequency domain radar cube. This will be explained using Fig. 16.7, which shows the outline of a target vehicle together with the corresponding range-angle cells. Each cell is associated with so-called cell frequencies corresponding to range and angle, respectively. The grey scale represents the values of (16.10) for the respective cell frequencies. In the vicinity of the car, the cell with index zero is referred to as main cell and the cells with indices from one to four as adjacent cells. The corresponding cell frequencies are denoted by

$$\left\{\rho_c^{(i)}\right\}_{i=0}^{4}, \left\{\nu_c^{(i)}\right\}_{i=0}^{4},$$

where

$$\rho_c^{(0)} = \rho_c^{(2)} = \rho_c^{(4)}$$

and

$$\nu_c^{(0)} = \nu_c^{(1)} = \nu_c^{(3)}.$$

The cell power values

$$\begin{aligned} \gamma_k^{(i)} &= \left| a_k S_L\left(\rho_k - \rho_c^{(0)}\right) S_N\left(\nu_k - \nu_c^{(i)}\right)\right|^2, \quad i = 0, 2, 4 \\ \delta_k^{(i)} &= \left| a_k S_N\left(\nu_k - \nu_c^{(0)}\right) S_L\left(\rho_k - \rho_c^{(i)}\right)\right|^2, \quad i = 0, 1, 3 \end{aligned} \tag{16.19}$$

are defined as the absolute value squared of (16.14) for the respective cell frequencies. By comparing (16.19) with

$$\begin{aligned} |S_L(\rho_k - \rho)|^2 &= \left| \frac{\sin\left((\rho_k - \rho)L/2\right)}{\sin\left((\rho_k - \rho)/2\right)} \right| \\ |S_N(\nu_k - \nu)|^2 &= \left| \frac{\sin\left((\nu_k - \nu)N/2\right)}{\sin\left((\nu_k - \nu)/2\right)} \right| \end{aligned} \tag{16.20}$$

it is found that (16.19) represents scaled samples of (16.20). This enables the use of local interpolation techniques as described in [8] for estimating the true frequencies, which are related to (16.20) by

$$\begin{aligned} \rho_k &= \arg\max_{\rho} |S_L(\rho_k - \rho)|^2 \\ \nu_k &= \arg\max_{\nu} |S_N(\nu_k - \nu)|^2. \end{aligned}$$

As an example local parabolic interpolation is considered, which leads to the following frequency estimates [8]

$$\begin{aligned} \hat{\rho}_k &= \rho_c^{(0)} + \Delta\rho_k \\ \hat{\nu}_k &= \nu_c^{(0)} + \Delta\nu_k, \end{aligned} \tag{16.21}$$

where

$$\begin{aligned} \Delta\rho_k &\cong \frac{1}{2} \frac{\gamma_k^{(4)} - \gamma_k^{(2)}}{\gamma_k^{(4)} + \gamma_k^{(2)} - 2\gamma_k^{(0)}} \\ \Delta\nu_k &\cong \frac{1}{2} \frac{\delta_k^{(3)} - \delta_k^{(1)}}{\delta_k^{(3)} + \delta_k^{(1)} - 2\delta_k^{(0)}}. \end{aligned} \tag{16.22}$$

In (16.22) the cell power values given by (16.19) are used and

$$\begin{aligned} \max_i \gamma_k^{(i)} &= 0, \quad i = 0, 2, 4 \\ \max_i \delta_k^{(i)} &= 0, \quad i = 0, 1, 3 \end{aligned}$$

is assumed.

Applying (16.18) in each cell and taking the absolute value squared provide estimates for

$$\begin{aligned} \gamma_i &= \left(\gamma_0^{(i)} \cdots \gamma_{K-1}^{(i)}\right)^T, \quad i = 0, 2, 4 \\ \delta_i &= \left(\delta_0^{(i)} \cdots \delta_{K-1}^{(i)}\right)^T, \quad i = 0, 1, 3. \end{aligned} \tag{16.23}$$

Employing (16.21) and (16.22) to the estimates of (16.23) leads to the desired range and angle-dependant frequency estimates. The required frequencies μ_k, $k = 0, \cdots, K - 1$ in (16.17) are estimated in the main cell using high-resolution techniques in the range-rate dimension.

Note that even though cell interpolation was derived for high-resolution range-rate estimation, it can also be applied for high-resolution range or angle estimation. Cell interpolation can also be extended to the two-dimensional case by applying stacked vectors and matrices in (16.16).

16.6 Experimental Results

In this section two-stage processing is applied to experimental data recorded with an automotive radar sensor. The sensor is described in Sect. 16.3, and relevant system parameters are listed in Table 16.1. High-resolution estimation is applied in the range-rate and angle dimensions of the radar cube and leads to cells parameterized by range. For high-resolution estimation the two-dimensional spectral RELAX algorithm is used, which is described in detail in [4].

The experimental setup is depicted in Fig. 16.8. The target vehicle is approaching the radar sensor with a constant velocity of 14 m/s and a constant lateral offset of 1 m. Starting at a longitudinal distance of 20 m, the target car drives past the radar sensor, which is not moving.

Also shown in Fig. 16.8 are two sample scattering contributions denoted by S_0 and S_1 at the front of the car. The longitudinal distance of the car to the radar sensor is 8 m, and its lateral offset as well as its velocity is according to the experimental setup described before. The angle and range-rate differences of S_0 and S_1 can be converted to frequency differences by (16.4) and the system parameters listed in Table 16.1. Comparing the frequency differences with the corresponding Rayleigh

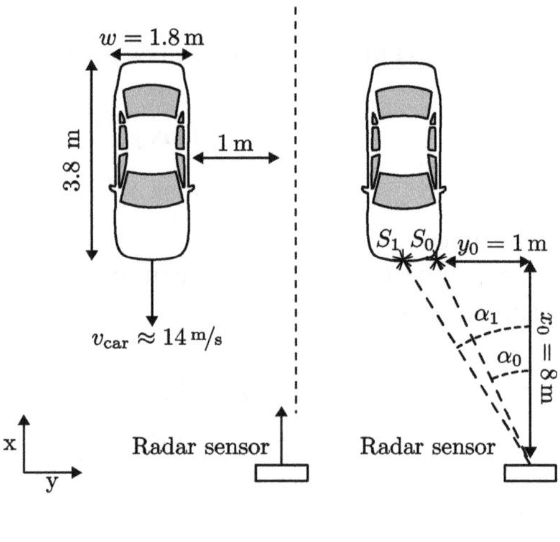

Fig. 16.8 Experimental setup and sample scattering contributions at the car's front

limits gives an approximate measure of the periodogram's expected resolution performance. Using the parameters from Table 16.1 in (16.8) and (16.9) yields

$$\Delta\nu_R = \frac{2\pi}{M} = 0.024\text{rad}$$

$$\Delta\mu_R = \frac{2\pi}{N} = 0.897\text{rad}.$$

The azimuth angles of S_0 and S_1 are given as:

$$\alpha_0 = \tan^{-1}\frac{y_0}{x_0} = 7.1^\circ$$

$$\alpha_1 = \tan^{-1}\frac{y_0 + 0.5w}{x_0} = 13.3^\circ.$$

The angular difference of 62° can be converted to a frequency difference of

$$0.34\text{rad} < \Delta\mu_R,$$

which is smaller than the respective Rayleigh limit given by (16.9). The range-rate of S_0 and S_1 can be calculated by [8]

$$v_0 = v_t \cos\alpha_0 = 13.99m/s$$
$$v_1 = v_t \cos\alpha_1 = 13.71m/s$$

which results in a range-rate difference of 0.28 ∘ m/s. The corresponding frequency difference

$$0.022\text{rad} < \Delta\nu_R$$

is again smaller as the respective Rayleigh limit given by (16.8). Therefore the periodogram will most likely not resolve scattering contributions with a lateral difference of less than half the car's width. In contrast, the second processing stage using RELAX does not share the resolution limitations of the periodogram and is therefore expected to resolve closely spaced scattering contributions. With this example in mind, the experimental results will be discussed.

Figure 16.9 shows detections of the first periodogram processing stage in a birds' eye view for one CPI. Detections corresponding to the target vehicle can be distinguished by marker size, which is proportional to range-rate. This means that the greater the marker size, the greater the range-rate of the corresponding detection. The target car's approximate position is outlined for reference. It can be observed that one detection is located at the front of the car and several others at the rear. The rear detections originate from the wheelhouse and the wheel, respectively. The area around the car's front is also shown in Fig. 16.9 in

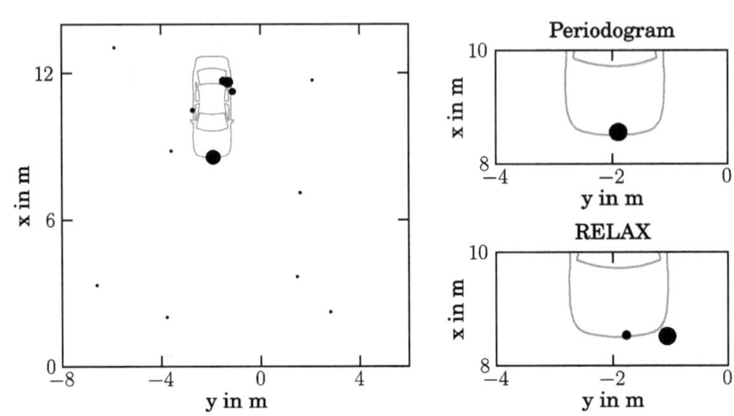

Fig. 16.9 Birds' eye view for one CPI

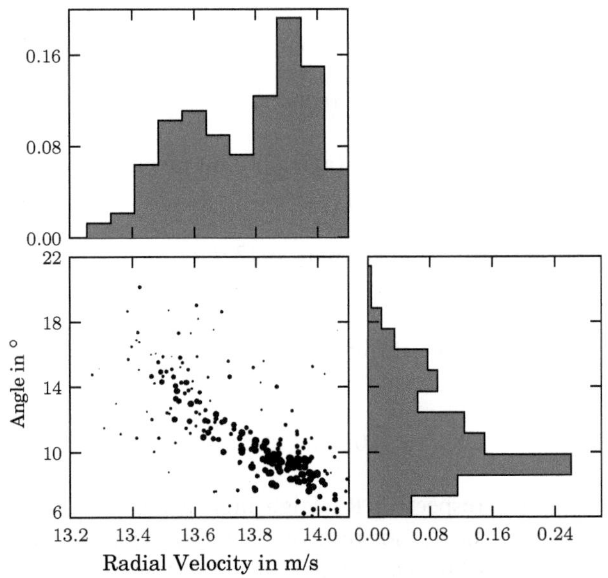

Fig. 16.10 Angle and range-rate estimates of the second processing stage using RELAX

greater detail. In addition to the first-stage detections, the second-stage detections obtained with RELAX are shown. It can be observed that one first-stage detection is obtained which is used to select the range-cell needed for the second-stage processing. The RELAX-based second processing stage is able to resolve two scattering contributions at the car's front. The periodogram cannot resolve multiple contributions but can be used for cell selection.

To see how reliably RELAX yields multiple detections at the car's front, the same experiment was repeated 37 times. Figure 16.10 shows the second-stage angle and range-rate estimates from all experiments in distances from 7.5 to 8.5 m in a scatter plot. Also shown are the relative frequencies for angle and range-rate, respectively. The angle estimates range from approximately 7° to 18°. This corresponds to contributions from the right corner to approximately the middle of the front. The range-rate estimates are consistent with angle estimates as

$$v = \cos\alpha.$$

16.7 Conclusions

A two-stage approach to improve the angular resolution of automotive radar sensors has been discussed. It was shown how one- and two-dimensional high-resolution parameter estimation can be incorporated in a Fourier-based signal processing chain. The concept of cell interpolation was presented, which yields angle estimates through the use of high-resolution range or range-rate estimates. Therefore cell interpolation allows one to choose the high-resolution data dimension according to the radar sensor design. In particular low-cost automotive radar sensors with small antenna arrays could benefit from the use of cell interpolation.

Two-stage processing was applied to experimental data and was shown to successfully resolve closely spaced target parts. High-resolution estimation was employed in two data dimensions and implemented using the spectral RELAX algorithm. The resulting angle and range-rate estimates were observed to be in good agreement with the expected results.

References

1. M. Andres, P. Feil, W. Menzel, H.L. Bloecher, J. Dickmann, Analysis of automobile scattering center locations by SAR measurements, in *Proceedings of the Radar Conference (RADAR)*, Kansas City, USA, 23–27 May 2011
2. J. Van Bladel, *Electromagnetic Fields* (John Wiley & Sons, New York, 2007)
3. H. Buddendiek, Streuzentrenmodelle zur Simulation der Wellenausbreitung für automobile Radar- und Funksysteme, PhD Thesis, Technische Universität München, 2011
4. F. Engels, Target shape estimation using an automotive radar, in *Proceedings of the 5th Biennial Workshop on DSP for In-Vehicle Systems*, Kiel, Germany, September 2011
5. J. Gunnarsson, L. Svensson, L. Danielsson, F. Bengtsson, Tracking vehicles using radar detections, in *Proceedings of the IEEE Intelligent Vehicles Symposium*, Istanbul, Turkey, 13–15 June 2007
6. F. Gustafsson, Automotive safety systems. IEEE Signal Processing Magazine **26**(4), 32–47 (Jul 2009)
7. P. Heidenreich, Antenna array processing: autocalibration and fast high-resolution methods for automotive radar, PhD Thesis, Technische Universität Darmstadt, 2012
8. E. Jacobson, P. Kootsookos, Fast, accurate frequency estimators. IEEE Signal Processing Magazine **24**(3), 32–47 (2007)

9. R. Klemm, *Principles of Space-Time Adaptive Processing* (The Institution of Engineering and Technology, London, 2006)
10. E.F. Knott, J.F. Shaeffer, M.T. Tuley, *Radar Cross Section* (Artech House, Boston, 1993)
11. N. Levanon, E. Mozeson, *Radar Signals* (John Wiley & Sons, New York, 2004)
12. D.A. McNamara, C.W.I. Pistorius, J.A.G. Malherbe, *Introduction to the Uniform Geometrical Theory of Diffraction* (Artech House, Boston, 1989)
13. M. Pesavento, Fast algorithms for multidimensional harmonic retrieval, PhD Thesis, Ruhr-Universität Bochum, 2005
14. R.H. Rasshofer, K. Naab, 77 GHz long range radar systems status, ongoing developments and future challenges, in *Proceedings of the European Radar Conference (EURAD)*, Paris, France, 6–7 October 2005
15. A.W. Rihaczek, *Principles of High Resolution Radar* (Artech House, Boston, 1969)
16. R. Schneider, Modellierung der Wellenausbreitung für ein bildgebendes KFZ-Radar, PhD Thesis, Universität Karlsruhe, 1998
17. M. Schoor, Hochauflösende Winkelschätzung für automobile Radarsysteme, PhD Thesis, Universität Stuttgart, 2010
18. P. Stoica, R. Moses, *Spectral Analysis of Signals* (Prentice Hall, Upper Saddle River, NJ, 2005)
19. J.A. Stratton, *Electromagnetic Theory* (John Wiley & Sons, New York, 2007)
20. H. Winner, Radarsensorik, in *Handbuch Fahrerassitenzsysteme*, ed. by H. Winner, S. Hakuli, G. Wolf (Vieweg + Teubner, Wiesbaden, Germany, 2009)

Index

G. Schmidt et al. (eds.), *Smart Mobile In-Vehicle Systems: Next Generation Advancements*, DOI 10.1007/978-1-4614-9120-0,

MIX
Papier aus verantwortungsvollen Quellen
Paper from responsible sources
FSC® C105338

If you have any concerns about our products,
you can contact us on
ProductSafety@springernature.com

In case Publisher is established outside the EU,
the EU authorized representative is:
Springer Nature Customer Service Center GmbH
Europaplatz 3, 69115 Heidelberg, Germany

Printed by Libri Plureos GmbH
in Hamburg, Germany